Remote Sensing for Biodiversity and Wildlife Management

About the Author

Steven E. Franklin specializes in the application of satellite and aerial remote sensing in environmental science and management. He studied in the School of Forestry at Lakehead University and the Faculty of Environment at University of Waterloo, where he received his B.E.S. (Honours), M.A. (Physical Geography), and Ph.D. (Remote Sensing) degrees. He has held academic appointments at University of Waterloo (1984–1985), Memorial University of Newfoundland (1985–1988), University of Calgary (1988–2003), and University of Saskatchewan (2003–2009), where he served as Vice-President Research. Dr. Franklin was appointed President and Vice-Chancellor of Trent University in Peterborough, Ontario, Canada in 2009, where he also holds a joint appointment in the Department of Geography and the Environmental and Resource Science/Studies program.

Dr. Franklin has been a visiting professor in the College of Forestry at Oregon State University (1994), the Department of Geography at University of California (Santa Barbara) (2000), and the Department of Spatial Sciences at Curtin University of Technology (2008). His research and consultancy work have taken him to Argentina, India, Norway, Colombia, France, China, and numerous field sites in Australia, Canada, and the United States. His research team has been recognized by the Canadian Forest Service (2006 Merit Award) and the Alberta Emerald Foundation (2005 Excellence Award). Dr. Franklin has authored more than 125 peer-reviewed articles and several books, and has supervised more than 50 senior undergraduate and graduate students and six postdoctoral fellows. In October 2007, he received the Canadian Remote Sensing Society Gold Medal Award.

Remote Sensing for Biodiversity and Wildlife Management

Synthesis and Applications

Steven E. Franklin

New York Chicago San Francisco
Lisbon London Madrid Mexico City
Milan New Delhi San Juan
Seoul Singapore Sydney Toronto

The McGraw-Hill Companies

Cataloging-in-Publication Data is on file with the Library of Congress.

McGraw-Hill books are available at special quantity discounts to use as premiums and sales promotions, or for use in corporate training programs. To contact a representative please e-mail us at bulksales@mcgraw-hill.com.

Remote Sensing for Biodiversity and Wildlife Management

1 2 3 4 5 6 7 8 9 0 DOC/DOC 0 1 4 3 2 1 0 9

ISBN 978- 0-07-162247-9
MHID 0-07-162247-0

The pages within this book were printed on acid-free paper.

Sponsoring Editor
Taisuke Soda

Acquisitions Coordinator
Michael Mulcahy

Editorial Supervisor
David E. Fogarty

Project Manager
Gita Raman

Copy Editor
Shivani Arora

Proofreader
Eina Malik

Indexer
Edwin Durbin

Production Supervisor
Richard C. Ruzycka

Composition
Glyph International

Art Director, Cover
Jeff Weeks

For Gillian

In all things of nature there is something of the marvelous
—Aristotle, 384-322 BC

To take away the dignity
of any living thing
even tho it cannot understand
the scornful words
is to make life itself trivial
and yourself the Pontifex Maximus
of nullity
—Al Purdy, 1918-2000

Contents

Preface

Remote sensing has long been considered an important source of environmental information that is capable of supporting a deeper understanding of trends and clarifying management strategies in a wide range of ecological applications. Biodiversity and wildlife management are of global significance and increasing concern as current environmental trends intensify. These three broad areas—remote sensing, biodiversity, and wildlife management—intersect to create a dynamic *transdisciplinary* area of research with practical applications that are based on key new methods, concepts, and insights acquired in a wide variety of field studies and management situations. The urgency of many environmental challenges and the fast-moving technological world of remote sensing provide a stimulating context for innovative applications and research in environmental management. A full implementation of a remote-sensing approach may make new solutions to demanding tasks in biodiversity and wildlife management possible.

The initial promise of remote sensing originates in the ability to deliver high-quality scientific observations at many scales, over large areas, with regular revisit intervals, in a format that is both well-suited for ecological interpretation and readily integrated with data from other components of a modern spatial information and environmental management system. Many introductory and specialized remote-sensing books provide a rich technical compendia and survey of developments, but few focus on the ecological remote-sensing applications that are now possible. None of the current literature specifically highlights and expands on the critical remote-sensing contributions to biodiversity and wildlife management. Several recent journal article reviews document a growing success in applying remote-sensing technology and methods to the analysis and modeling of habitat, species-environment relationships, and biodiversity. Recent commentary suggests that what is needed is more effective collaboration among all those working in remote sensing, wildlife management, and biodiversity science and conservation. The intent of this book is to bring a remote-sensing approach closer to many readers and environmental managers—to synthesize the relevant aspects of this expanding scientific world in an informative, timely, and concise fashion.

A number of related themes are addressed in this book. In the opening chapters, the information needs—the critical targets that remote sensing can potentially address—are reviewed in support of environmental management applications. The broad and growing area of *biodiversity indicators* and its detailed treatment by other researchers are included in this review, while the basic operation of certain remote-sensing instruments is explained. These chapters are followed by a review of image analysis and related interpretation techniques of interest in biodiversity assessment and wildlife management (notably GIS, GPS, and spatial modeling). The heart of the book presents examples and case studies that were carefully selected to emphasize key insights and methods that have worked in actual conservation and management projects.

The information contained in this book will be useful to a full range of researchers and management professionals in the wide range of natural resource disciplines and subfields. Increasingly, traditional disciplinary-based or expertise-based divisions are fading as solutions are developed by researchers, managers, policy-makers and all manner of resource professionals who are working together in multidisciplinary, interdisciplinary, or transdisciplinary teams. Such teams bear front-line responsibility for environmental planning, management, assessment, and recovery planning for general environmental impacts and conservation in some of the most important work of our time. These teams, as well as senior undergraduate and graduate courses in the broad areas of biodiversity and wildlife management and remote sensing, will be able to draw on the material as a textbook.

The book was completed while the author had the good fortune to experience administrative leave at Curtin University of Technology in Perth, Australia. Among the many unique features of Curtin is a strong Department of Spatial Sciences which is adjacent to a strong Department of Environmental Biology and located within a dynamic intellectual setting. Colleagues at the Western Australia Department of Environmental Conservation, Murdoch University, University of Western Australia, and CSIRO shared generously of their knowledge and expertise—their outstanding institutions are among the world leaders in ecological aspects of remote sensing and spatial sciences and are located in the spectacular environmental setting of Western Australia, a biodiversity hotspot of international significance. While exploring this interesting and beautiful part of the world, I learned a great deal and was stimulated to gain a deeper understanding and to communicate better the contribution of remote sensing in biodiversity and wildlife management.

Steven E. Franklin

Acknowledgments

Special thanks are due all those who made this book possible: in particular, Peter Mackinnon (University of Saskatchewan) and Bert Veenendaal, Graeme Wright, and Jonathan Majer (Curtin University). New colleagues and friends helped in many ways: Giles Hardy (Murdoch University); Cecilia Xia, Grant Wardell-Johnston, Tom Schut, and Ahmed El-Mowafy (Curtin University); Graeme Behn and Richard Robinson (Department of Environmental Conservation); Rochelle Pilon, and David Brereton. The preparation of the book was supported by my current research collaborators: Gordon Stenhouse, Marc Cattet, Greg McDermid, John Boulanger, Matt Vijayan, David Janz, Xulin Guo, and Scott Nielsen. Outstanding help with the appendices, tables, figures, and plates was provided by Dr. Yuhong He (University of Toronto Mississauga). Careful reviews, color plate or other materials were provided by Nicholas Coops, A. Gasiewski, Rob Gutro, Ron Hall, Yuhong He, Doug King, Ellsworth LeDrew, Tom McCaffrey, Greg McDermid, Jon Pasher, Bryan Schreiner, Gordon Stenhouse, Ralph Tiner, Lee Vierling, Michael Wulder, Cecilia Xia, and Rommel Zulueta. A sincere thank you to Sharon Munger-Osborne (University of Saskatchewan); Caroline Rockcliffe (Curtin University); Lauralee Proudfoot (Trent University); and Gita Raman (Glyph International) for their administrative and editorial assistance. Cover photos were provided by Tom McCaffrey (University of Calgary) and Gordon Stenhouse (Foothills Research Institute). The author photo was by Wayne Eardley (Brookside Studios). Thanks are due to Taisuke Soda, Michael Mulcahy, and the very efficient and professional team at McGraw-Hill Professional.

Support during the writing of this book was provided by the University of Saskatchewan and Curtin University of Technology. My research program in the past two decades has been funded by the Natural Sciences and Engineering Research Council of Canada, the Canadian Forest Service, Parks Canada, and Sustainable Resource Development of the Province of Alberta. Partnership support through the Foothills Research Institute, Hinton, Alberta is gratefully acknowledged.

I would like to express my gratitude to my family: Dawn, Meghan, and Heather, my parents Jean and Eric, and my sisters Gillian and Barb for their essential, patient, and understanding support of my academic work.

Remote Sensing for Biodiversity and Wildlife Management

CHAPTER 1
Introduction

And the dramatically successful Earth Resources Satellites, Landsat 1 and 2, may be the precursors of a sort of global inventory, not only surveying all our planet's resources, but also monitoring their misuse through pollution and environmental degradation.

—*Arthur C. Clarke, 1975*

Remote Sensing for Biodiversity and Wildlife Management

A Remote-Sensing Approach Across Disciplines

Remote sensing has developed over the past five decades to form an important part of the foundation for a *transdisciplinary approach* to biodiversity and wildlife management. This new approach is based on the capability to derive multispectral views of environment at multiple spatial and temporal scales, which are readily integrated with other forms of data, including global positioning system (GPS), geographical information system (GIS), and observational data essential to provide a foundation for species-specific resource selection models, a wide range of wildlife habitat suitability maps, testable predictions of population dynamics, and the development and testing of biodiversity indicators and species-environment hypotheses. New ideas in a transdisciplinary remote-sensing approach have emerged for the detection, evaluation, and mapping of wildlife habitat resources, and the identification of factors determining species richness and wildlife health. Such applications have helped generate an enormous expansion of knowledge and insight as environmental scientists and remote-sensing scientists, sometimes independently, but increasingly together in teams, explore solutions to previously intractable environmental science and management problems. High-level skills in remote sensing, and in GPS, GIS, and spatial modeling, are increasingly recognized as valuable, and should be possessed by those involved in environmental science, conservation, monitoring, and assessment.

Transdisciplinarity represents one of several powerful themes that facilitate "bringing multiple disciplines together" (Rosenfeld 1992,

Lattuca 2001, Naveh 2007). Transdisciplinarity seeks to apply concepts, theories, and methods *across disciplines* dealing with urgent scientific and management questions that transcend the disciplinary perspective. A new form of scientific progress is made possible, and new management options are developed, partly because the endeavor is accomplished by muting the sources of theories and disciplinary methods that cannot, alone, provide solutions. Wilson (1999), in a grand vision of the unity of knowledge, has called "the jumping together of knowledge" across disciplines, *consilience*. In a sense, the discipline or field becomes a research setting, and of secondary importance, relative to the desired advancement of knowledge or "the common groundwork of explanation," which is thought possible in no other way and will be judged independently of the discipline (Stokols et al. 2003, Rhoten 2004). There is a growing consensus that transdisciplinary approaches, and other forms of interdisciplinarity spanning the sciences, social sciences, and humanities, present remarkable opportunities for progress (Heberlein 1988, Brewer 1995, Clark et al. 2001), and are at the frontier of science and knowledge (Smith and Carey 2007, Payton and Zoback 2007, Hirsch-Hadden et al. 2008).

Transdisciplinary remote sensing is comprised of elements of conceptual, technological, and methodological innovation, which when integrated appear to provide an excellent opportunity to implement creative solutions to problems that exist at the leading edge of environmental science and management. Such problems are among the most important and complex issues of our time.

Significant disruptions to natural ecosystems are widely expected as a result of global climate change (Saxon et al. 2005, Metzger et al. 2005). Habitat loss, degradation, and fragmentation have reached unprecedented levels (Myers 1997). Pollution of land, water, and air has led to severe and lasting environmental and health impacts, and there is increasing agreement on the evidence that the health of wild species is adversely affected by human activities and landscape change (Daszak et al. 2001, Majumdar et al. 2005, Botkin et al. 2007). Virtually every ecosystem on earth has recent and damaging experience with invasive alien species (Mooney et al. 2005), and solutions have not kept pace with the growing number of management challenges (Underwood and Ustin 2007). Biodiversity is in decline across the planet as a direct result of human activity and inadequate management strategies focused mainly on single species and protected areas. Evidence continues to accumulate that complex social, cultural, and ecological processes operating at multiple scales but particularly, the crucial *landscape scale*—are among the primary factors influencing conservation and environmental management success or failure.

Species extirpation and, ultimately, species extinction, are a final and overwhelming loss that must be avoided. Some observers have concluded the situation is beyond dire; we are already too little and too late (Lovelock 2007). Even if all human activity were to cease today, many species would continue on their human-assisted path to extinction, a consequence of irreversible environmental damage already done and processes previously, though perhaps inadvertently, set in motion. Others read the evidence to suggest we may be improving in our ability to recognize and reverse negative environmental outcomes, and that there is time—though time is of the essence, and incisive and urgent action is at a premium (Wilson 2006). Some point to the emergence of a new understanding of diversity, multiple-species complexity, and the need to preserve wholly functioning ecosystems (Shugart 1998). These ideas appear to be gaining widespread scientific and, increasingly, public acceptance (Waring and Running 1998, Chen and Saunders 2006). Perhaps it is not too much to hope that we can turn around the critical resource management situation of the planet. This book is written with this hope in mind.

Many scientists and resource managers already recognize the importance of adopting an *ecosystem approach* in managing natural resources (Boyce and Haney 1997, Payne and Bryant 1998, Ward 2008), and in parallel with this recognition, the use of remote sensing and GIS approaches have rapidly increased (Miller 1994, Turner et al. 2003). In many situations, only by capitalizing on the enormous *informational strengths* of these powerful technologies can complex environmental management succeed. For example, current research has begun to confirm a link between human-induced habitat changes and long-term physiological stress, leading to damaging health consequences in individual animals (i.e., impaired reproduction, diminished growth, suppressed immune function), and subsequent negative effects at the population level (i.e., low natality and survival, diminished abundance). A transdisciplinary approach to understanding these relationships, based on sensitive and reliable measures of health, stress, and landscape change, is both urgently needed and impossible to conceive without remote sensing.

At the same time, no mainstream remote-sensing research text has dealt with this emerging approach in sufficient detail to help the growing collaboration among those specializing in remote sensing, wildlife management, and ecosystem science, in the critical decision making and efforts on the ground. Although many individual accomplishments and achievements signal the developing consensus, what may be useful, then, is a *synthesis* to highlight this new approach. The term synthesis is used in the broadest sense of the word: that of *selecting crucial information from the copius amounts available; arraying that information in ways that make sense* (Gardner 2006). In a way, synthesizing may be the type of *thinking* that is needed to stimulate new ideas and approaches in the increasingly multidisciplinary, interdisciplinary, and transdisciplinary work of applying remote sensing in conservation biology and wildlife management—and, in effect, provide practical guidance to those interested in environmental questions.

This synthesis will be a companion contribution to earlier presentations focusing on introducing and explaining the remote-sensing approach and documenting a long history of successful remote-sensing applications. A select few of these earlier contributions are summarized briefly in the annotated bibilography of App. I. These books established the *fundamental quantitative remote sensing approach* (Swain and Davis 1978) based on digital imagery and computer processing that has since proven so effective in supporting environmental research, monitoring, and management (Townshend 1981, Howard 1991, Sample 1994, Gholz et al. 1997, Skidmore 2002). The present volume acknowledges and builds on these early efforts by the many developers and practitioners of remote sensing and environmental science; their work has provided the solid foundation from which many of the current initiatives are launched.

Remote Sensing and Wildlife

The earliest technological remote-sensing wildlife applications were photography experiments conducted to count or study individual animal movement and behavior (Altmann and Altmann 2005). Initial success was quickly followed by aerial surveys and simple map overlays to reveal or explain known species range and distribution. Aerial survey and detection of individual animals proved beneficial as inputs to management and for scientific purposes, but with satellite sensors, the initial emphasis was often on creation of a vegetation or land-cover map based principally on *spectral response patterns* (Robinove 1981), or an augmented spectral response pattern which incorporated terrain variables (Hutchinson 1982). Such maps were increasingly developed by integrating remote sensing and GIS data (Estes 1985).

Few successful satellite remote-sensing studies of individual animal detection or species distribution mapping were reported initially because of the relatively coarse spatial resolution and long revisit times of the available imagery (De Leeuw et al. 2002). The earliest civilian satellites were of limited use in many wildlife applications where the goal was to detect individuals, but were more applicable in certain situations, such as studies of population-level changes (e. g., habitat loss). In the past decade, though, purpose-designed aerial applications, and new satellite sensor technology, have overcome many of the early challenges with new forms of imagery. The key characteristics of imagery continue to improve; for example, higher image quality (decreased signal to noise ratio), and ever increasing spatial, spectral, radiometric, and temporal resolutions (described in Chap. 3), with significant implications in wildlife studies (described in Chaps. 4 and 5 of this book).

The early successful satellite remote-sensing applications in wildlife management often focused on well-understood species-environment relationships, particularly where a species of interest engaged in environmental modifications that were highly visible and quantifiable on available imagery. Mapping forest defoliation by insects using satellite sensor data is one of the earliest and more obvious success stories (Hall et al. 2007). The essential approach in studying a particular species-environment relationship of interest is to develop a remote-sensing vegetation map, water resources map, or land-cover map, and rely on expert opinion to interpret or infer the impact of conditions and change over time on habitat, food resources, or shelter/connectivity, or individual or population behavior or characteristics, that could be used in specific management situations. This idea is based on the general concept that an organism's characteristics and behaviors at both the individual and population levels are inextricably linked to the physical habitat in which it occurs (Boyce and McDonald 1999).

This approach has now evolved into a rich array of applications and enhancements to help address wildlife management questions about food quantity and quality, and habitat associated with specific behaviors including reproduction, nesting, security, and shelter (McDermid et al. 2009). Those engaged in wildlife management typically express great interest not only in species distribution and behavior, but in many aspects of wildlife habitat condition. However, *habitat* is a concept not without some challenges, including ongoing ambiguity and even confusion over what is meant by habitat and habitat *niche* (Morrison et al. 2006), discussed in more detail in later sections and in Chap. 2. Vegetation structure and land-cover maps produced by remote sensing are not habitat maps (Corsi et al. 2000, McDermid et al. 2005). Generally, remote sensing is an excellent way to produce an image or map product that may be used to help create a wildlife habitat map—or perhaps, a wildlife suitability map. The suitability of habitat for a given species is typically judged according to a recognized source of expertise.

Remote sensing-derived habitat or wildlife suitability maps provide a base for *wildlife suitability models* designed to predict suitability of a given set of land attributes, again for a particular species of interest. The habitat suitability index (HSI) (United States Fish and Wildlife Service 1981), for example, is a hypothesis about species-environment relationships. A common approach has been to build an empirical HSI model to work with field observations associated with locations on the remote-sensing land-cover map, or other remotely sensed input layer, and create a generalized set of statements about the suitability of the landscape, and potential distribution and abundance, for a species of interest. Human activity factors are often thought to be a strong influence on habitat suitability for many species, and certain human activity factors can be incorporated into HSI maps and both species and habitat distribution models with current remote-sensing technology.

While many habitat suitability and species distribution prediction models are based on general land-cover maps, such models can be expanded spatially or otherwise strengthened with access to good quality remote-sensing input variables such as detailed land cover, specific resources of interest (e.g., water), fragmentation and connectivity, or vegetation structure and productivity estimates (Jacquin et al. 2005). The temporal or dynamic nature of both the resource base and the wildlife can be addressed using remote sensing, although the transferability of any developed models to new areas and multiple species remains a significant challenge and an important research topic (Hollander et al. 1994, De Leeuw et al. 2002, Huettmann and Cushman 2009).

Remote Sensing and Biodiversity

Biodiversity management and monitoring represents a relatively new challenge, but again significant progress has occurred in applying remote-sensing solutions (Muchoney 2008). An important approach has been to identify *biodiversity indicators*, which are measures that allow the determination of the degree of biological or environmental changes within ecosystems, populations, or groups (Strand et al. 2007, Leyequien et al. 2007). A remote-sensing approach can lead to new ways of biodiversity monitoring through an understanding of linkages to spatial heterogeneity and net primary productivity (Fuller et al. 1998, Turner et al. 2003, Balmford et al. 2006, Sarkar et al. 2006, Coops et al. 2008c).

New ideas are constantly emerging given the rapid development and complexity of remote-sensing technology, methods, and interrelated developments in other geospatial analysis areas (e.g., GPS). Chapter 3 of this book contains a review of these developments, and the implications in biodiversity studies are considered in Chaps. 4 and 6. For example, early in the development of remote sensing, significant contributions were anticipated in mapping *species richness*, one reasonably well-understood indicator of biodiversity (Scott et al. 1987, Stoms and Estes 1993, Wang et al. 1998). The research community responded with significant initiatives to develop new methods of mapping both animal and plant species richness, examination and testing of hypotheses concerning the biophysical factors believed to be affecting species richness, and the anthropogenic processes causing the accelerating rate of species extinctions. Of particular interest was the relationship between remotely sensed productivity and species richness, and explanations of spatial heterogeneity, landscape structure and climate influences, which can now feasibly be addressed with a remote-sensing approach.

Recent developments have helped to place the appropriate geospatial and remote-sensing approaches in the right hands (Nohr and Jorgensen 1997, Nagendra and Gadgil 1999a) to assist in the increasingly urgent task of biodiversity monitoring at all relevant spatial and temporal scales (Nagendra 2001). This work has only just begun to inform larger initiatives, such as those designed to monitor national and international biodiversity status, described in Chap. 6.

As in all ventures across disciplines, it may be beneficial to first pay some attention to the terminology and basic concepts that have developed. The following sections provide some of the context which may be useful in developing a common understanding of the scope of the environmental issues to be addressed, and the role of remote sensing in possible solutions. Further elaboration of these initial concepts and methods are provided in later chapters.

Definitions and Context

Remote-Sensing Science

Remote sensing is the "detection, recognition, or evaluation of objects by means of distant sensing or recording devices" (Avery 1968). Remote sensing is both technology (sensors, platforms, transmission and storage devices, and so on) and methodology (radiometry, geometry, image analysis, data fusion, and so on). There are numerous standard textbooks available to provide an introduction to the rapidly maturing science of remote sensing (Lillesand et al. 2004, Aronoff 2005, Jensen 2007). It is now five decades since the term *remote sensing* was applied by Evelyn Pruitt of the Office of Naval Research, and a new field of science was born (Risley 1967, Weaver 1969). Pruitt later wrote about the excitement and satisfaction of launching the new remote-sensing initiative in those early days (1979: p. 106):

> ...increasing interest was being focused on the nonconventional photography (multispectral, infrared, and ultraviolet), and even on nonphotographic sensors. The whole field was in flux...it was finally decided to take the problem to the Advisory Committee...Walter H Bailey and I pondered a long time on how to present the situation and on what to call the broader field that we felt should be encompassed in a program to replace the aerial-photointerpretation project. The term "photograph" was too limited because it did not cover the regions in the electromagnetic spectrum beyond the "visible," and it was in these nonvisible frequencies that the future of interpretation seemed to lie. "Aerial" was also too limited in view of the potential for seeing the earth from space. A new term was needed, so "remote sensing" was invented, and I am generally credited as its coiner.

That remote sensing is a defined field of science with significant questions and applications based on theory was quickly established (Gregory 1994, Kondratyev et al. 1974, Bowden and Pruitt 1975, Colwell 1983, Curran 1987), and is not seriously debated or challenged today (Ryerson 1998, Campbell 2002, Tatum et al. 2008).

Remote-sensing data collection from field, aerial, and satellite sensor platforms will accommodate analog (or visual-interpretation) and digital uses of imagery. The majority of remote-sensing applications are now computer-based (Aronoff 1989, Buiten and Clevers 1993, D'Erchia 1997, Aronoff 2005). By far the most widely used remote sensors operate in the relatively familiar optical portion of the electromagnetic spectrum (visible and near-infrared light), but RADAR (RAdio Detection and Ranging), LiDAR (Light Detection and Ranging), and hyperspectral instruments represent a growing presence and will be increasingly deployed in many kinds of environmental applications. Remote sensing has matured, in the eyes of many practitioners and developers, to become an interdisciplinary endeavor moving beyond mainly empirical and descriptive studies to more comprehensive analyses that extend over multiple spatiotemporal scales.

Wildlife Management

A current and widely accepted definition of wildlife management is the scientific process of identifying appropriate management principles and practices, developing and implementing management plans, and evaluating success of treatments or actions under the plan (Giles 1978, Ward 2008). There is a long-standing tradition that wildlife management can be considered in two integrated components, which cannot easily be separated—species management (game or endangered species, for example), and the management of habitats of targeted species. Many trace the origin of wildlife management in the United States to the influential contributions of Aldo Leopold (1933), and suggest that the trajectory of wildlife management is now clearly established as an integral part of the larger paradigm of ecosystem management (Boyce and Haney 1997).

Ecosystem functioning is the dominant value to be managed and balanced between ecological and social dimensions in an ecosystem management approach. Consequently, a significant contemporary influence on wildlife management practices in many jurisdictions is the multitude of regulatory and statutory requirements that provide a decision making context in which human activities occur or are prevented.

Biodiversity

Biodiversity is defined as "all hereditarily based variation at all levels of organization, from the genes within a single local population or species, to the species composing all or part of a local community, and finally to the communities themselves that compose the living parts of the multifarious ecosystems of the world" (Wilson 1996). This definition has posed an interesting challenge in scientific work because it appears to mean that biodiversity is everything—the whole expression of life on earth, ranging from genes to life zones sensu (Holdridge 1967). And, in fact, this very idea is encountered not infrequently in the literature. For example: "Biodiversity, as such, is impossible to define precisely; it refers to diversity at every level of the taxonomic, structural, and functional organization of life" (Sarkar et al. 2006).

Defining biodiversity has been a major task, but what many have tried to do is place biodiversity among the primary *values* with which society regards the environment (Lugo 1998, Faith 2007). Such values may originate in any number of theories and philosophies, including a clear and compelling association of biodiversity with aesthetics and environmental services (Sarkar 2005). Instead of focusing on how many species exist in the world and their distribution, or accumulating more data on some megafauna or management issue with one or two species, a *change in biodiversity* and the *rates of decline in biodiversity* are of primary concern (Janzen 1997). Biodiversity indicators represent an attempt to develop simple and multidimensional measures of biodiversity that can be quantified and, therefore, monitored over time (Magurran 2004, Strand et al. 2007).

Habitat

By far the most significant problem in protecting the world's biodiversity is habitat destruction (Lovejoy 1997). But what is habitat? Every wildlife species has its own habitat needs, and every wildlife species occurs in particular places; habitat is especially (perhaps only) relevant when attached to a species. Habitat is usually expressed as conditions necessary to support a particular wildlife species or guild, but it can be used to mean both *where* a population lives, and *the type of environment* where a species lives:

> ...habitat is an area with a combination of resources (like food, cover, water) and environmental conditions (temperature, precipitation, presence or absence of predators and competitors) that promotes occupancy by individuals of a given species (or population) and allows these individuals to survive and reproduce.
>
> *(Morrison et al. 2006: p. 10)*

Most definitions of habitat encompass the set of resources necessary to support a population through space and time (Corsi et al. 2000, McComb 2008). But some prefer to avoid the word habitat all together, instead using wildlife in its place (thus, not *habitat suitability map*, but *wildlife suitability map*) (De Leeuw et al. 2002). The concept of habitat is sometimes discussed in terms of *macrohabitat* (e.g., broad vegetation cover types) and *microhabitat* (based on the perceptual world of the organism, and related to behavior and evolutionary determinants of survival and fitness). For effective mapping and remote sensing, it is necessary to know the habitat resource needs for the species of interest, and to carefully develop an approach to match those resources to the information available for the area under consideration at the appropriate scale. Since wildlife encompasses many species, including humans, a study of habitat will always need to consider habitat function, including ecosystem energy (Phillips et al. 2008).

Environmental conditions include a host of other factors (such as hunting pressure, weather, interspecies dynamics) (Morrison et al. 2006), and suggest a multidimensional understanding of the *niche* concept in understanding habitat and the distribution of organisms. One interpretation of niche focuses on the attributes of the species, and on how these attributes determine geographical distribution; the *community-centric niche* concept, on the other hand, identifies function and structure (competition) performed by individual species (Shugart 1998). Kolaso and Waltho (1998) helpfully distinguish between species attributes (such as movement and dispersal, habitat specialization, demographics, genetics) and habitat attributes (such as quality, size, connectivity, availability of food). This approach provides credible value to managers and scientists when habitat maps are developed as attributes of a specific species (e.g., Reading et al. 1996).

Landscape Level or Landscape Scale

Scale is a complex concept with significant potential for definitional and interdisciplinary confusion (Peterson and Parker 1998, Linke et al. 2007). Scale refers to measurement characteristics—generally, scale means the "size" of something; thus, a large-scale phenomena is simply a large phenomenon (Atkinson et al. 2005). In ecology, a large-scale ecological study will suggest a large area is covered, and this usually implies a low level of detail. A small-scale ecological study will suggest a small area is covered with a high level of detail. Terms such as *extent* and *grain* are used to express ecological scale, and the general term is interpreted to mean both an area of coverage and a particular level of detail (or an ecological hierarchical level) (Wiens 1989, Shugart 1998). The terms resolution and grain are often used interchangeably in the ecological literature.

Landscape scale is then used as a kind of shorthand to refer to a specific and intermediate position in an ecological hierarchy (a level), and at the same time to refer to a general notion of reasonably large spatial extent (Fig. 1.1; Shugart 1998, Bolliger et al. 2007).

Typically, landscape scale is used to signify a regional study; one in which the scale of the analysis, and the scale at which the processes

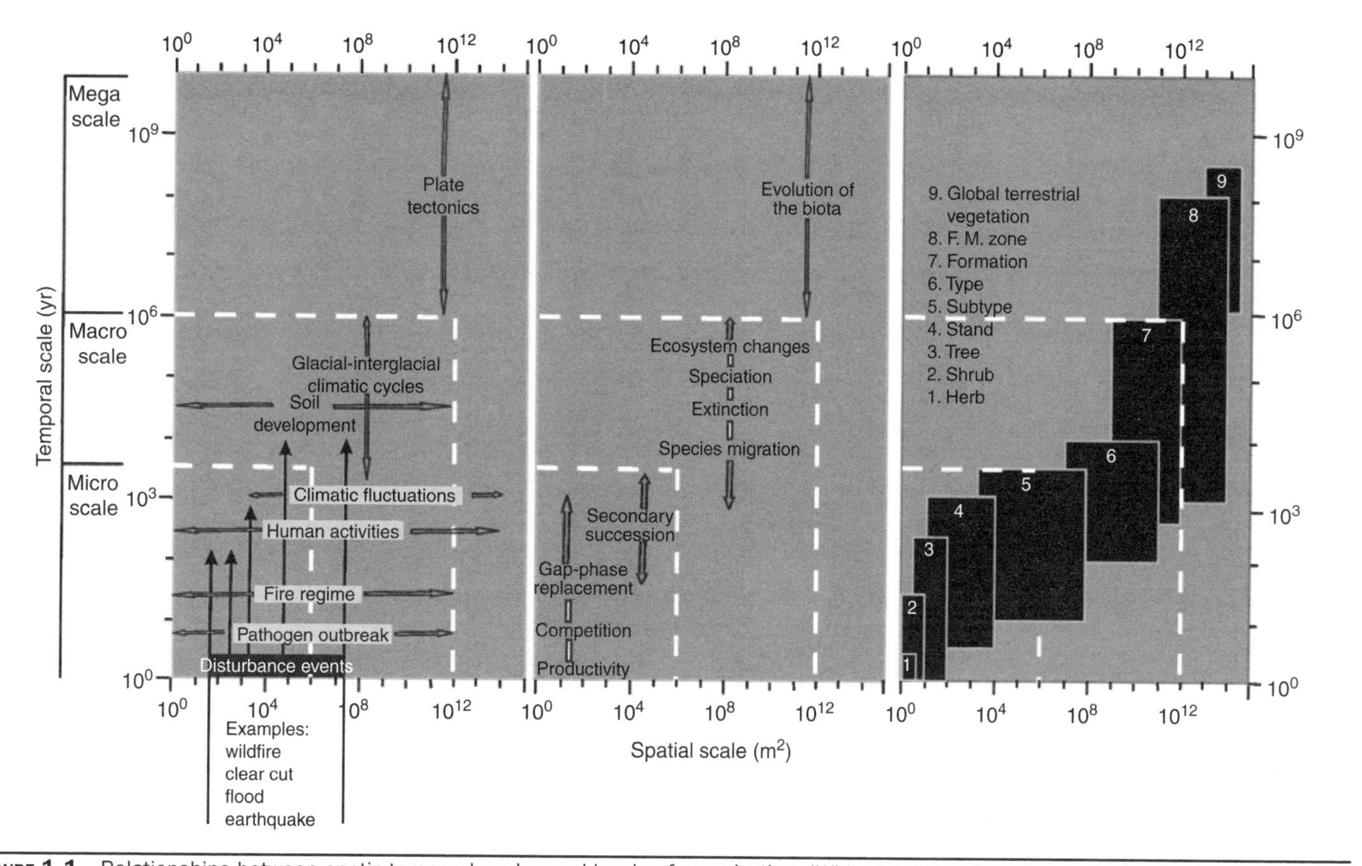

FIGURE 1.1 Relationships between spatio-temporal scales and levels of organization. (*With permission from* Turner, M.G., R.H. Gardner, and R.V. O'Neill. 2001. *Landscape ecology in theory and practice: pattern and process*. Springer-Verlag, New York.)

of interest operate, can be expressed at a position between large spatial *map scales* (e.g., 1:1000) and small spatial map scales (e.g., 1:1 000 000) studies. This notion of landscape-scale or -level can then usually be interpreted to occur over a particular region or geographic area of study at a particular mapping scale (Zheng et al. 2006). However, without qualification or context, the use of the terms landscape-scale and -level are meaningless generalities (Allen 1998).

An intermingling of the meaning of ecological scale and cartographic or map scale may be an unfamiliar characterization and confusing treatment of scale for some. But much of the success in using remote sensing in wildlife management and biodiversity applications has come at specific large-, medium-, or small-mapping scales, which may roughly (though inversely) coincide with an ecological interpretation of the levels and processes operating within a landscape hierarchy. With appropriate context, the idea of multiple scales of ecosystem structure, process, and function has been used with a knowledge of mapping scale to develop essential insights to be expanded and built-upon in terms of resource management, and in research and monitoring applications (Silbernagel et al. 2006).

Interface of Management and Research

There is often a creative tension and occasional conflict that can develop at the interface of management and research (Hobbs 1998). Environmental management typically requires an intense commitment to key information gathering endeavors including mapping, monitoring, and modeling (Phinn et al. 2006), prior to grappling with solutions (Green 1999). *Applied remote-sensing research*, on the one hand, may be designed to achieve a more complete response to management questions or objectives. Applied remote-sensing research may have little or nothing to contribute to theory, or even to continued improvements in the general understanding of environment, but may be essential in helping management arrive at a decision-point in the face of ongoing uncertainty, lack of public policy, and conflicting evidence. *Pure remote-sensing research*, on the other hand, may be more focused on activities that seem likely to lead to the acquisition of fundamental knowledge through the development and testing of theory. Remote-sensing scientists engaged in these pure research activities are acutely aware of the limits of their knowledge—of the error, uncertainty, and lack of precision that exist in any study of complex environmental phenomena. Such awareness can create a reluctance to commit to the type of concrete statements or choices favored in management decision making. Pure remote-sensing research will typically have little immediate application to management problems, but can lead to insights and knowledge that provide the foundation for all management.

The early promise of remote sensing was that this approach would lead to significant contributions to both pure and applied research questions in environmental management. Such promise has been overwhelmingly confirmed, as the material presented in this book, and elsewhere, can attest. And it is in this tradition—remote sensing is both a pure and an applied science—that this book considers remote-sensing contributions in biodiversity and wildlife management. Something to be kept in mind, however, is the idea that environmental management problems for which remote-sensing contributions are sought are actually *human management problems* that arise from human values, goals, plans, and activities (Clark et al. 2000). Remote sensing, then, even with an ambitious and successful pure and applied scientific agenda—*to deliver the information required to improve management*—may make only a modest contribution to the more fundamental management issues of human decision making and environmental sustainability.

Early Developments

Field-Based Remote Sensing

Remote sensing has been used in field-based animal studies at least since 1872 when E. J. Muybridge settled a bet, for railroad tycoon and then-governor of California Leland Stanford, that a running horse had all four feet in the air at some point (Altmann and Altmann 2005). Muybridge (1887) had set up one of the first trip-camera studies of locomotion, and apparently, the idea inspired the development of the *zoopraxiscope* for viewing animal photographs acquired of specimens in the Philadelphia Zoo (Fig. 1.2).

These animal locomotion photographs were published widely to great acclaim in Scientific American and other respected journals.

In the late 1890s and early 1900s, pioneer wildlife photographer George Shiras began experimenting with cameras and trip wires to capture nighttime images of deer in their natural environment (Sunquist 1997, Sanderson and Trolle 2005, Bray 2006). Using a glass-plate style camera and a tray of magnesium powder for the flash

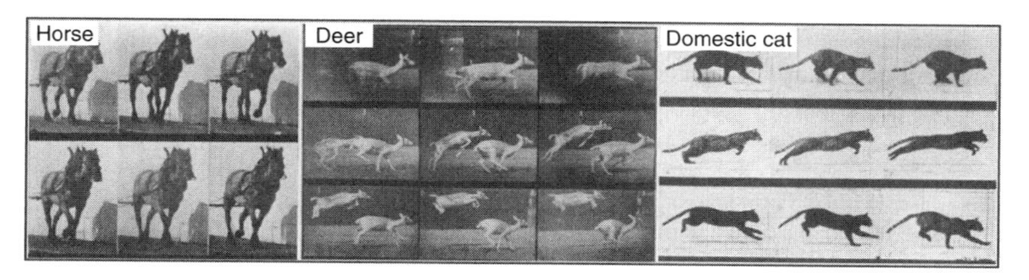

Figure 1.2 Three early examples of animal locomotion photography. (*With permission from* Muybridge, E. 1887. *Animal locomotion: an Electro-photographic Investigation of Consecutive Phases of Animal Movement, 1872–1884.* University of Pennsylvania Press, Philadelphia.)

FIGURE 1.3 Early auto-trigger camera photography of wild deer. (*With permission from* Shiras, G. 1906. *Hunting Wild Game with Camera and Flashlight.* National Geographic Society. Washington.)

(which exploded in a blinding, brilliant white light that startled the animals and sometimes set the surrounding forest on fire), the photos were unprecedented in their beauty and scientific value (Fig. 1.3).

The photos were originally displayed at the Paris Exposition in 1900, featured in what remains one of the most popular issues of National Geographic Magazine (July 1906), and then published in a two-volume book (Shiras 1906). Independently, and approximately over the same time period using similar technology, Indian forest officer F. W. Champion produced some astonishing pictures of wild tigers that captured the imagination of the public (Fig. 1.4, Champion 1928).

In the 1940s, British natural history photographers Hosking and Newberry (1944) began to photograph birds in flight, and exceptional

FIGURE 1.4 Early auto-trigger camera photography of wild tigers. (*With permission from* Champion, F. W. 1928. *With a Camera in Tiger-Land.* Doubleday Doran, New York.)

improvements in electric flash, shutter speed, film technology, and auto-trigger systems continued in subsequent years as the new approach became more widely adopted (Gysel and Davis 1956). A dual-camera 16 mm movie unit was used in early work on mouse runways in California grasslands by Pearson (1960). Among many interesting findings, Pearson recorded 26 different species using runways apparently "maintained" by one type of mouse. Some of the photographs showed species that were completely unexpected (e.g., snakes, weasels, insects not previously observed in the area). Since this pioneering work, camera traps have been used frequently in biodiversity surveys and in wildlife management applications to, among other things, document species presence, understand current range and distribution, and isolate individual behavior of animals (Lee et al. 2003, Stein et al. 2008). Periodic evaluations and new applications of ever-improving auto-trigger camera technology (Jones et al. 1993, Yashuda and Kawakami 2002) have highlighted their considerable value in tropical forest environments (Griffiths and van Schalk 1993), mountainous terrain (Pei 1998), and other challenging environments (McCullough et al. 2000, Liebezeit and Zack 2008).

In recent decades, optical and opto-electronic devices, videography, thermal imaging, radio frequency identification (RFID), radiotelemetry, X-ray radiography and computed tomography, sodar, sonar, Doppler radar, and ground-based VHF-radio tracking have been used to study bird, flying mammal, or insect movement and migration (Eastwood 1967, Kenward 1987, Riley 1989, Reynolds and Riley 2002). These technologies have been gradually improved and augmented with satellite-tracking systems and the advent of the global positioning system (GPS), which is now essential in many behavioral and resource analysis studies of larger animals (Gillespie 2001, Reynolds and Riley 2002). The advantages of VHF-radio tracking and GPS satellite techniques are considerable (Mech and Barber 2002) particularly when compared to sole reliance on traditional field observation and intensive mark-recapture activities. Typical trade-offs in accuracy and data reliability depend on the species and environment of interest. Since the mid-1990s satellite-tracking GPS data, and other location-sensing data, increasingly are being stored and analyzed in a GIS with other remote-sensing information (Kaplan 2005).

Aerial Remote Sensing

The tremendous potential of remote sensing was recognized early for mapping individual animals and animal distribution from aerial platforms (Hay 1958, McLaren 1966, Graham and Read 1986). Balloons, kites, and even purpose-built launch towers and floating booms were all deployed to raise observers, cameras, and other sensing devices to various heights to record particular species' behavior or habitats of interest. The goals of early fixed-wing airplane and helicopter aerial camera surveys resonate today (Burns and Harbo 1972: p. 280):

> An aerial census of ringed seals in areas of land-fast ice along the north coast of Alaska was undertaken during June 1970 in order to: (1) develop and document a method for census of ringed seals along the northern coast of Alaska, the procedures for which could be duplicated in future years; (2) establish a base line index of ringed seal abundance during late spring, prior to the seasonal influx of migrating seals; (3) determine comparative density and distribution of ringed seals in different areas along the northern coast; (4) determine the effects of human activity in areas of land-fast ice occupied by ringed seals.

Many examples of successful aerial image census of wildlife were included in the Manual of Color Aerial Photography originally published in 1968 by the American Society for Photogrammetry. A comprehensive listing of such work has not been compiled recently, but some representative operational examples of animal census work using a variety of aerial sensors includes surveys of whales (Cliff et al. 2006), waterfowl (Bajzak and Piatt 1990), ungulates (Otten et al. 1993), saltwater crocodiles (*Crocodylus porosus*) (Harvey and Hill 2003), walrus (*Obenidae*) (Barber et al. 1991), and beaver (*Castoridae*) (e.g., detection of animals and their works on aerial photographs, and detection of individual animals in water or inside lodges using thermal sensors) (Broschart et al. 1989, Hood and Bayley 2008).

Satellite Remote Sensing

The first images of Earth from space were photographs acquired in 1946 over White Sands, New Mexico, using a camera attached to a captured V-2 rocket. Satellite remote sensing of the atmosphere and meteorological conditions from space began in 1959 with the launch of the Explorer VI satellite (Hall 1992).

In the subsequent decades, enormous improvements in satellite platform capability and image quality have occurred; for example, recently, high spatial detail satellite imagery have been used to support individual animal surveys. In one western United States study, Laliberte and Ripple (2003) counted cattle on Ikonos satellite imagery with satisfactory results, suggesting that the method would work well with elk (*Cervus elaphus*) and American bison (*Bison bison*). In another study, in India, Sasamal et al. (2008) identified flamingos (*Phoenicopteridae*) on QuickBird imagery, and identified potential improvements in bird census confidence and reliability compared to traditional individual bird counting field methods. S. Bergen et al. (2008) experimented with target fur placement and animal counting at the Bronx Zoo during satellite sensor-data acquisition. Successful examples were presented of high-contrast animal detection with African elephant (*Loxodonta africana*), giraffe (*Giraffa camelopardalis*), elephant seal (*Mirounga*), bison,

Arctic musk ox (*Ovibos moschatus*), black wildebeest (*Connochaetes gnou*), and caribou (*Rangifer tarandus*) using QuickBird and other high spatial detail satellite sensors (Color Plate 1.1).

Early satellite mapping of habitat modification caused by animals was widely attempted. Terrestrial ecosystem habitat modification studies using relatively coarse spatial resolution Landsat imagery have included hairy-nosed wombats (*Lasiorhinus latifrons*) in Australia (Loeffler and Margules 1980), lesser snow geese (*Anser caerulescens*) in the Canadian subarctic (Alisauskas et al. 2006), and Rocky mountain ecoregion elk habitat (Huber and Casler 1990). Other early studies focused on mapping known and distinctive habitat requirements associated with endangered species or species of interest, including wetland habitat of wood stork (*Mycteria americana*) (Coulter et al. 1987), old-growth forest habitat of spotted owl (*Strix occidentalis*) (Ripple et al. 1991), and grassland habitats of black-tailed prairie dogs (*Cynomys ludovicianus*) (Gustafson 2002) and lesser prairie chicken (*Tympanuchus pallidicinctus*) (Cannon et al. 1982).

Growing Confidence in Remote-Sensing applications

There have been increasingly confident statements in the literature that suggest a strong role for remote sensing in applications of wildlife and biodiversity management (Table 1.1).

The initial animal locomotion studies, then aerial census with cameras, and now an impressive array of aerial and satellite sensors from which to choose, show that some traditional information needs of environmental managers and scientists can be addressed reliably and innovatively using remote-sensing technology. Current information needs are increasingly complex (Chap. 2), and may include (Kushwaha 2002, Asner et al. 2008; *see* Fig. 1.5):

1. Inventorying patterns of species richness and other biodiversity indicators
2. Helping determine the factors controlling species presence and abundance, particularly net primary productivity, landscape physiology, chemistry, health, and structure
3. Determining anthropogenic effects (e.g., fragmentation) and natural disturbances
4. Mapping wildlife resource requirements, habitat suitability, quality and spatial heterogeneity, and multiscale structural habitat properties
5. Linking remotely sensed data with ground-plot information to provide spatial context and to complement other data types
6. Land cover mapping, land-use planning, and functional terrain mapping

Value	Observation	Reference
Unique temporal and spatial perspective on processes and conditions	"Use of satellite imagery from the habitat of the giant panda proved to be of vital importance for understanding the crux of the real problems affecting the giant panda... two time series of satellite images showed the rapid reduction and rapid fragmentation of the giant panda's habitat in China. More than any other factor, it was this perspective provided by satellite imagery that changed the manager's views about the main threats to panda survival."	Mackinnon and de Wulf 1994: p. 128–130
Synoptic landscape view 'from above'	"Images from satellites have revolutionized our perception and approaches to understanding landscapes and regions."	Forman 1995: p. 35
Integration with other data and technology	"Perhaps the most significant recent advance in our ability to quantitatively evaluate primate habitat is the combination of remote sensing technology and geographic information system analysis."	Dietz 1997: p. 345
Powerful inference and habitat interpretation capability	"... for coverage of animal species of concern, large-scale reserve design schemes are usually based on inferences about presence or absence from habitat characteristics that can be determined by remote sensing."	Case and Fisher 2001: p. 67
Powerful model-building capability	"New imagery and data sets are now enabling remote sensing, in conjunction with ecological models, to shed more light on some of the fundamental questions regarding biodiversity. These tools should prove useful to those seeking to generate basic knowledge about why organisms are found where they are, as well as those asking the more applied question of where to invest conservation funds."	Turner et al. 2003: p. 308

Table 1.1 Selected Characteristics of Remote Sensing That Add Value in Wildlife and Biodiversity Applications

Value	Observation	Reference
Multiple scales of analysis	"Remote sensing tools will continue to increase in accuracy, decrease in cost, and become more accessible to the conservation community... we believe that there is an opportunity to utilize these systems to measure the status of marine habitats on a national, regional and global scale of effort."	Crosse 2005: p. 848
Long term monitoring of environmental correlates and fundamental processes	"As ever more detailed and sophisticated remotely sensed data and analytical methods become available for assessing spatial correlates of disease risk and incidence, it will become increasingly important to root explorations in biologically meaningful hypotheses... long-term and spatially distributed data sets that provide wide coverage of variable biotic and abiotic conditions...at multiple spatial scales."	Killilea et al. 2008: p. 5

Table 1.1 (*Continued*)

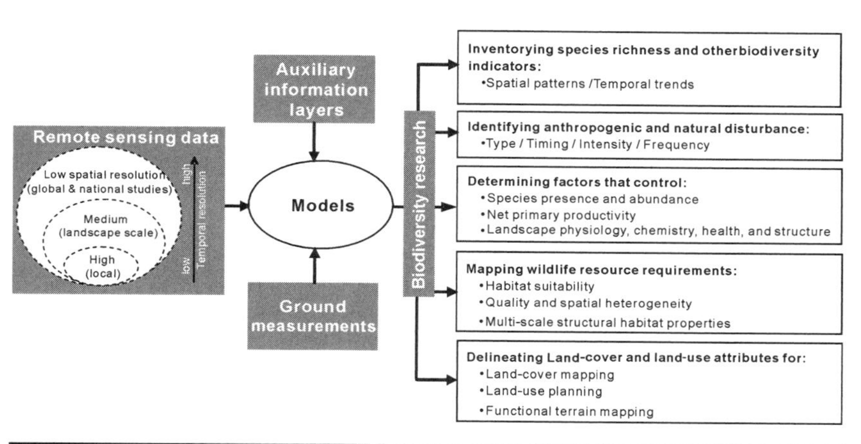

Figure 1.5 Conceptual schematic of major areas for remote-sensing contributions to biodiversity assessment.

Early work typically addressed species distributions in studies of individual plant or animal species of economic or conservation importance, and rarely considered biodiversity as a whole (Stoms and Estes 1993). But almost as soon as the term biodiversity gained widespread acceptance (originally *BioDiversity*—Wilson and Peters 1988), remote sensing was considered a prime candidate to provide critical information of great value to biodiversity inventory, conservation, and management objectives (Scott et al. 1989, Iverson et al. 1989, Stoms 1991, Miller 1994, Frohn 1998, Gould 2000).

Current Status

Critical Remote-Sensing Contributions

The situation today is that remote sensing is a critical component in biodiversity and wildlife management and will play an even more significant role in future (Nagendra 2001, Turner et al. 2003). Achieving the goal of effective remote-sensing contributions will require considerable coordination of effort and development of common approaches (Muchoney 2008, McDermid et al. 2009). Significant, integrated, and transdisciplinary effort is required to:

1. Support continued remote-sensing research to understand spectral response patterns and effective data processing methods (Chaps. 3 and 4 of this book)
2. Identify habitat preferences and applications, and test biodiversity hypotheses (Chaps. 5 and 6 of this book).

One review of three decades of remote sensing in avian-habitat relationship research and management work concluded that satellite-based remote sensing had greatly increased the ability to assess *causal effects* in species-environment relationships (Gottschalk et al. 2007), but that more must be done to build on the capacity of remote sensing to support fundamental ecological science, including inventory of species distributions (e.g., Avery and Haines-Young 1990, Fuller et al. 1998, Levin et al. 2007, Coops et al. 2008a,c). Measuring, mapping, and modeling biophysical factors, such as forest structure, disturbance, and physiology, which are related to changing biodiversity and habitat, represent significant areas of remote-sensing contributions, particularly over the past decade (Wulder and Franklin 2007). Continued rapid progress is widely anticipated in such applications in a wide variety of ecological settings and over the short- and long-term (McDermid et al. 2005).

Biodiversity Indicators

The use of indicators has emerged as an essential approach in environmental monitoring and management. The broad goal of development of habitat and biodiversity indicators is to create purpose-specific, hierarchical, flexible, relevant, and consistent measures that allow comparisons over time. Typically, monitoring systems require a baseline or reference point, something that may be difficult to determine post-hoc. One approach is based on indicator design to achieve identified standards of performance in comparative studies (e.g., Frohn 1998, Saura et al. 2008). Considerable national and international effort has occurred to create a comprehensive and common suite of indicators based on existing databases and supplemented with remote sensing (Strand et al. 2007).

The overall role that remote sensing can play in developing and monitoring biodiversity indicators has been related to various strategic components of the *Convention on Biological Diversity (CBD) COP8 (Decision VIII/15)* and are reviewed in detail in Chap. 2 of this book. However, one general trend is that while environmental management situations at many scales increasingly require objectives related to biodiversity, in many such situations the field or archived information available does not support the subsequent decision making at the appropriate level (Clark et al. 2000, Pretziger et al. 2001). Remote sensing is then properly viewed as one source of the required information, together with field, GPS and GIS data, that may allow much more specific solutions to be developed.

The development of biodiversity indicators and a system to monitor biodiversity requires multiple perspectives and represent inherently transdisciplinary problems. Transdisciplinarity as a concept and a suite of methods—or *a mode of knowledge production* (Lattuca 2001)—can be considered a broad and developing way to consider complex scientific problems, which depend on, for example, understanding fundamental underlying relationships, patterns and processes. Transdisciplinarity may be comprised of specific interdisciplinary studies and diverse approaches that may appear at first glance to lack a coherent philosophical structure. However, for the purposes of remote sensing in wildlife and biodiversity management, the overriding goal would be to organize practical activities that will lead to better understanding of how to identify and successfully secure the benefits that remote-sensing methods can bring to the research and conservation of biodiversity and wildlife.

This approach is described more fully in Chap. 2 and informs much of the discussion in later chapters of this book. Significant transdisciplinary effort has been made to document success in applying remote-sensing technology to the analysis and modeling of biodiversity indicators and wildlife-habitat relationships (Cohen and Goward 2004, Leimgruber et al. 2005). These are reviewed in this book to highlight the common themes and areas of understanding in specific areas, for example, the use of remote sensing to estimate primary productivity in support of ecosystem services (such as carbon sequestration) (Naidoo et al. 2008), biodiversity and species richness monitoring and mapping. Such efforts are critical new developments that may lead to more successful implementation of biodiversity indicator development and monitoring systems (Turner et al. 2003, Leyequien et al. 2007).

Linking to Animal-Based Data Sets

Critical remote-sensing wildlife applications emphasize the growing capacity to link animal-based data sets, such as observations related to a species' distribution, abundance, health, or genetics, and a variety of spatially explicit environmental variables obtained by remote sensing or through GIS database development efforts (McDermid et al. 2009). While the investigation of these links must be well-grounded by solid field observations, the multiple scales and extent over which information is compiled suggests a key role for remote-sensing instruments and related technologies. Initially, habitat suitability maps and models of species distribution are significantly improved with remote-sensing inputs (such as general land cover or other macrohabitat attribute); later development of more complex models (such as HSI or resource selection functions) is feasible with a strong link between increasingly more detailed (in both spatial characteristics and information content) remote sensing and other observational data (e.g., data collected using GPS collars, and detailed modeling of microhabitat attributes).

Technological innovation occurs very rapidly, and at an increasing rate in the past few years. As a data source and analysis tool in animal ecology, radiocollar- and GPS-tracking and remote sensing are still relatively new and have not yet attained the status of a user-friendly application. Even the basic land cover and change-detection analyses, which are two instances of remote-sensing products that have become widely accepted in many environmental user communities based on decades of experience and success, are, unfortunately, not yet routine. Such work must be made more readily available to a wider group of practitioners and scientists to secure even greater success and benefits to environmental management. This theme—*the operational use of remote sensing*—represents a significant challenge and will recur throughout this book.

Assessing the benefits of the decision to adopt a remote-sensing approach and the resulting investment should take into account, among other things, the accuracy of the information products generated, the value of the resulting information (e.g., spatially explicit landscape scale habitat maps for a species of interest or for multiple species known to occur in a region), and the utility of the developed database for other resource management applications. Typically, a remote-sensing approach will recommend itself when it is clear that no other way exists to resolve the information management and environmental decision making task at hand.

Purpose and Organization of the Book

A rich and varied literature has developed to form an excellent foundation for further work in remote sensing for biodiversity (Miller 1994, Turner et al. 2003, Strand et al. 2007, Muchoney 2008) and wildlife management (De Leeuw et al. 2002, Leyequien et al. 2007, McDermid et al. 2009). What is attempted here is a transdisciplinary *synthesis* that will be of interest to those addressing some of the following questions:

- What are the key background concepts and definitions in scientific remote sensing, wildlife management, habitat studies, and biodiversity assessment and monitoring?
- What are the important geospatial tools available to biodiversity and wildlife practitioners? How do remote sensing, GPS, geostatistics, modeling approaches, and GIS components complement field observations and other data, and fit together to comprise a new approach to environmental issues? What are the science and management implications of the integration of remote sensing and related geospatial technologies (GPS, modeling, GIS, geostatistics)?
- What are the important characteristics of remote-sensing data and methods that can be understood to provide a foundation for a remote-sensing approach to these types of environmental applications?
- Is there a framework for applications in environmental science that can help resource managers and scientists obtain the best value from a remote-sensing approach?
- What are the landmark case studies and illustrative examples of best-practices in biodiversity and wildlife management remote sensing? Will success in studies of landscape fragmentation, multiscale vegetation mapping, structural assessment, and other remote-sensing outputs be applicable in new environments?
- How should multidisciplinary, interdisciplinary, and transdisciplinary teams organize themselves and their selected remote-sensing activities to help wildlife projects succeed?
- What are the key new insights and research ideas in remote sensing applied to environmental questions?

This book is focused on the actual and potential remote-sensing contributions to biodiversity and wildlife management, and therefore does not address completely, or even approach more than superficially, either the complete biodiversity and wildlife management agenda, or the remote-sensing research agenda. Instead, the book is explicitly focused on presenting a transdisciplinary remote-sensing approach which has already led to success and insights that support ongoing environmental management and interdisciplinary science applied to wildlife mapping, modeling, and biodiversity assessment.

The book is organized into seven chapters. Chapters 1, 2, and 3 serve to provide introductory material, definitions, and background disciplinary material that may help promote fundamental understanding of the remote-sensing approach in environmental applications. A second objective is to lead the interested reader to more detailed sources. Chapter 4 introduces recent proven methods for remote-sensing applications in land cover and habitat mapping. This is followed by two chapters presenting a compendium of successful remote-sensing approaches and examples in wildlife management and biodiversity studies and projects. Chapter 7 is the final and concluding chapter containing a summary of three prominent themes highlighted throughout this book—*transdisciplinarity, technological innovation in methods,* and *new applications*. This final chapter contains some thoughts on management implications, which may lead to better use of remote sensing in future and new ideas for wildlife management and biodiversity monitoring.

Chapter Summaries

Chapter 1. Introduction

The discussion of remote sensing in biodiversity and wildlife management begins by introducing emerging interdisciplinary concepts of a transdisciplinary remote-sensing approach, highlighting the potential in "bringing together the disciplines" to resolve significant environmental challenges. Clarification of important terms and concepts are provided, and a brief review of the history of remote sensing in studies of wildlife and species-environment relationships associated with habitat and biodiversity is presented. Field-based, aerial, and satellite remote-sensing perspectives are introduced and appropriate environmental questions for which remote-sensing contributions and solutions are possible are outlined.

Chapter 2. Management Information Needs and Remote Sensing

As science and management identifies new goals and objectives, the types, amount, and quality of information required increases. New knowledge is essential in many pressing environmental situations. A complete understanding of the ways in which remote sensing can support science and management is not attempted; instead, considerable attention is paid to the ways in which interdisciplinary and transdisciplinary teams might work together to achieve significant progress. Conceptual and application frameworks are described to help in the process of identifying information needs and appropriate ways to address those needs. Certain information needs for which remote sensing may have a particularly valuable contribution to make (e.g., land cover and change mapping, habitat conditions, spatial heterogeneity, water resources and quality) are introduced and the fundamental limitations of a remote-sensing approach are considered.

Chapter 3. Remote Sensing Data Collection and Processing

Image acquisition, data characteristics, and analysis procedures form the heart of any remote-sensing endeavor. Five types of remote sensing are highlighted in this chapter. Thermal, passive and active microwave, and passive and active optical sensing are considered from the platform and sensor point-of-view, and the common image processing steps that are typically necessary as part of the implementation of a remote-sensing approach are introduced. This chapter also introduces three critical complementary geospatial technologies—GPS, GIS, and spatial models—which, together with remote sensing, constitute a powerful and compelling integrated approach for use in biodiversity and wildlife management applications.

Chapter 4. Remote Sensing of Ecosystem Process and Structure

This chapter reviews selected well-developed methods of interpreting and analyzing remotely sensed data. Since both visual and digital analysis methods have been applied to remotely sensed data in a wide variety of applications yielding new information of enormous value, both are introduced. However, there is a growing trend to digital remote-sensing applications, and therefore the quantitative approach is the focus of the presentation. Examples of remote sensing in ecosystem processes mapping (e.g., landscape fragmentation) and structural analysis (e.g., invasive alien species mapping) are highlighted, but a wide range of terrestrial and coastal habitats are considered. The emphasis is on the choice of methods and evaluation protocols.

Chapter 5. Remote Sensing in Wildlife Management

This chapter introduces a review of remote sensing in wildlife management, and highlights case studies and specific examples of habitat suitability modeling, structural habitat analysis, detection of wildlife resources, and environmental modeling approaches based on remote sensing. Key aspects of successful use of remote sensing in wildlife management are introduced and used to support a broader presentation of habitat suitability and quality analysis, direct animal sensing, linking animal data sets to habitat models, and mapping anthropogenic disturbances. The examples and case studies are drawn from the recent literature and together provide a foundation from which important applications' lessons are drawn.

Chapter 6. Remote Sensing in Biodiversity Monitoring

The biodiversity-indicator approach, including species richness mapping and productivity modeling, have benefitted through development of remote sensing. The emerging approaches are reviewed and highlighted here in a series of examples and case studies from the recent literature. New programming to ensure adequate biodiversity monitoring at local, regional, and global scales is reviewed; the remote-sensing contributions are significant and of increasing importance, for example, in species richness prediction, invasive alien species monitoring, and special species habitat analysis. This chapter introduces the key design principles—a focus on indicators, thematic habitat, and ecological processes—typically invoked in current or planned biodiversity monitoring programs; many more such programs are anticipated in the near future to support biodiversity management and research.

Chapter 7. Conclusion

This chapter concludes the book with a review of the remote-sensing approach, highlighting three major themes—*transdisciplinarity, technological innovation in methods, and new applications*. Some of the key principles of successful environmental research and management applications are reviewed. Issues relating to new research and future developments are clarified; expectations for an emerging research agenda based on the growing awareness and appreciation of remote sensing and transdisciplinary collaboration are considered. Finally, the book concludes with a review of the larger, perhaps less tangible benefit of remote sensing—a view from above, which reinforces a deep knowledge of the importance of natural resource management and environmental conservation.

CHAPTER 2

Management Information Needs and Remote Sensing

My own suspicion is that the universe is not only queerer than we suppose, but queerer than we can suppose.

—J. B. S. Haldane, 1927

Information Requirements for Management

Types of Information Required

Wildlife and biodiversity management practices are increasingly, and firmly, embedded within the larger management paradigm of *ecosystem management* with significantly large and challenging information needs (Boyce and Haney 1997, Chen and Saunders 2006). Ecosystem management, and specific environmental projects, will have information needs associated with operational activities, decision making, strategy definition, governance, and risk management, among others (Canter 1993). Some of these information needs flow from various ecological hypotheses concerning the geographic distribution of habitat and biodiversity, and specific indicators, such as measures of rarity, endemism, and species richness (Franklin and Dickson 2001). A quick glance through the literature reveals a vast array of individual projects and studies, and also suggests that arriving at a conclusive and comprehensive list of detailed information needs will be difficult and, perhaps, not particularly useful. Each study or management situation will have a myriad of information needs and specific requirements. A more useful approach may be to consider information needs in broad categories, which naturally incorporate different axes or dimensions, including spatio-temporal resolution and scale (Aldrich 1979, Hollander et al. 1994, Leyequien et al. 2007).

Ground-Based Observations

One obvious broad category of information needs is focused on the direct observation of individuals of the species or guilds of interest. Much of the information required for wildlife management and biodiversity studies must be derived from individual observations or species-specific distribution or habitat assessment studies that can only be acquired in the field by trained personnel. Included here are the detailed observational data on individuals and populations, habitat, and behavior, including purpose-acquired locational information and data from existing range maps, specimen collections, museum, or other data sets. Such data may lead to species models, population monitoring, and population viability analysis, for example, or be part of a genetic population assessment. Remote sensing can complement these observational data; there is almost never a replacement for such traditional field work. There will always be the need—no, the privilege!—to go into the field.

Maps and Models

Other necessary information for management is developed through a broad category of information needs associated with models of all types, including broadly based conceptual models, biological models, behavioral models, and Bayesian models (Milner-Gulland and Rowcliffe 2007). Shugart (1998) provided an introduction and review of terrestrial ecosystem models useful in understanding climate change impacts (Table 2.1); five categories were outlined, illustrating

Type of Model	Model Elements
Conceptual	Diagrams or descriptions of the important connections among the components of an ecosystem
Microcosm	Small (lab bench) physical and biological analogues to a larger ecosystem
Population and community	Systems of differential or difference equations that compute the change in the number of individuals in a population or interactions with other populations
Compartment and multiple commodity	Equations that follow the transfer of one or more elements, energy or other material through an ecosystem
Individual-based	Changes in individuals are used to understand or predict larger ecosystem dynamics

Source: With permission from Shugart, H. H. 1998. Terrestrial Ecosystems in Changing Environments. Cambridge University Press, Cambridge.

TABLE 2.1 Examples of Different Types of Models Commonly Used in Ecosystem Analysis

the enormous variety of models for questions of ecological significance related to the climate-change theme. Data requirements for habitat or biodiversity modeling are exceedingly diverse. Species distribution and abundance models, for example, may rely principally on climatic data obtained in a variety of ways, included by model prediction (John et al. 2008). For such species distribution models to work well, typical information needs will often include biogeoclimatic map and topographic data sets, which provide a number of model parameters. These, in turn, might be used to exclude certain areas from the habitat analysis (e.g., temperature or elevation ranges that are known to be unsuitable habitat).

Remote Sensing

Critical to the appropriate use of both individual animal observations from the field, and types of information required for models, will be the principal contribution of remote sensing—spatially explicit and multitemporal information on key ecosystem structure and processes. Included in this broad category of information needs are the structure and composition of vegetation, disturbances and historical vegetation patterns, water resources, vertical complexity, horizontal patchiness, dead and damaged trees and nonphotosynthetic material, food availability and quality, productivity (e.g., net primary production and specific ecosystem components, such as browse and fruit production), forest floor and soil conditions, physical conditions (geology, topography, hydrology, climate), and cultural data (e.g., pollutants, land use, road network, human activity) (McComb 2008). A list of such information needs is virtually endless, with many more relevant information needs that might not yet be known or even anticipated. Clearly, not all of these information needs—or even most of them—can be obtained by remote sensing at the required scales, levels of accuracies, or acceptable cost (Strand et al. 2007).

Remote sensing represents a powerful suite of data sets and methodological procedures for direct use, and as a contribution capable of *complementing and extending* ground observations and existing map databases, and models, accurately and efficiently over large geographic areas. The work required in identifying and testing a remote-sensing approach is important, but has only just begun, and needs to be considered within the larger context of monitoring; for example, Balmford et al. (2006 : p. 227) reviewed the biodiversity monitoring information needs when considering the Convention on Biological Diversity 2010 targets and concluded:

> ...much good work has been done and is in progress, most notably by the nongovernmental and intergovernmental sectors, and using data collected largely for other purposes. However, a great deal still needs to be done if we are to deliver robust and timely measures of progress against what is probably the most significant conservation agreement of

the decade. Established indicators need to be subjected to ongoing scrutiny; promising indicators need development and expansion; new measures and models, especially those addressing ecosystem services, need to be conceived; and many new partnerships must be forged. This will not be easy, and it will not be cheap. However, we contend that it is essential, it is achievable, and it will prove less costly and more rewarding than looking for life on a lifeless planet.

Sampling Strategy

One of the greatest strengths of remote sensing is to provide an effective foundation for sampling landscapes at multiple scales (Environmental Remote Sensing Group, 2003). For large areas, and the landscape-level analysis (which implies a regional study), an information strategy to support biodiversity or species-specific management will often require a sampling approach. Both low and high spatial detail imagery may be required, although many other factors will be considered and are discussed in detail in Chap. 3 of this book (*see also* Coops et al. 2007). A multiscale, hierarchical sampling design, however, will often be a decisive component. This design typically embeds observations at one scale into sampling protocols that expand spatial and temporal data acquisition at succeeding scales in a hierarchy.

One such design was developed in a recent study with MODIS and Landsat imagery to estimate loss of forest cover in the humid tropics (Hansen et al. 2008). Initially, MODIS satellite imagery covering a very large area at low spatial resolution were used to map broad land-cover conditions, and then individual Landsat image scenes were acquired for specific areas within this larger area, but at much higher spatial resolution. Field data were also collected at specific locations within each Landsat image. Because the Landsat imagery could detect changes in land cover at a much higher spatial resolution than the MODIS imagery (*see also* Pape and Franklin 2008), the acquisition of the Landsat imagery was designed to sample areas within the larger map region that were likely to experience different rates of change. Similarly, field observations were collected in areas where the sample variance could be expected to be high (e.g., near roads, and in areas with higher probability of human activity, such as valleys or alluvial terraces). Overall, forest clearing rates were calculated to be 1.39 percent of the total biome area, with much higher regional variation in forest clearing than had been previously documented in analyses at only one scale of observation (e.g., large-area MODIS mapping, or in comparison to estimates based only on field data).

This multiscale analysis approach may be particularly valuable at the landscape- or waterscape-level and in situations where it is possible to acquire data with increasingly higher spatial detail *in situ*, aerial and satellite imagery nested hierarchically. The *in situ* observations or field imagery, or the aerial image data, will represent the highest spatial detail at the highest hierarchical (or ecological scale) level distributed using a sampling strategy within a larger area mapped using coarser resolution imagery. Together with existing thematic map data (biophysical, geophysical, climatic, cultural), acquisition of such multiscale field observations and remotely sensed data are increasingly the preferred source of regional local habitat information (including human activity) and fragmentation (landscape structure) data (Weiers et al. 2004, Estes et al. 2008, Hill et al. 2008). In such sampling designs, the prospect of developing a complete understanding of the remote-sensing information needs for wildlife studies and biodiversity (and particularly, its global measurement) remains a significant undertaking, but nonetheless, likely more practical than efforts to map larger and larger areas at the highest spatial resolution available (McDermid et al. 2009).

Science Team Dynamics

The challenges and rewards of bringing together multiple disciplinary perspectives and approaches in environmental science, and the differences in teamwork and models of disciplinary, multidisciplinary, and interdisciplinary science leadership that may be necessary, have been partially described (Rhoten 2004), but are generally only poorly understood and implemented (Naiman 1999). Working across disciplines presents significant new challenges to science and applications' or management teams, demanding personal and institutional commitment (Stokols et al. 2003, Payton and Zoback 2007). *Multidisciplinarity* refers to a process whereby researchers or practitioners in different disciplines work independently or sequentially, each from his or her own discipline-specific perspective. *Interdisciplinarity* is a process in which researchers or practitioners work jointly from each of their respective disciplinary perspectives. Interdisciplinary questions are usually recognizably distinct from clear disciplinary-based scientific or management questions because they require the integration of different epistemological perspectives.

One innovation in addressing this integration challenge presented in complex interdisciplinary work has been termed *transdisciplinarity*. Transdisciplinary approaches begin with a process by which people work jointly to develop and use a shared conceptual framework that draws together discipline-specific theories, concepts, and methods (Hirsch-Hadden et al. 2008). While some have suggested that transdisciplinarity is a powerful approach—perhaps, even, the ultimate form of interdisciplinarity—that involves coordination among the disciplines, others have focused more intensely on identifying a transdisciplinary mode of knowledge production (Lattuca 2001). In practical terms, the types of questions that are of interest in transdisciplinary work imply models of natural systems with characteristics or features that are likely to be revealed through a comprehensive effort to identify common structures or relationships.

Experience suggests that the creative potential increases as one moves from disciplinary to multidisciplinary to interdisciplinary and transdisciplinary approaches, since the latter entail more extensive dialogue and collaboration among scholars and practitioners from different fields. This process is more likely to yield conceptual integrations of broader scope and significance (Stokols et al. 2003); such an outcome was anticipated by Zaveh (2007) in a powerful essay on the meaning and importance of transdisciplinarity and the role of new types of integrated thinking in both landscape- and restoration-ecology. In Zaveh's view, the science and management of landscapes in every part of the world requires a holistic approach, one in which the frontiers of disciplines are challenged and transcended, leading to a radical innovation for the dominant paradigm. Perhaps the end point of this epistemological transformation and evolution is a *transdisciplinary ecology,* which will ultimately assume the central position in the ongoing scientific revolution toward integrative environmental theory and practice.

One early task in any science team effort is to define the terms of the collaboration in addressing a common problem. Recently, impressive web resources have been made available to help those embarking on multidisciplinary, interdisciplinary, and transdisciplinary remote-sensing projects in environmental studies (e.g., American Museum of Natural History's Center for Biodiversity and Conservation, Center for Biodiversity Informatics http://biodiversityinformatics.amnh.org/ web site accessed August 11, 2008). Some key questions might include:

1. What are the particular needs in remote-sensing project management?
2. Is there an appropriate balance between the use of consultants and the development of in-house expertise?
3. What type and amount of training opportunities should be provided?
4. Is there an effective approach to writing technical specifications for remote-sensing applications?

Even preliminary answers to such questions can help build a strong foundation and help ensure successful team and project outcomes. Additional helpful material has been assembled on project and individual team member selection and evaluation, team building, infrastructure and space considerations, project communications, and development and implementation of successful publication strategies (Caenepeel and Wyrick 2001, Tompkins 2005).

To succeed, teams need to set aside resources to facilitate their work *as a team*; individuals should understand that those resources are available (for example, to support an integrated work plan), and that there are a few reasonably straightforward ways to make effective use of these resources and help promote personal growth in the complex multidisciplinary, interdisciplinary, or transdisciplinary research or management system that typically develops around significant environmental management projects. Multidisciplinary, interdisciplinary, and transdisciplinary teams need to work respectfully together to avoid some common frustrations. For example, perhaps an environmental applications' challenge has been insufficiently understood or poorly communicated (Landgrebe 1978). In 1989, Woodham (p. 5) wrote: "You know that problem we've been working on for the last 10 years? Well, it's a lot more difficult than we thought!" Two decades after Woodham's statement, the problem referred to—modeling radiative transfer using optical imagery and DEMs in mountain areas, to estimate the topographic effect, for example—continues to provide an interesting intellectual challenge across disciplines, and has not yet been completely resolved.

Another common source of frustration originates in an overzealous early promotion of remote sensing in many scientific disciplines. Almost 30 years ago, in 1979, Aldrich (p. 4) commented that:

> ...much of the oversell of remote sensing in recent years has come from overly enthusiastic individuals and their agencies. Many times speculative statements at the end of inconclusive studies have been quoted out of context... and blown out of proportion, resulting in a credibility gap between the remote sensing community and the user community.

Clearly, a remote-sensing credibility gap has persisted in certain disciplines (Kummer 1992, Green et al. 1996, Wynne and Carter 1997). Any multidisciplinary, interdisciplinary, or transdisciplinary team needs to work through this challenge, which can become an early and potentially damaging obstacle to a full and productive collaboration. A critical objective is to avoid unrealistic expectations—for example, "...the methods have not yet demonstrated their expected uses as wonder tools" (Herbreteau et al. 2007). No team should develop or expect a remote-sensing approach or specific product to become a *panacea* to resolve all problems (Landgrebe 1983, Meyers and Werth 1990, McDermid et al. 2005).

Institutional obstacles to the adoption of technology (Lauer et al. 1991), and the lack of recognition of meritorious contributions across disciplines (the reward structure) (Smith and Carey 2007), are often important in determining ultimate project success. The influence of these and related institutional factors is often greater than perhaps will be initially realized, although relatively simple methods exist to identify and overcome such obstacles. For example, some observers have noticed that the overly complex technical aspects of remote sensing can be intimidating, and diminish team coherence, as different team members (perhaps represented by researchers from different disciplines) seek to maximize different performance goals. Some of the team scientists may be primarily interested in accomplishing disciplinary science, and publishing papers in familiar journals, but others on the team may need to have concrete results in specific operational tasks before progress can be made in their science or application area (Andréfouët 2008). Some team scientists might prefer to use methods that are interesting and novel, and worthy of publication, but are not particularly interesting or understandable to other team scientists, nor necessary to achieve management goals—simpler approaches may work well and may have several added advantages valued by management and scientists in other areas; for instance, reduced cost, less risk, and a greater likelihood of being understood.

In a straightforward disciplinary or multidisciplinary collaboration and an application of limited complexity, the situation of working within one's discipline could still constitute a workable-team approach to accomplish a common goal. Multidisciplinarity suggests that individuals can approach the common problem from their own disciplinary perspective without significant modification of their working models and without adopting new integrated models of team behavior. Perhaps each team member pursues or contributes to the common challenge, but with differing intensity and "risk-taking" from an individual disciplinary perspective. However, in more complex situations, such a multidisciplinary approach may be less effective in organizing team activities, and may threaten the overall coherence of the team and diminish the likelihood of achieving a common goal. Instead, a carefully planned and agreed upon interdisciplinary or transdisciplinary approach to this common goal will likely more readily accommodate multiple outcomes, and reduce or completely avoid individual scientist performance optimization at the expense of a team effort.

An additional, and possibly even more challenging, obstacle to complex interdisciplinary or transdisciplinary remote-sensing projects may originate in the continued assumption of the existence of two separate and distinct communities: (i) the remote sensing, or producer community, and (ii) the user community. Sometimes, these two communities will appear to coincide or partially overlap with a disciplinary scientist dualism (e.g., remote-sensing scientists/ecologists) (Andréfouët 2008), or worse, a technician/scientist dualism, or perhaps still worse, a scientist/manager dualism (Hobbs 1998). In a true interdisciplinary or transdisciplinary team collaboration, the existence of such dualism—any dualism—is false and counterproductive.

The simple questions posed in the "information needs definition circle" outlined by Hoffer (1994) and shown in Fig. 2.1 may help to visualize the problem. The question apparently posed by resource managers (*What can the tools do to help me?*) is often shaped by the environmental context, the indicator selected to map or monitor, available finances, and infrastructure, including image processing

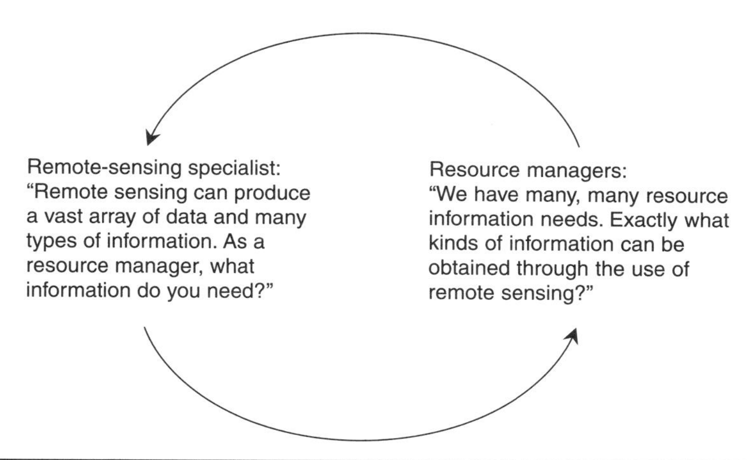

Figure 2.1 Information needs circle defining early collaborations of multidisciplinary and interdisciplinary remote sensing. (*With permission from* Hoffer, R. M., 1994. In Sample, V. A., ed., *Remote Sensing and GIS in Ecosystem Management.* Island Press, Washington. pp. 25–40.)

system capability and human expertise (Phinn et al. 2006). The other side of the circle (*What information do you need?*) is shaped partly by the remote-sensing specialists' understanding of the environmental problem at hand. These two questions promote a kind of "division of labor" in a problem-solving research and management environment. The existence of two separate and distinct communities—the producers and the users—will, paradoxically, constitute a constraint to effective interdisciplinary and transdisciplinary collaboration (Pohl 2005).

Instead, what is needed is a lucid sharing of information, the development of a joint commitment to identify the common approach, and a shared *conceptual framework*, which integrates and transcends respective disciplinary or producer/user dualistic (or nonintegrated, multiple) perspectives. A shared conceptual framework may evolve from current discipline-specific theories, concepts, and methods, or represent a completely new approach designed to address a common problem (Clark et al. 2000, Stokols et al. 2003). The idea is to focus on integration of knowledge and transcendence of disciplinary boundaries. Many such examples exist—for instance, Rapport et al. (1998) emphasized that only this type of transdisciplinary approach was likely to succeed in pursuit of complex ecological concepts, such as ecosystem health and integrity, because an integrated approach will lead to mutual enlightenment and a clear strategy to accomplish goals. In other settings, such as in the policy sciences approach to environment (e.g., Brewer 1988), or in the human health sciences, for example, genuine and comprehensive integration is considered a stringent criterion of all successful interdisciplinary or transdisciplinary scientific collaboration and operational practice (Clark et al. 2000, Hirsch-Hadden et al. 2008).

To put it bluntly, organizing the team may require some differentiation of tasks and activities, but team members with expertise in remote sensing are not simply providing a "service" to a larger, and possibly only vaguely understood, goal. Instead, there is full participation in defining the science to be achieved. Similarly, the team members with ecological expertise are not simply the "receivers" or users of data products, and therefore, excluded from executive decisions concerning necessary information development and delivery. Instead, there is collegial discussion and agreement on approaches and specifications. Certainly, in some situations, individuals may be free to pursue disciplinary goals that will provide significant outcomes (and perhaps an overall project *gestalt*), but individual disciplinary pursuits or activities that might reduce the effectiveness of the team to achieve progress on the larger transdisciplinary goals must be carefully evaluated within an overall agreement on effort and intensity of contributions.

A number of characteristics have been identified to define and clarify the transdisciplinary approach (Table 2.2). The debate on this

Aspects of Transdisciplinary Approaches	Defining Characteristics
Scope, process, and outcomes	Definition—across disciplines Scope and relevance Recursive processes Recognition of multiple knowledge forms Contextuality and generality Specialization and innovation
Transdisciplinary practice	Participatory processes Mutual learning, relearning Integration and collaboration Values and uncertainties Management and leadership Education and career building Evaluation and quality control
Fundamental nature of issues addressed	Scientific challenges Institutional challenges Societal challenges

Source: With permission from Hirsch-Hadden, G., H. Hoffmann-Riem, S. Biber-Klemm, W. Grossenbacher-Mansuy, D. Joye, C. Pohl, U. Wiesmann, and E. Zemp, eds. 2008. Handbook of Transdisciplinary Research. Springer, Dordrecht.

Table 2.2 Selected Characteristics That Help Define Transdisciplinarity and the Transdisciplinary Research and Practical Approach

approach is young and far from complete, but some interesting themes have emerged; for example, transdisciplinarity will recognize many forms of knowledge and knowledge-creation processes (e.g., scientific and traditional or Aboriginal knowledge and "ways of knowing"). Transdisciplinary scientific problem-solving is well-suited to an environment in which significant uncertainty exists, where solutions are not predetermined, but the outcomes have great meaning or importance to people. Transdisciplinarity will include discussion about not only facts and practices but *values*, typically requiring an extended discussion between managers, researchers, and the society at large.

Rhoten (2004) compared the attributes of a traditional disciplinary-project approach (largely homogeneous, disciplinary or multidisciplinary, and hierarchical), and new transdisciplinary approaches (often heterogeneous, interdisciplinary, horizontal, and fluid). Clearly, these approaches are effective in tackling different types of questions. Distinguishing interdisciplinary and transdisciplinary *processes* from disciplinary and multidisciplinary processes is similar to distinguishing the types of questions that are posed; in interdisciplinarity, the emphasis is on unified and unifying problem definition *and* project direction. The skills to participate in the new approach include the methods, techniques and knowledge of a chosen discipline, but also the broader problem-solving skills that require learning, unlearning, and relearning across disciplines. In successful interdisciplinary and transdisciplinary collaborations, the team works together to understand what is possible and can be achieved with available resources. Interactions in the field and in the laboratory are of enormous importance (Read et al. 2003). The discussions include references to both management and research objectives, and clarity is achieved on the appropriate scales of investigation. Reciprocal use of tools and analytical techniques, and the development of common models, is always encouraged (Chen et al. 2006).

Experience suggests the importance of ensuring that any group working across disciplines understands and appreciates the differences in the types of questions, the shared nature of the team challenges, and perhaps even more importantly, the constraints or limits introduced by the interdisciplinary and transdisciplinary methods themselves. When evaluating and then adopting a remote-sensing approach, some examples of these methodological challenges might include:

- *Maps.* Maps produced for general purposes rarely fulfill different and specific management purposes equally well (Robinove 1981); this issue frequently arises as great effort is expended to generate a general information product, for example, a land-cover map for an area of interest. But then a realization sets in that *other* attributes or aspects of the land are equally, if not more important, and this general land-cover map does not depict those well, and new investments are needed.

- *Specifications*. The specifications for the ideal remote-sensing data can vary, depending on vegetation conditions, study area size, and available image processing techniques (Phinn et al. 2003); this challenge can preoccupy analysis teams as they search, obtain, analyze, and ultimately accept or compromise on nonoptimal remote-sensing data sets for reasons of availability, accuracy and precision, complexity or rigidity, and ease-of-use.
- *Data handling*. Data preparation, management, and interoperability issues in remote sensing and spatial data information processing can easily consume vast amounts of human and capital resources, even before any of the interesting analysis questions are tackled (Green 1999).
- *Costs*. There are at least four main types of costs encountered in environmental remote-sensing applications: project setup, field surveys, image acquisition, and data analysis (Mumby et al. 1999, Steininger and Horning 2007). The reality of these costs can sometimes be surprising—and such surprises are rarely a good thing!

Even with an awareness and effective management of these largely process challenges, beginning an interdisciplinary or transdisciplinary remote-sensing endeavor can be a daunting prospect, but ultimately can be expected to return greater rewards to individual participants and to overall project success (Turner et al. 2003).

The Role of Leadership

One aspect of environmental science and management—including the environmental remote-sensing multidisciplinary, interdisciplinary, and transdisciplinary work of primary interest here—that has attracted recent commentary is the role of *leadership*. The idea of leadership is fascinating and elusive, and one of enormous societal interest. A rich assortment of current and recent publications on business and organizational leadership has developed. The curriculum of recent environmental management programs often now includes explicit leadership training. The diversity of views and insights on leadership are far-reaching, confirming both the importance and complexity of leadership across the entire spectrum of human organizational activity. The issue suggests some attention should be paid to aspects of leadership within complex science and management team construction and operation.

Without referencing any particular contribution or leadership philosophy here, it is useful to consider the function and dynamics of leadership in complex interdisciplinary and transdisciplinary remote-sensing projects. The science or management focused on a true interdisciplinary or transdisciplinary accomplishment will flow from a goal that cannot be reached in other ways but which is synthesized from different viewpoints; the leader must explain this synthesis through multiple perspectives, and reinforce cultural practices that break down artificial disciplinary distinctions among team members. Genuine interdisciplinary thought may, in fact, be a relatively rare individual achievement (Gardner 2006), but a successful synthesis by fully participating team members may be transformed into a highly prized and motivating *project vision* (Rhoten 2004, Hirsch-Hadden et al. 2008).

Leadership in an endeavor using remote sensing across disciplines can be decisive because leaders are in a position to set the direction and tone. The interdisciplinary or transdisciplinary team leader does not need to have all the answers—in fact, *cannot* have all the answers! Instead, team leaders must work to create opportunities for information exchange and mutual learning; ensure collective sharing of data and the responsibility to collect data; and work diligently to foster mutual respect and team members' appropriate involvement in the diverse research cultures and disciplinary traditions represented. Leaders are responsible for decision making and will construct appropriate processes leading to common agreement on the direction of the project. The leadership should foster understanding and communicate frequently—and work to ensure that all members of the team are valued and understand the importance of their individual contributions as well as the overall outcome. Effective leadership is context-specific; in other words, specific leadership behaviors are a part of the broader reflective understanding about what is needed according to the particular situation.

Leadership is a profoundly human experience about *values*—influencing people and work through individually held values—and comes from within individuals, not imposed or artificially constructed. Leading is not managing; management and leadership responsibilities, and the tasks required to fulfill them, are in many ways at opposite ends of the spectrum, and are readily differentiated (Strang 2007). Leadership across disciplines requires a significant and lasting commitment to listening and viewing issues from a variety of perspectives. The purpose of interdisciplinary and transdisciplinary leadership is not to create followers, or certainty, or a rigid and narrow focus for the team, but to create *more leaders*, more people working across disciplines, making connections, being innovative, and deriving new insights. Successful interdisciplinary and transdisciplinary teams work to ensure that such leadership extends to, and exists within, all team members—not only the principal investigator and the project manager—but all participants. This should be fully apparent in the creation of a shared project vision. The vision will ensure project goals reach across disciplines, and all project members will focus intensely on fostering common understanding of these goals and how to achieve them. Having fewer, more focused, jointly developed, and well-understood goals means the team will more likely have the energy required to sustain the effort to achieve them.

Application Framework

The previous discussion has emphasized the importance of the *conceptual framework* within which the interdisciplinary team works jointly to develop and use a shared approach that draws together and transcends discipline-specific theories, concepts, and methods to address a common problem. This jointly developed, commonly understood, conceptual framework will then lead naturally to an *application framework* to optimize the choice and use of appropriate remote-sensing data and methods in an ecological context (Glenn and Ripple 2004, Phinn et al. 2006). The application framework will effectively align the information content of the imagery with the data needs of the project (Phinn 1997, McDermid et al. 2005).

A well-designed application framework will potentially reduce the sometimes overwhelming and intimidating technological dimensions of an interdisciplinary or transdisciplinary remote-sensing approach. Typically, the application framework will organize team activities in a logical sequence of steps (Phinn et al. 2003):

1. Identify the information requirements for the project; this might include the type of information required, acceptable error levels, scale(s) of the information, available time and finances, and whether a prescribed analytical technique is to be used.
2. Link the information needs organized into an ecological hierarchy with the characteristics of the remote-sensing image; this step may involve a preliminary matching of the different types of image resolution to the desired ecological information (for example, avoiding the use of a particular sensor if the desired information is represented by features that are too small to be detected by that sensor).
3. Conduct an exploratory analysis using existing digital data; this may be as simple as a visual inspection of imagery to confirm quality, or as complex as implementing a geospatial transformation to understand image variance.
4. Identify the ideal remote-sensing data, considering spatial, spectral, radiometric, and temporal resolutions; this step may be reasonably straightforward, resulting in a selection from available image sources, or perhaps a specific remote-sensing acquisition mission will need to be designed to acquire the necessary data.

5. Select and apply a suitable set of processing strategies to extract the required information; generally, the choice of data should provide strong cues about the subsequent image processing techniques to be pursued. Then, this step involves evaluating the available image processing techniques, their cost and implementation, and expected accuracy levels to meet the project needs.
6. Conduct a cost-benefit analysis; by evaluating the success of the project based on the selections in earlier steps, or perhaps comparing accuracy or cost of output products from different sources that theoretically could satisfy the same information need.

A simplified flowchart for an application framework developed in a coastal marine environment is included in Fig. 2.2 (Phinn et al. 2006). Initial specification is required of the ecosystem or habitat indicator of interest, the area to be covered and the timeframe, and the financial resources. The application framework guides the project team to select the appropriate image data set and to evaluate the image processing methods and associated cost with appropriate tools.

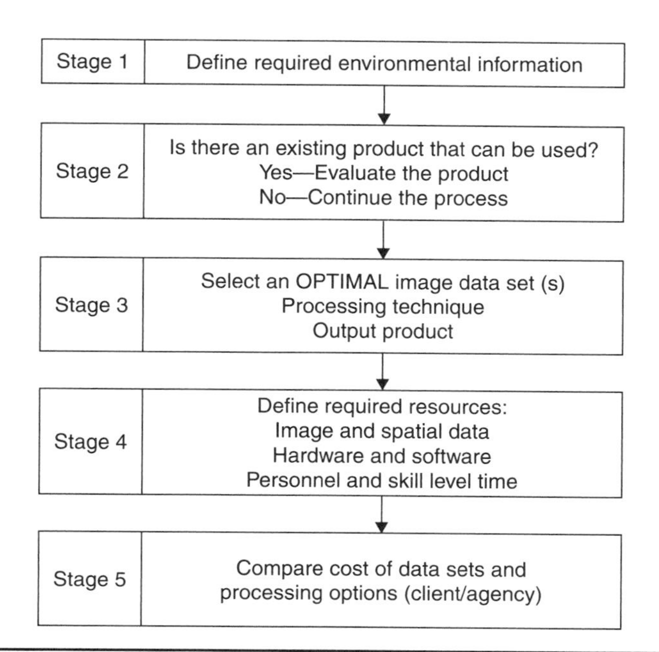

Figure 2.2 Conceptual framework for integrating remote sensing with environmental monitoring programs. (*With permission from* Phinn, S. R., K. Joyce, P. Scarth, and C. Roelfsema. 2006. In Richardson, L. L. and E. F. LeDrew, eds. *Remote Sensing of Aquatic Coastal Ecosystem Processes.* Springer, Dordrecht.)

The application framework may include additional steps. For example, there may be a need to address possible gaps in understanding through the establishment of widely accepted standards. Depending on the maturity of the application, additional consideration might be needed of the full range of geospatial technologies to be accessed in the project. Without this or a similar structural approach, keeping on top of a wide range of such technological developments in GIS and GPS, and the science of remote sensing, could quickly become an impossible task. The complexity of working across disciplines, in an application often concerned with pressing and public environmental management issues, suggests that the information needs situation may well become unmanageable if not thoroughly analyzed and explained.

There may be differences in emphasis for different steps in the application framework developed for any particular project. Typically, map accuracy is thought to be the key feature of remotely sensed land cover and habitat maps, particularly those produced quantitatively by satellite remote-sensing methods over large areas and intended to serve general purposes. For example, Throgmartin et al. (2004) noted land-cover mapping errors had influenced the accuracy of avian habitat models in a regional study based on the National Land Cover Data set, a reasonably current and widely available representation of land cover in the conterminous United States. In another remote-sensing habitat mapping application framework, this time presented for a large area, multijurisdictional grizzly bear (*Ursus arturus*) management project in Alberta, McDermid (2005) identified five map attributes, including accuracy, as essential for project success. The five attributes of map scale, robustness, accuracy, consistency, and flexibility were derived from more comprehensive discussions of spatial data quality [*see* for example, Wiens (1989) on spatial data *scale*, Worboys (1998) on spatial data *precision*, and Stine and Hunsaker (2001) on spatial data *uncertainty*].

In the grizzly bear project, an empirical *map quality index* was developed as a balance among the desired attributes of map scale, robustness, accuracy, consistency, and flexibility. McDermid (2005) had noted that the success of the grizzly bear project was strongly dependent not only on the remote-sensing map accuracy; at least as important, or perhaps even more so in particular applications, were other map attributes, such as map consistency, data precision, and scale. Certainly, map accuracy needed to be sufficiently high to provide a convincing demonstration for the whole team to move more confidently in the direction of a satellite remote-sensing approach as the project continued to expand in scope (Franklin et al. 2002b). But while certain parts of the study area *did* have access to higher accuracy maps at more detailed mapping scales produced from aerial photointerpretation methods, because those maps were not available for the whole study area, the subsequent grizzly bear resource selection functions (RSFs) based on them could not be confidently extended to the larger study region of interest. An important characteristic of map consistency and robustness, then, was map *completeness*. In this example, the satellite-based remote-sensing habitat maps produced over the whole study area (because of near-synoptic, large-area image acquisition) generated a higher level of consistency at the appropriate scale in the grizzly bear RSF model development and application.

In essence, the interplay among the five remote-sensing habitat map attributes of interest in the grizzly bear application (scale, robustness, accuracy, consistency, and flexibility) was balanced using the map quality index based on the intended purpose of the mapping products (McDermid 2005). All aspects of the five map attributes were clearly articulated within an effective application framework which involved discussion and agreement of all team members. This meant, for example, that the detailed land-cover map legend was developed in a collaborative way with all team members participating in the discussion. Specific targets for attributes such as map accuracy (e.g., greater than 80 percent overall accuracy), and map robustness (e.g., 100 percent wall-to-wall coverage with map legend consistency), and development costs, were then more readily managed. Overall team satisfaction with the finished remote-sensing products (land-cover maps, and estimates of vegetation structure, such as LAI and species composition), were traced back to a combination of (i) a shared understanding of product development strategies and characteristics, and (ii) realistic expectations.

Many of the relevant issues associated with the steps to be completed in an application framework are discussed in the remaining sections of this chapter and in Chaps. 3 and 4. The following sections outline the specific details of information resources that are typically required in wildlife management and biodiversity applications, which are then discussed in detail in Chaps. 5 and 6, respectively. Not all information needs are addressed; such a comprehensive assessment would be impossible to complete. Instead, the focus is on a sample of those information needs that have potential to be supported in the remote-sensing approach to address environmental science and management. The chapter concludes with a synopsis of the biodiversity and habitat indicators where remote sensing has been identified as an appropriate contributor, and the general limitations that may be encountered in any remote-sensing application.

Ecosystem Structure and Process

Land Cover

Land cover is an essential information need in biodiversity and wildlife management. Land cover is typically defined simply as the vegetation and artificial coverings of the land surface as represented by the *attributes* of the surface (Robinove 1981, Townshend 1981). Land-cover attributes typically are based on physiognomic or vegetation structure characteristics or relationships. By considering differences among these attributes, land-cover classes and areal units of land that are "reasonably homogeneous" in those attributes can be produced, such as a class or unit of "trees" versus a class or unit of "shrubs" versus a class or unit of "herbs." Some differences in interpretation of which attributes to use, and their various constraints, have naturally led to varying land-cover map outcomes for ostensibly similar purposes. One jurisdiction may consider a parcel of land to be "trees" or "forest" land cover with a minimum of 6 percent coverage by tree species; another jurisdiction may use a different percent, say 10 percent, or another attribute altogether, for the same decision. This has led some observers to recommend functional mapping of land cover based not on vegetation structure, floristic, or physiognomic attributes, but on characteristics more closely related to ecosystem processes, such as leaf longevity and leaf type (Running et al. 1995).

Presently, no standardized hierarchy of land-cover classes has been developed for global use although there is increasing agreement that a standard approach is required (Friedl et al. 2002). The broadest or highest level land-cover classifications are biome or continental in extent, and are now based on global remote-sensing databases (e.g., Friedl et al. 1999, Huang et al. 2003, Justice et al. 2008). Their antecedents can be found in the broad climatic and physiognomic classifications of early climatologists and biogeographers (e.g., Peel et al. 2007). Sometimes these early classifications reflected well the *potential* vegetation and land cover, rather than actual land cover, which is more readily obtained by remote sensing (Franklin 2001).

There are numerous regional and national land-cover classification systems, which serve to clarify the main elements of a potential (though yet to be consistently implemented) global-mapping approach (Loveland et al. 1991, Fuller et al. 1994, Wulder et al. 2003, Wallace et al. 2004). For example, two critical requirements are: (i) data consistency and (ii) *hierarchical class structures* across the full range of phenomena that are to be classified. The early development of most regional and national land-cover classification systems incorporated standardized land-cover class hierarchies that originated in land-systems mapping (Story et al. 1976), metric aerial photointerpretation exercises (Heath 1956, Kreig 1970), and existing practical vegetation inventory classifications, such as forest resource or rangeland inventories (Bailey et al. 1978, Pregizter et al. 2001). Perhaps the most influential land-cover class hierarchy for use with aerial photography and digital satellite imagery continues to be the three- and four-level land-cover USGS system described by Anderson et al. (1976). Current classification systems often incorporate available legacy data and the best of these systems as needed (Wulder et al. 2003).

Land-Cover Change

Land-cover change is a key component of environmental applications and as a result has received an enormous amount of attention in biodiversity monitoring and habitat mapping. An essential information requirement is land-cover change *at the scale of interest*. The main processes or flows generating landscape structure formation and landscape change over time can often be expressed as changes in land cover. The processes responsible for these changes are varied, operate at multiple scales, and may include natural and anthropogenic disturbances (e.g., wildfire, insect infestation, harvesting), biotic processes (e.g., succession, birth, death and dispersal), and environmental conditions (e.g., soil quality, terrain, climate) (Linke et al. 2007). Land-cover change detected by satellite sensors has quickly emerged as the basis for many national and regional environmental monitoring programs (Franklin and Dickson 2001); changes in land cover often have reasonably well-known and predictable effects on wildlife habitat and biodiversity.

There is a tendency in the literature—often pointed out and frequently lamented, but, nevertheless, still common—to confuse land cover and land use. Land use and land-use changes should be considered different dimensions from land cover and land-cover change. Land use is a *functional interpretation* of land cover—it is obviously an important attribute of land—and may also be required to support land capability or wildlife suitability mapping. Certain land-cover conditions may be associated with a specific and known land use and the relationships are straightforward—then, a change in land cover can be safely predicted to denote a change in land use (Emch et al. 2005). Agricultural land cover and deforestation are two obvious examples. Some changes in land use will not cause any changes in land cover. Others may instead cause a change only in one or more attributes of the land cover, influencing land use, while the land cover remains the same. Partial forest harvesting or selective "high grading" may change relative forest density, for example.

Habitat Attributes

The concept that every species has a set of environmental requirements for life and reproduction is generally known as the species-habitat or *niche hypothesis* (Shugart 1998). Species richness in a community or landscape may be a function of the number of distinct habitats or niches. Spatial heterogeneity in habitat conditions or attributes may be one of the driving factors in the explanation of species richness partly because of increased niche differentiation (Leyequien et al. 2007). Managers have adapted these relatively simple ideas to provide insight into complex management requirements to support the notion of *indicator species*—indicators of environmental conditions, or perhaps indicators of guilds, that are believed to respond to changing environmental conditions in a similar fashion (Innes and Koch 1998, Hager 1998, Betts et al. 2003). *Umbrella species* are species with large area requirements that will encompass a variety of species with smaller area requirements, and *keystone species* are those species whose activities are known or believed to determine the well-being of a number of other species (Simberloff 1998).

These insights into species-habitat relationships have been extremely useful and appreciated by those working with species at risk, vulnerable species, and endangered species. For example, White et al. (2006), working to support a recovery plan for endangered Puerto Rican parrots (*Amazona vittata*), highlighted the importance of spatial-scale and species-habitat interactions. In their view, understanding nest–site selection at multiple spatial scales provided insights into how specific environmental components interacted to influence reproductive outcomes. For example, habitat- and spatial-characteristics were used to differentiate between used and unused nest sites because "identifying biologically relevant variables that influence nest-site selection is an essential step toward developing effective management strategies." By integrating area-specific information on fruiting phenologies, microclimate, topography, and forest structure, the placement of artificial cavities was optimized, which resulted in improved nesting success, and which, in turn, yielded dividends in lower extinction probabilities (White et al. 2006). Increased understanding of required treatments or actions and policy development in these kinds of situations has stimulated increasingly urgent demands for greater and more detailed information on specific habitat attributes or conditions.

Consequently, there are many different habitat attributes that are of interest to wildlife managers and in biodiversity assessment—and, in fact, a virtually unlimited number of habitat conditions and resources could be of interest. However, certain characteristics—or attributes—of a particular land cover will be of interest in virtually any environment, because they are known to be, or are considered to be, important macrohabitat variables. Such variables may be decisive in understanding species-environment relationships or have strong predictive relationships to microhabitat or other environmental conditions of interest. In vegetated ecosystems, percent canopy cover, mean plant height, density, and so on, represents basic, widely applicable, and measurable macrohabitat conditions of interest in many species management and inventory situations (McComb 2008). For a particular species of interest, there will likely be, in addition to such general macrohabitat conditions, a very specific microhabitat information need. For example, in forests, cover and density may be macrohabitat conditions important to understand invertebrate species distribution. Such information can be combined with estimates of microhabitat associated with invertebrate forest floor resources, such as litter depth and decomposition processes (Hattenschwiler et al. 2005).

In another example, in estimating grassland forage biomass and quality for ungulates, density and cover are of interest (macrohabitat), together with microhabitat variables, such as concentrations of nitrogen (N) and neutral- and acid-detergent fibre (Starks et al. 2004, Zhao et al. 2008).

To illustrate the level of detail and type of understanding that is typically sought in wildlife management, two detailed information needs, as examples of specific microhabitat attributes, are reviewed in the remainder of this section.

Different types of *dead and downed trees* provide a variety of habitats and resources for numerous species. The lack of dead wood as a habitat resource is considered a possible threat to certain wildlife, such as the American marten *(Martes americana)* (Bunnell et al. 2002), and in some environments, to biodiversity (Bütler and Schlaepfer 2004). Various snag/downed wood sizes and states of decay produce different cover and nesting sites for many species. Therefore, effective wildlife and biodiversity management may require an inventory of dead wood habitat conditions and access to management strategies that respond to the needs of cavity nesters, foragers, and wood decay processes. Tree mortality rates, snag abundance, dead-limb branching characteristics, size distribution, age, and characteristics of downed woody debris are of interest, though often very difficult to obtain; estimating cavity abundance, in particular, is very challenging (Read et al. 2003, Clark et al. 2004, McComb 2008).

Trees that fall into streams play a role in habitat quality for aquatic and semiaquatic species (McComb and Lindenmayer 1999, Gregory et al. 2003). Forest floor and soil characteristics influence community structure and abundance of invertebrates (Burke and Nol 1998). Type and depth of litter will be of interest since below-ground conditions represent significant resources but are difficult to assess (Hattenschwiler et al. 2005). Litter and soil organic matter will affect soil fertility and ecosystem processes (e.g., carbon and nutrient supply and turnover, infiltration, storage of water) (Smith and Raison 1998).

Coastal aquatic ecosystems contain critical habitat for many species, and specific microhabitat attributes are of interest in assessing biodiversity potential and impacts (Table 2.3, Holden and LeDrew 1998, Purkis et al. 2008). Macrohabitat components of such ecosystems include substrate, benthos, water column, and water surface features extending from the mean-high water mark to the edge of the continental shelf (Phinn et al. 2006). Coral reef macrohabitat conditions must be mapped including composition, reef geomorphology, and a host of dynamic elements. Vertical orientation, branching structure, location of photosynthetic and productive components, color, live versus dead coral, coral density (sparse vs dense), debris, sand, seagrass, and algal definition are possibly of interest or are required to understand coral microhabitat and specific relationships with biota (Newman et al. 2006).

Habitat Variables
• Depth
• Structural complexity
• Live coral cover
• Branching coral cover
• Number of holes
• Total hole volume
• Mean hole volume
• Distance to reef edge
• Distance to river mouth
• Total area

Source: With permission from Knudby, A., E. F. LeDrew, and C. Newman. 2007. Progess in the use of remote sensing for coral reef biodiversity studies. Progress in Physical Geography 31: 421–434.

Table 2.3 Habitat Variables Associated with Coral Reef Biodiversity Measures of Fish/Gastropod/Sea Urchin Species Richness, Diversity, or Abundance

A knowledge of *benthic change* is a significant information need in many marine environments. Species of coral may be of interest (Tsai and Philpot 1998), and the important sediment processes influencing reef development include spatial distribution, transport patterns, and amount of sediment introduced onto reefs. Potential biodiversity impacts and reef degradation are known to be influenced by pollution and sediment vulnerability (Storlazzi et al. 2003, Knudby et al. 2007).

Habitat Patches

Together with an understanding of the processes operating to produce spatial patterns in the ecosystem, the *hierarchical patch dynamics paradigm* can be considered a central organizing principle of considerable value at the landscape level (Hessburg et al. 2004). The patch-corridor-matrix model (Forman 1995) has been the focus of much attention in landscape ecology (Chen et al. 2006, Linke et al. 2007):

- *Patch*. A patch is a homogeneous area that differs from its surroundings; patches occur at various scales, and can be defined using attributes of land cover, vegetation composition and structure (Jorgensen and Norh 1986), geomorphology (Burnett et al. 1998, Nichols et al. 1998), ecoclimatic stability (Fjeldså et al. 1997), and levels of photosynthetic activity (Walker et al. 1992), among others.

- *Corridors.* Corridors are a form of patch with a particular shape and function and may provide some form of connectivity between patches (Haddad 1999) though different corridor effects have been documented (e.g., fewer specialists, impacts on nesting parasitism and predation) (Hilty et al. 2006).
- *Matrix.* The matrix is the most extensive component of the landscape, is highly connected, and controls regional dynamics (Forman 1995).

The hierarchy imposes certain constraints; for example, on energy flows within individual levels of the hierarchy and between levels (Hessburg et al. 2004). *Area-of-edge influence* (AEI) includes transitional zones between patches and may account for different types of edges (inherent edge vs created edge, for example) (Chen et al. 2006). The edge amount and pattern are also relevant—edge orientation, age, shape, and contrast introduce differences in habitat and ecosystem processes. The shape of landscape patterns may be linked to the underlying processes that have configured boundaries and affected the diversity of patches in certain landscapes (Saura et al. 2008). Currently, process-type questions are prominent in attempts to understand the links between pattern and process, and how processes create, maintain, modify, and destroy patterns, across scales, and in long-term monitoring experimental designs (Franklin and Dickson 2001, Hessburg et al. 2004, Linke et al. 2007).

Metapopulation theory has been used to help explain the importance of patch dynamics. A metapopulation is a group of discrete localized subpopulations with dispersal among them. Most population dynamics occur within the subpopulation; however, dispersal between populations on a regular basis promotes gene flow and helps to decrease the probability of population extinction and fluctuations in population size. In order for a metapopulation to function, habitat patches must be accessible; otherwise, isolation of subpopulations will occur. The composition, structure, and quality of corridors affect the connectivity; that is, the degree to which organisms can move between subpopulations through the landscape matrix, and influence the size of the metapopulation (Anderson and Danielson 1997). To maintain subpopulation balance, the number of migrating individuals should not exceed that of those emigrating. Potential breeding habitat must exist and reproduction must successfully replace losses due to mortality; otherwise, local extinctions will occur. The size and shape of habitat patches are important because area-related edge effects may influence, for example, predation and reproduction.

The theory of island biogeography suggests island area and colonization rate may determine species richness (MacArthur and Wilson 1967). At equilibrium, the local-extinction rate is inversely related to area; that is, higher rates of extinction occur in smaller areas. Rates of immigration to islands decline with distance from the mainland colony. Therefore, larger islands closer to the mainland will have greater species richness than smaller, more distant ones. Similar relationships between species richness and patch size, distance between patches, and colonizing populations have been attributed to terrestrial systems. Patterns of species richness may be explained by structural variables related to different processes. Forest landscapes have been described as being oceans with islands (patches) of habitat; as forests become fragmented by disturbance, patches become smaller and more distant from one another (Harris 1984). This model may have less application in terrestrial systems in which patch area has a limited relationship to number of species. Species richness will be influenced by the edge/area ratios of patches within landscapes. The direction of the effect will depend on species habitat requirements; for example, whether a species is an edge- or core-dwelling. Factors such as the patch size and diversity, the distance and connectivity between patches, and the edge/area ratios may be important (Noss 1990).

Spatial Heterogeneity and Fragmentation

Spatial heterogeneity may be positively correlated with species richness of an area, and in such areas biodiversity may be largely a function of patch diversity within a landscape (Wickham et al. 1997). Landscapes, having corridors, that act as conduits and connected patches that act as stepping stones that facilitate movement are likely to have greater species diversity. Increased spatial heterogeneity, or a trend in specific variance in spatial pattern, has sometimes been interpreted to mean *habitat fragmentation* (Fig. 2.3; Fahrig 2003). In forests, disturbance size, frequency, and severity influence stand dynamics (e.g., forest initiation and stem exclusion, understory characteristics) (Oliver and Larson 1996), successional pathways (McComb 2008), and landscape structure (Linke et al. 2007). Thus, habitat fragmentation may occur at different rates, and through different processes,

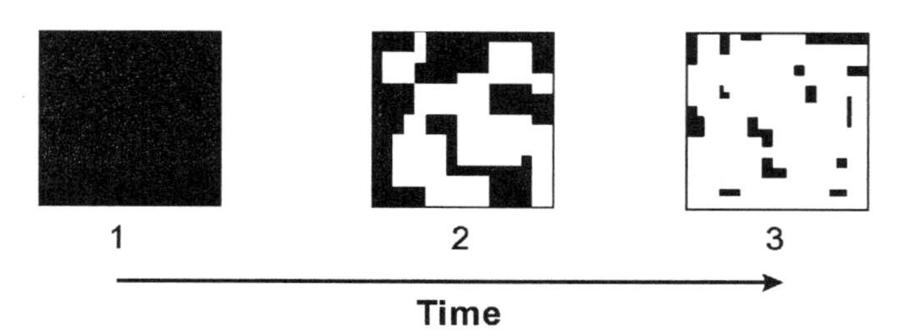

Figure 2.3 Habitat fragmentation defined as the "breaking apart of habitat." (*With permission from* Fahrig, L. 2003. *Annual Review of Ecology, Evolution, and Systematics* 34: 487–515.)

resulting in widely varying spatial patterns overlain on a mosaic of changing landforms and natural disturbances. Several types and phases of habitat fragmentation have been identified, and two recent reviews have considered the fragmentation issue in all its complexity (Fahrig 2003, Gergel 2007).

Habitat fragmentation and habitat loss should be described separately (Fahrig 2003), since, although a fragmented landscape may result in loss of habitat, fragmentation does not always imply habitat loss. Habitat fragmentation can be considered a change in habitat configuration, for which there are many possible measures and causes. Fragmentation of habitat is species-specific; for example, Letcher et al. (2007) found that fragmentation, independent of habitat loss, increased extinction risk for trout in a stream network. Fragmenting habitat for some species of birds may be caused by reducing the size and connectedness of a mature eastern hardwood forest, but fragmenting habitat for other species [e.g., grassland bobolink (*Dolichonyx oryzivorus*)] may be caused by forest encroachment on grasslands (McComb 2008). Species-specific habitat fragmentation and overall landscape fragmentation (the "breaking-apart" of land cover or habitat) together may be of interest in wildlife and biodiversity management, and measuring these effects can be challenging (Franklin and Dickson 2001). The literature provides strong evidence that habitat fragmentation may have no effect, positive effects, or negative effects on biodiversity (Fahrig 2003). Habitat loss, on the other hand, has large, consistently negative effects on biodiversity.

Energy and Productivity

The productivity of environments may be positively correlated with species richness and biodiversity at continental scales (Wickham et al. 1997, Phillips et al. 2008). Areas of high net primary productivity have more resources to partition among competing species, and can support a greater number of species and larger populations than areas with low net productivity (Walker et al. 1992). Increased niche differentiation for species to exploit may be directly influenced by productivity, one of the driving conditions in spatial heterogeneity and temporal variability. However, the relationship between species richness and productivity is not completely straightforward, depending on the level of productivity, the particular species, and the balance between available nutrients and light (Rosenzweig and Abramsky 1993, Tilman and Pacala 1993). Bass et al. (1998) hypothesized that energy-efficient specialized species replace more generalized, less energy-efficient species with time. Consequently, ecosystems develop toward more efficient utilization of energy and greater species diversity; less thermal energy will be released from systems that more efficiently utilize incoming energy. More mature forest ecosystems with higher levels of species diversity will emit less thermal energy than immature forest ecosystems.

Recent research has examined this concept of energy efficiency with landscape-scale thermal remote-sensing data. For example, Fraser and Kay (2004: 333–334) have observed that:

> ...the one unmistakable trend is that the more developed ecosystem is cooler...a grassland is warmer than an adjacent forest...a lawn had the warmest temperature, an undisturbed hay field was cooler, and a field which has been naturally regenerating for 20 years was coldest...in conclusion, ecosystem surface temperature appears to correlate with ecosystem maturity...the Second Law of Thermodynamics has an important role to play in understanding ecosystem phenomena.

The relationships between thermal condition and plant-ecosystem stress (photosynthetic stress or water stress), and ecosystem health have been identified as one way that remote sensing could contribute to an understanding of habitat distribution and biodiversity (Innes and Koch 1998, Moran 2004).

Species richness patterns may be explained by some measure of available energy, but the actual mechanism is not yet clear. Rates of evolution, thermoregulatory load, and population size may be constrained by energy (Phillips et al. 2008). Initial studies of energy relationships have used a number of climate variables to represent energy including potential and actual evapotranspiration, ambient temperature, precipitation, and water energy balance. For example, rainforest frog-species richness in 22 subregions of the Australia Wet Tropics was explained with different categories of variables by Williams and Hero (2001), who emphasized that it is not sufficient to examine patterns based purely on species richness within a broad taxonomic group, but instead it is necessary to use meaningful and objective groups based on functional ecology in order to understand the determinants of biodiversity:

- *Habitat generalists.* The species richness of habitat generalists was largely unaffected by rainforest variables and was primarily related to broad habitat diversity and climate.
- *Non-microhylid rainforest frogs.* The spatial patterns of species richness of non-microhylid rainforest frogs were the result of processes associated with historical biogeography, especially extinctions and subsequent recolonisations in those subregions most affected by Quaternary fluctuations in rainforest area.
- *Microhylid species.* In contrast, the most significant influence on spatial patterns of microhylid species richness may have been *in situ* speciation in areas of consistent rainfall, driven by altitudinal gradients, isolation, and low dispersal ability.

Estimates of available energy using the satellite-derived terrestrial gross and net primary production global data sets have significantly improved (Zhao et al. 2005), and operational net primary productivity models which rely on remote-sensing input (Running and Hunt 1993, Chen et al. 1999, Running et al. 2004) have been implemented at global, regional, and site-specific scales. Researchers working with these data, together with improved capability to use medium- and high-spatial resolution imagery to model NPP, showing interannual variability, and seasonality (Schmidt et al. 2008), indicate that many more such studies correlating specific habitat conditions and biodiversity with net primary production or other measures of energy will be forthcoming (Turner et al. 2003, John et al. 2008).

Water Resources and Quality

Ecosystem water balance may be of particular interest in many wildlife management situations and as a strong influence on biodiversity (Currie 1991). An inventory of intermittent and permanent streams, seeps, springs, vernal pools, ponds, swamps, marshes, and lakes will, in many situations, be necessary to determine critical habitat (temperature, sediment load, chemical constituents, nutrient flux) and proximity relations (Phinn et al. 2006, McComb 2008).

The hydrological cycle will influence habitat and biodiversity (Fig. 2.4; Waring and Running 1998, Franklin 2001).The most important effects may be as a result of the evapotranspiration flux, comprised of

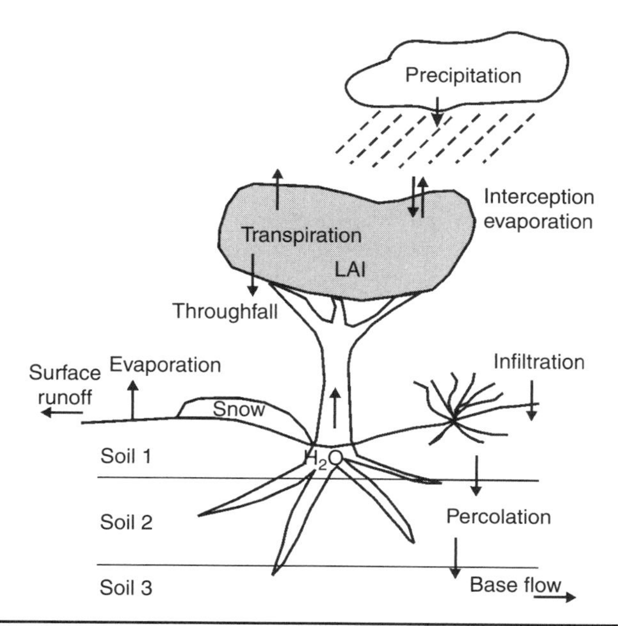

Figure 2.4 The hydrological cycle represented by a forest water balance. (*With permission from* Waring, R. H., and S. W. Running. 1998. *Forest Ecosystems: Analysis at Multiple Scales*. 2d ed. Academic Press, New York.)

plant transpiration, canopy and litter interception, and evaporation from soil, water and vegetation surfaces. Water, in the form of precipitation, is intercepted and evaporated from the surface of vegetation and the litter layer, and infiltrates the soil. Extraction by roots, surface runoff, and percolation through to the water table is affected by the density and depth of root channels and soil organic matter (Waring and Running 1998). Water taken up by plants may be stored temporarily within the stem, branches, and foliage, but most is transpired from the leaves to the atmosphere through leaf stomata. As soils dry out, resistance to water flow into tree roots increases, and stomatal conductance is reduced (Peterson and Waring 1994). Heat and water vapor transport processes, such as transpiration, are strongly influenced by LAI and by vegetation roughness.

Estimates of these processes may be developed through leaf-area-index and branch structural characterization by field or remote-sensing methods (Running et al. 1989). Runoff, surface and subsurface flows, and storage components can be modeled with access to topographic data and estimates of biophysical condition (Band et al. 1991, Minar and Evans 2008). Mechanistic ecosystem process models have advanced considerably in recent years, and can be used to identify the flux and mechanisms influencing water, carbon, and nitrogen movements. Models that predict spatial pattern and processes in ecosystems are increasingly relying on remote-sensing information in all aspects as necessary inputs (Franklin 2001, Zheng et al. 2006). Typically, model simulations are used to explore the sensitivity of patterns and processes to land-use changes, energy, and disturbances related to landscape heterogeneity, and multiple scale effects.

Ecosystem Health and Integrity

Earlier sections have introduced some research and management information needs in the broad areas of ecosystem processes (e.g., photosynthesis, nutrient cycling), and ecosystem configuration (e.g., distribution of patches, patterns of disturbance). Processes and configuration have strong relationships to emerging concepts of ecosystem integrity and sustainability (health) (Rapport et al. 1998, Robinson 2008). Ecosystem or habitat integrity suggests the provision of *ecosystem services* and a self-correcting capacity when subject to disturbance. Integrity is typically defined relative to a natural state or historical baseline—for example, *ecosystems have integrity when they have their native components (plants, animals and other organisms) and processes (such as growth and reproduction) intact*. The measurement or monitoring of ecological integrity shares many information needs with those associated with specific wildlife management and biodiversity issues (Woodley and Kay 1993, Crabbe et al. 2000). A key requirement is an understanding of reference conditions against which the current situation can be compared and contrasted (theoretically or practically) (Hessburg et al. 2004, Tiner 2004).

Using a freshwater coastal wetland to illustrate, Grabas et al. (2006) outlined an approach to meet specific information needs to understand ecosystem services:

> Freshwater coastal wetlands provide extensive ecological services to the Great Lakes basin. They cleanse water, mitigate effects of weather events, provide diverse biological resources, generate primary production, provide food, cover, and breeding/rearing habitat for a multitude of fauna, and offer excellent recreational opportunities. Threats to this natural capital include watershed development, adjacent urbanization, and lake-level water regulation....(what is needed is) the development and application of an interdisciplinary regional monitoring framework with biotic community assessments of vegetation, fish, birds, amphibians, and macroinvertebrates.... supported by GIS monitoring at the watershed scale (land use, vegetation cover) and by field surveys at the site level (water/sediment chemistry, bathymetry, water levels). Results have a range of applications, including helping to set ecological thresholds and refining delisting criteria in the Lake Ontario Area of Concern and monitoring the effectiveness of wetland restoration efforts to aid in adaptive management.

Ecosystem health and integrity monitoring as a direct application of remote sensing is in its infancy—indeed, definition and use of the concepts of health and integrity, sampling issues, cumulative impacts, ecosystem services, and measurement of reference conditions, are all still very much a subject of debate and refinement (Polls 1994, Quigley et al. 2001). Two examples are used here to illustrate the state of maturity in the conceptual development of ecosystem health and integrity and practical use of the idea in management:

1. *River system health.* Typically, measures of river health or river system integrity are developed from a wide range of input variables; for example, Amis et al. (2007) used a weighted function of measures such as fish assemblages, macrophytes, bank erosion, water quality, and flow modification to rate river health. Fourteen land-use, settlement and infrastructure variables available in the GIS database quantified river integrity across watersheds in South Africa (Table 2.4). Results emphasized the critical role of field data collection, but also suggested that riparian and instream habitat integrity of river systems could be predicted in the GIS model with reasonable accuracy, leading to a monitoring component for the state of rivers in South Africa.
2. *Forest ecosystem health.* Generally thought to be a crucial component of ecosystem health, and identified as one of the sustainability indicators as part of the Montreal Process, forest health tends to be poorly defined and approaches are consequently highly varied (Stone et al. 2001, Robinson 2008).

Selected Measures for River Habitat Integrity Prediction
• Natural vegetation cover (forest and grassland)
• Plantations
• Degraded areas (low vegetation cover)
• Cultivated areas
• Rural clusters (dwellings)
• Residential/commercial areas
• Industrial
• Mines
• Eroded areas
• Roads
• Dams
• Population density

Source: With permission from Amis, M. A., M. Rouget, A. Balmford, W. Thuiller, C. J. Kleynhans, J. Day, and J. Nel. 2007. Water SA 33: 215–221.

TABLE 2.4 A Selection of Measures Derived from a GIS Database to Predict River Habitat Integrity in South Africa

Remote-sensing approaches include detecting indicators or markers of physiological or structural response to stress (e.g., derived from leaf reflectance, canopy chemistry, or bioindicator plants), and capturing long-term changes in health and vigor by classification and measuring characteristics of vegetation development (Dendron Resources Inc. 1997, Olthof and King 2000, Xiao and McPherson 2005, Liu et al. 2006). Resource managers may be interested in vegetation chemistry and pigment concentrations, for example, for process modeling purposes and directly estimating forage quality influenced by the chemical constituents. For example, some plant species contain high levels of phenols, which may influence their digestibility for herbivores, some of whom, in turn, produce a substance called prolene in their saliva that binds with phenols and reduces their effectiveness (McComb 2008).

Habitat and Biodiversity Indicators

Habitat Suitability

Habitat suitability is typically defined as the ability of the habitat in its current condition to provide the life requisites of a species (British Columbia Wildlife Habitat Rating Standards 1999). A habitat suitability index (HSI) is useful for representing in a simple, understandable, and deterministic form, the major environmental factors thought to most influence species occurrence and abundance (Morrison et al. 2006). *Habitat capability* is an ideal or comparative habitat situation based on assumed or measured optimal conditions. Thus, habitat capability can exceed habitat suitability for a wide variety of reasons, including seral stage of vegetation, adjacent disturbance, or human use factors. A habitat integrity indicator may be derived by a comparison of the current habitat conditions to the assumed or ideal capability or to a reference condition (Tiner 2004, Juntii and Rumble 2006).

Habitat quality is sometimes defined separately as a condition considered to exist when the habitat capability and suitability are identical. *Habitat effectiveness* is typically defined separately from habitat quality, capability, and suitability, as a proportion of suitable habitat available to a species (Kansas 2003). Habitat effectiveness and quality, in particular, are subjective interpretations based on a range of existing conditions, including in some (but not all) instances, species abundance or density (Van Horn 1983)—clearly, suitable habitat may be unoccupied for a large number of reasons. Weisberg et al. (1997) and Diaz et al. (2004) provide examples of habitat integrity and quality in a benthic environment, and those presentations demonstrate a continuing evolution in the use of such terms.

All of these habitat concepts can be viewed as species-environment hypotheses rather than as representative of causal relationships, since information on population trends, and known behavioral responses of individuals, are only infrequently included. Interaction terms and statements of uncertainty or variability are also uncommon in development of such habitat indicators and relative measures of habitat quality (Morrison et al. 2006). Some observers have remarked on these limitations, and the preponderance of habitat terms and definitions, suggesting these may be a factor in continuing confusion and lack of clarity in the pursuit of appropriate habitat methods and analysis approaches (e.g., Bissonette and Storch 2003).

Monitoring Programs

Most biodiversity monitoring programs contain a common core or suite of measures of biodiversity components, including habitats, for defined units of space and time (Henry et al. 2008). The measures may be organized into a framework that identifies trophic levels and scale or hierarchical levels of interest.

Species monitoring, combined with habitat (or general environmental conditions) monitoring, provides the most important opportunity to assess biodiversity change (Henry et al. 2008). In species monitoring, typically four main data types are used: presence/absence, counts (of individuals, including vegetation coverage), individual follow-up (capture-mark-recapture data), and measures of individual traits (e.g., age, size, health of individuals). Multispecies indicators may be combined from single-species protocols, and used to construct biodiversity indicators. An ideal, though rare, situation is to include in such indicators habitat specialization, functional traits, trophic levels, or other species traits. Generally, traditional habitat and biodiversity indicators, such as species counts or total area under different management planning or protection, are reasonably familiar to many environmental managers and decision makers. Remote sensing-based habitat and biodiversity indicators are less well-known (Tiner 2004, Strand et al. 2007).

An example of a list of coastal indicators designed to reveal changes in Australian coastal systems is part of the *Coastal Indicator and Knowledge System* (Table 2.5, OzEstuaries 2003, Robinson and Culley 2003). Indicators are defined for their ability to represent key elements of an important environmental issue or ecosystem function. The developers of this system provide some cautionary remarks that apply to all indicator-based monitoring programs:

- Indicators may reveal trends, but do not show cause and effect
- Many indicators have not been validated
- Interpretations can easily become overly simplified, misleading, and erroneous
- Monitoring is not a substitute for a comprehensive research program

Issues of scale are of central importance when adopting an indicator approach in monitoring programs. National assessments of biodiversity typically rely on field measures and methods of *scaling-up* from plots or sites to larger and larger areas, culminating in global assessments. Many such measures currently under development as part of the CBD 2010 Target require a remote-sensing component (Strand et al. 2007). A variety of regional biodiversity monitoring programs based on selected indicators are already in place, but with flexibility according to the particular region and stage of development. To cite only one example described in more detail in later sections of this book, the *Alberta Biodiversity Monitoring Institute* (ABMI) program (Stadt et al. 2006) considers the organization of biodiversity in an hierarchy of four levels (regional landscapes, communities-ecosystems, species-populations, and genes), each with three primary ecosystem attributes (composition, structure, and function). Indicators, such as amounts and patterns of landscape vegetation types (from remote sensing), local habitat structure, and species counts are sampled in specific monitoring protocols. The goal of this regional biodiversity monitoring program is to ensure that biota are responding as predicted by models of landscape and habitat change associated with human use:

> The ABMI focuses on monitoring the structure of ecosystems and species. The ABMI also monitors human footprint to correlate with biodiversity

Biophysical Indicators	Pressure Indicators
Water quality • Chlorophyll a • CO_2 partial pressure • Dissolved oxygen • Marine pathogens • Metal contaminants • pH • Salinity • Turbidity • Water column nutrients • Water temperature *Sediment quality* • Benthic CO_2 flux • Denitrification efficiency • Organic matter & nutrients • Sediment P/R ratio • Sedimentation rates • Sediment TOC:TS ratios • Toxicants *Habitat extent & quality* • Beach & dune indicators • Changes in mangrove areas • Changes in salt marsh areas • Changes in seagrass areas • Changes in wetland coverage • Index of habitat variability • Isohaline position • Maturity index & lifespan *Biotic indicators* • Benthic invertebrates • Diatom species composition • Fish assemblages • Hermit crabs • Imposex frequency • Intertidal invertebrates • Macroalgal indicators • Seagrass species • Shorebird counts	*Agricultural* • Impoundment density • Percent of agriculture on steep slopes • Percent of agricultural land area • Pesticide hazard • Rivers through forest *Industrial* • Aquaculture pressure • Coastal tourism • Fishing pressure • Industrial point source hazard • Shipping, boating, and yachting *Urban* • Coastal discharges • Coastal population • Nutrient point source hazard • Stormwater discharges *Catchment condition* • Catchment condition index • Native vegetation extent • Rivers in acid hazard **Coastal Management Indicators** • Coastal care community groups • Marine network participation • Marine protected areas

Source: Adapted from OzEstuaries (2003) and Robinson and Culley (2003).

TABLE 2.5 Coastal Indicators; Biophysical, Habitat, and Human Pressure Indicators Required in a Comprehensive Coastal Management Information System for Australian Coasts

information. Specifically, the ABMI implements protocols to sample the state of vascular plants, mosses, fungi, lichens, phytoplankton, birds, mammals, fish, springtails, mites, zooplankton, benthic invertebrates, live and dead trees, down logs, soil, shrub cover, vegetative litter, water physiochemistry, water basin characteristics, and landscape characteristics. Ultimately, the ABMI monitors the status and trends of more than 2000 species, more than 200 habitat elements, and more than 40 human footprint variables. http://www.abmi.ca/abmi/home/home.jsp; accessed *20 July 2008.*

Clearly, the list of biodiversity indicators that might be of interest in a regional or national biodiversity monitoring program is large and still growing. In Canada alone, for example, at least ten such initiatives to monitor different aspects of national biodiversity have been proposed or partially developed in the past decade (Ahern 2007). For example, low spatial resolution landscape change indicators obtained by satellite remote sensing and intended as broad biodiversity or macrohabitat indicators would include land-cover change, burn-scar maps and other disturbances (e.g., defoliation by insects), leaf-area index, length of growing season, indicators based on photosynthesis, and net primary productivity, potential vegetation and variations from potential vegetation, and snow cover and ice formation on lakes (as an indicator of climate change). Many of these are described in greater detail in later sections.

Globally, the Convention on Biological Diversity (Decision VIII/15) includes a recommended process in which specific biodiversity and habitat indicators are designed and selected for use in any particular region or ecological setting (de Sherbinin 2005, Strand et al. 2007). Overall, indicators should be designed to provide information on the *rate of decline in biodiversity*—and, therefore, three data points through time are required. Remote sensing has been identified as an immediate and important contributor for the following broad categories of indicators:

- Trends in extent of selected biomes, ecosystems, and habitat
- Trends in abundance and distribution of selected species
- Change in status of threatened species
- Trends in invasive alien species
- Water quality of freshwater ecosystems
- Connectivity or fragmentation of ecosystems

Within each of these areas, many possible indicators have already been, or will be soon, developed for use (Noss 1990, Franklin and Dickson 2001). Species richness, historically an important biodiversity indicator, is increasingly well-established (Nagendra 2001), but for many ecosystems, simple richness measures must be supplemented with biodiversity indicators of habitat integrity (Tiner 2004), floristics (Egler 1954, Wardell-Johnson et al. 2004), productivity and phenology (Bradley and Fleishman 2008), and alternative measures based on structure and community ecology (Wardell-Johnson and Horwitz 1996, Ellner et al. 2001), to ensure credibility in monitoring and conservation programs (Turner et al. 2003, Ahern 2007).

A selection of the many possible biodiversity indicators is listed in Table 2.6 to illustrate the wide range and diversity of ideas in this

Forests	Marine and Coastal Habitats
• Extent of component ecosystems • Forest change • Rate of deforestation/reforestation • Forest intactness • Area and number of large forest blocks • Forest fragmentation • Carbon storage • Area and location of old-growth forest • Area and location of plantations • Forest degradation • Area and location of sustainable forestry • Alien species • Fire occurrence	• Extent of coral reef ecosystems • Percent cover of living coral • Coral bleaching—direct observation • Coral bleaching—temperature proxy • Water quality—concentrations TSM/Tripton, Chla, CDOM • Algal blooms • Depth, substrate type • Extent and biomass of seagrass ecosystems • Extent and biomass of mangrove ecosystems • Mangrove habitat conversion (aquaculture, agriculture) • Changes in mangrove habitat extent resulting from natural hazard (e.g., tsunami) • Change in extent of mangroves, coral or seagrass as a result of regeneration or restoration • Biomass of mangroves • Connectivity between mangroves and their associated ecosystems
Dry and Subhumid Lands	**Protected Areas**
• Extent of grassland, desert and Mediterranean ecosystems • Intact biodiversity • Land degradation • Grazing pressure • Extent of alien species invasion • Climate change • Fire location and frequency	• Area of protected areas • Size distribution of protected areas • Representation of protected areas • Isolation of protected areas • Landscape condition adjacent to protected areas • Level of encroachment on, or degradation of, protected areas

Table 2.6 Habitat and Biodiversity Indicators Thought to be Amenable to a Remote-Sensing Approach Organized by Biome or Broad Environmental Theme (*Continued*)

Inland Waters, Riparian Areas, and Wetlands	Fragmentation and Connectivity
• Extent of inland waters, wetlands and riparian zones • Population trends of wetland taxa • Wetland restoration • Floodplain size and flooding • Riverbank components (slope, shape) • Distance from water edge to clear/developed land • Channel/floodplain width • Characteristics of pools/riffle/runs/backwater • Woody material (snags, type and sizes, recruitment) • Bed material/benthic types • Vegetated streamlength/connectivity • Variability of water levels and extent • Weir/number and distance to barriers/ • Coupling biological and physical assessments (e.g., pelagic Chla) • Water processes (dissolved organic matter, turbidity, temperature) • Changes in habitat quality and ecosystem quality	• Total number of land cover types • Patch size (largest, average) • Path density • Perimeter-to-area ratio • Core area index • Fractal dimension • Distance to nearest neighbor • Contagion • Juxtaposition index • Road length • Road density **Invasive Alien Species** • Area, distribution and trends in particular invasive alien species • Prediction of the distribution of invasive alien species • Indirect identification of areas vulnerable to invasion • Identification of potential sources of invasion and dispersal

Source: Adapted from Environmental Remote Sensing Group (2003), Petersen et al. (2004), Phinn et al. (2006), and Strand et al. (2007).

Table 2.6 Habitat and Biodiversity Indicators Thought to be Amenable to a Remote Sensing Approach Organized by Biome or Broad Environmental Theme (*Continued*)

increasingly urgent area of development. Some of these indicators are highlighted in later chapters of this book. Some additional more detailed sources of interest include the following resources:

1. The contribution of remote sensing to the development and use of indicators in four selected ecosystems (forests, dry and subhumid lands, inland waters, and marine and coastal

habitats) was summarized by Strand et al. (2007) with further discussion of remote-sensing indicators for species populations, protected areas, fragmentation, and invasive alien species.

2. Phinn et al. (2006) has provided a comprehensive summary of specific coastal aquatic ecosystem biodiversity indicators of interest in Australia.
3. Petersen et al. (2004) has provided some early thoughts on development of circumpolar biodiversity indicators designed to enhance conservation, monitor resource use, reduce significant loss, and convey the complexity of the Arctic biodiversity. The Circumpolar Biodiversity Monitoring Program is described in Chap. 6 of this book (*see also* CBMP 2007).
4. The RAMSAR Convention on Wetlands included recommended indicators for monitoring the biodiversity of wetlands.
5. Four continental-scale macroscale biodiversity indicators for the Canadian landscape were recommended by Duro et al. (2007). Topography, vegetation productivity, land-cover fragmentation, and disturbance were highlighted and form the basis of the Canadian BioSpace biodiversity monitoring program discussed in more detail in Chap. 6.

Limitations of Remote Sensing

Only a portion of the many existing and possible wildlife habitat variables and biodiversity indicators will be available through the use of a remote-sensing approach. Even with the remote-sensing indicators that have been proposed or developed, certain limitations may prevent successful application. Particular broad areas of concern have been noted repeatedly over the past few decades, as the remote-sensing approach has continued to develop and expand, and familiarity with remote sensing has increased. Notable are recurring problems with financial resources, data and methods availability, accuracy, validation, and human expertise. In the practical use of remote sensing in monitoring the tropical coastal zone, for example, Holden and LeDrew (1998) identified all of these broad constraints associated with remote-sensing data, and then highlighted specific challenges associated with image acquisition (cloud cover), lack of ground control, low accuracy, and inadequate user training in optimal ways to extract the information required.

Certain operational capability of remote sensing has not yet been developed despite "theoretical" advances that suggest feasibility (e.g., detecting river discharge). This is not only a function of the apparent divide that has occasionally been observed between pure and applied remote-sensing research (Andréfouët 2008). Even if applications are proven, aerial and satellite data are often not thoroughly integrated with other elements of environmental monitoring and observing systems [e.g., coastal water quality monitoring systems (Christian et al. 2006)]. Turner et al. (2003) considered the challenge of uncertain satellite data continuity to be significant; monitoring and research programs that are dependent on one or two systems can sometimes be threatened when unexpected technical problems arise (e.g., failure of Landsat 6 to reach operating orbit). Experience over many different remote-sensing projects suggests emphasizing one broad challenge frequently encountered in complex environmental monitoring applications; the need for expertise and software skills required to work with ecological models, which have a vital role in the process of converting remote-sensing data into ecological knowledge (e.g., predictions of species distributions and richness). This is part of the general problem of human resource development, which can take many forms and serve to undermine progress in remote-sensing monitoring systems and research initiatives.

Through proper development of the remote-sensing application framework and careful planning, implementation, and evaluation, many of the broad practical constraints will be avoided. However, certain technical constraints that flow from the fundamental characteristics of the remote-sensing endeavor—for example, image properties associated with the spatial and spectral resolving power of available sensors—may not be so easily managed.

A summary of the general limits in airborne and satellite remote sensing is provided in Table 2.7 (Landgrebe 1997, Franklin 2001). These are expressed as limitations that are frequently cited or encountered in remote-sensing projects. However, each limitation can just as easily be viewed as an opportunity to strengthen the remote-sensing approach. For example, entering a remote-sensing project with a lack of clear objectives has, without question, resulted in initially poor project design and inadequate results; conversely, a well-designed remote-sensing project will more readily flow from a statement of clear objectives.

Many of these limitations will be apparent in specific applications and are considered in detail in Chap. 3 and in later sections. The material presented in Chap. 3 will help explain the fundamental image properties, data sets and methods in remote sensing. The emphasis in the later chapters is on how to use remote sensing to derive appropriate habitat and biodiversity measures—including simple and complex indicators—for use in management situations. These presentations will help clarify the essential limitations and choices in developing a remote-sensing approach. Chapter 4, in particular, reviews appropriate methodology in the remote-sensing approach, with a few applications highlighted to illustrate where the approach has been applied successfully. The focus is on understanding

General Limits Encountered in Airborne and Satellite Environmental Remote Sensing

• Lack of clarity in statement of objectives
• Lack of appropriate remote-sensing data (e.g., cloud-free images, multipolarization SAR imagery)
• Lack of institutional commitment (management inertia); lack of a strong link between monitoring, research, and policy goals
• Lack of infrastructure to handle remote-sensing data
• Lack of availability of trained and experienced remote-sensing analysts
• Lack of an effective link between apparent "dualisms" (e.g., scientist/management, user/producer); continued existence of barriers to transdisciplinary practice
• Lack of expression of clear benefits compared to obvious and real costs of remote sensing
• Lack of congruence between the fundamental nature of remote sensing and the application; for example:
  • Spatial and radiometric data characteristics—the pixel as the fundamental unit of observation at specific scales
  • Temporal resolution—timing of image acquisition and phenomena of interest
  • Poor information extraction methods—lack of rigorous and standardized or acceptable analysis procedures

Source: Adapted from Landgrebe (1997) and Franklin (2001).

Table 2.7 General Limits Encountered Frequently in Airborne and Satellite Environmental Remote Sensing

the various degrees of success, and lessons learned, in extracting key information of value in biodiversity and wildlife management. Chapters 5 and 6 are designed to highlight specific examples and case studies of remote sensing in wildlife management and biodiversity research and monitoring applications, respectively.

CHAPTER 3

Remote-Sensing Data Collection and Processing

One fact was inescapable; with advances in computer technology and increasing abundance of imagery and other records from space, the earth sciences were in danger of being flooded with more data than could be sorted out and utilized.

—*E. Pruitt, 1979*

Image Characteristics

Image Formation

A wide variety of data and imagery from field, aerial, and satellite sensors may be required in any given wildlife or biodiversity management application. A basic understanding of the characteristics of these data and imagery is necessary to make appropriate decisions on the acquisition, use, and evaluation of remote-sensing approaches.

The image formation process is a good place to start to develop this understanding, and some excellent reference texts are available on this process, and related remote-sensing questions, for those seeking greater detail. For example, the first three editions (respectively, Reeves 1975, Colwell 1983, and Ryerson 1998) of the authoritative American Society for Photogrammetry and Remote Sensing *Manual of Remote Sensing* are an excellent source of information on the principles of image formation by instrumentation designed to operate in different parts of the electromagnetic spectrum. Three recent volumes in the *Manual of Remote Sensing* third edition series highlight advances in remote sensing for natural resources management (Ustin 2004), marine remote sensing (Gower 2006), and remote sensing of human settlements (Ridd and Hipple 2006). The accelerating pace of developments in remote-sensing instrument-and-platform design and

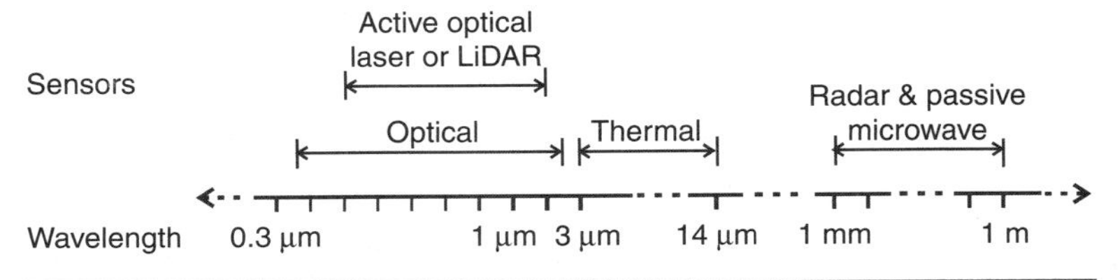

Figure 3.1 The electromagnetic spectrum showing principal remote-sensing bands discussed in this book.

capability over the past three and a half decades can be clearly discerned in these important references.

Image formation results from the detection of electromagnetic energy, generically referred to as *spectral response*, which is subsequently output in a spatially explicit digital (computer media) or chemical format (e.g., silver halide grains embedded in photographic emulsion). Briefly introduced here are five types of remote sensing image-formation-systems designed to detect spectral response patterns in different regions of the electromagnetic spectrum (see Fig. 3.1): thermal, passive microwave, active microwave (radar), optical, and active optical (laser or LiDAR). These regions have been selected for presentation here because of their applications in biodiversity and wildlife management, discussed in later chapters. Remote sensing in other regions of the spectrum—such as ultraviolet fluorescence, or acoustical and sonar sensing—may also have such applications, but are beyond the scope of this book. The reader is referred to Aronoff (2005), Jensen (2007), and Lillesand et al. (2004), among others, for further introduction and direction to the literature for remote sensing in the full range of the electromagnetic spectrum.

Thermal imagery is acquired in two separate wavelength regions, also known as bands, from 3 to 8 μm and from 8 to 14 μm. These bands occur in areas of the electromagnetic spectrum where atmospheric influences (primarily through water vapor and uniformly mixed gases) on spectral response patterns are relatively minor. The levels of energy measured remotely in these bands are closely related to the atmospheric water content and temperature profile, surface temperature, and emissivity of the surface. Emissivity is the radiating efficiency of the surface, and if this value is not known, remotely sensed thermal-image-values are interpreted as *radiometric temperatures* (sometimes called *brightness temperatures*) rather than true surface temperatures (Liang 2004, Quattrocci and Luvall 2004). More than a dozen thermal cameras are commercially available, from the consumer-product series of professional grade, hand-held infrared thermal imaging camera systems, such as the TVS-200EX and TVS-500EX manufactured by AVIO (Nippon Avionics), to a wide assortment of increasingly complex mounted systems designed to operate from aerial and satellite platforms, or in the field (Table 3.1).

Thermal Imagers
• FLIR Systems: ThermaCAM SC3000™
• IMSS—The mid-wave infrared IMSS
• Indigo Systems—The Merlin®
• Indigo Systems new Alpha NIR with VisGaAs sensor
• Infrared Solutions IR-320™—Thermal Imager
• IRCON quality real-time video temperature representation
• Radiance HS (RHS) Camera
• Sierra Pacific Innovations IR camera system
• VARIOSCAN Scanning
• ABS—Airborne Bispectral Scanner
• TASI 600—Pushbroom: Hyperspectral Thermal Sensor System

Source: Adapted from http://fullspectralimaging.net/LIAARS.aspx Accessed February 24, 2009.

Table 3.1 A Selection of Commercially Available Aerial Digital-Thermal Imagers and Cameras for Environmental Remote Sensing.

Passive microwave imagery is acquired by detecting weak microwave emissions from surfaces and the atmosphere. The levels of energy emitted in these regions are closely related to temperature and to moisture content (Sharkov 2003, Woodhouse 2006, Liang 2008). Typical environmental applications have included integrated cloud liquid and water vapor measurements, precipitation, temperature, and water vapor profiling, sea-surface wind speed, temperature and salinity, sea-ice concentration and age, soil moisture content and biophysical properties of snow/ice/water/soil features (e.g., glaciers). (Astronomers use passive microwaves to view pulsars, star forming regions, and black holes). Passive microwave instruments on aerial remote-sensing missions are primarily designed for research. Two examples are the polarimetric scanning radiometer (PSR) (Fig. 3.2) designed to obtain polarimetric microwave emission imagery of oceans, land, ice, clouds, and precipitation, and the airborne imaging microwave radar (AIMR) (Table 3.2). The AIMR detects surface and atmospheric microwave emissions at 37 and 90 GHz.

Active microwave imagery is produced by artificially illuminating a surface using relatively short microwave wavelengths (0.75–100 cm) and recording the reflected energy in specific wavebands typically coded using the original RAdio-Detection-and-Ranging (radar) military designations (*see* Table 3.3 which includes a selection of commercially available radar remote-sensing instruments). Radar remote sensing has a long and storied history of military and intelligence applications (Olsen 2007), and environmental work (Rees 2001). Atmospheric influences on active microwave spectral response

(a) PSR as installed on the NASA Orion P-3B aircraft

(b) PSR scanhead illustrating internal construction

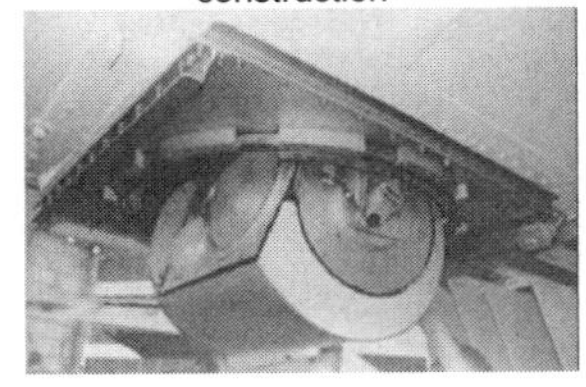

(c) PSR installation on the NASA DC-8

FIGURE 3.2 The polarimetric scanning radiometer, a passive microwave remote-sensing instrument used to obtain polarimetric microwave emission imagery of oceans, land, ice, clouds, and precipitation. (*Courtesy of* A. Gasiewski, University of Colorado at Boulder.)

Parameter	37 GHz	90 GHz
Frequency	37 GHz ± 0.1 GHz	90 GHz ± 0.3 GHz
Instantaneous Field of View (IFOV)	2.4°	1°
Dynamic Range of Apparent Brightness Temperature	0–350K (spec) 60–315K (display)	same
RMS Pixel Noise	1.3K max (spec) < 1.0K (acutal)	1.3K max (spec) < 1.5K (acutal)
Sampling	Every 0.5°	Every 0.5°
Integration Time	Adjusted in steps with V/H	Adjusted in steps with V/H
Typical IF Bandwidth	1.5 GHz	2 GHz
Antenna Beam Efficiency at 2.5 times half power angle	> 90%	> 90%
Operational Modes	Scanning and profile	
Scanning/Profile Angle Method	Rotating mirror	
Max Scan Rate	5.8 rps	
Scan/Profile Angle Range	± 60°	
Scan Line Advance (for V/H < 30 knots/k-foot)	0.5°	
Calibration Technique	Calibration loads viewed by scanner	

TABLE 3.2 Specifications for the (passive) Airborne Imaging Microwave Radiometer (AIMR)

Selected Operational and Commercial SAR Systems for Environmental Applications	Military Wavelength Code Designation	Web site (accessed 7 August 2008)
IFSAR STAR3i NextMap®	X-band	http://www.intermap.com/interior.php/pid/1/sid/317
Radarsat-2	C-band	http://www.space.gc.ca/asc/eng/satellites/radarsat2/default.asp
JERS-1 SAR	L-band	http://www.eorc.jaxa.jp/JERS-1/index.html
NASA/JPL AIRSAR/TOPSAR	Multifrequency	http://southport.jpl.nasa.gov/airsardesc.html
BioSAR™	Multifrequency	http://www.terresense.com/index.html
DLR F-SAR	Multifrequency	http://www.dlr.de/hr/en/desktopdefault.aspx/tabid-2326/3776_read-5691/
ERS-2 AMI	C-band	http://www.esa.int/esaEO/GGGWBR8RVDC_index_0.html
GEOSAR	P- and X-band	http://www.fugroearthdata.com/servicessubcat.php?subcat=radar-mapping
Orbisar-RFP	P- and X-band	http://www.orbisat.com.br/ingles/html/interna.php?chave=tecnologia
TerraSAR-X	X-band	http://www.dlr.de/tsx/start_en.htm
ALOS-PALSAR	L-band	http://www.eorc.jaxa.jp/ALOS/about/palsar.htm
ENVISAT ASAR	C-band	http://envisat.esa.int/instruments/asar/
YaoGan WeiXing 1/3 SAR (Military designation JianBing5)	L-band	http://www.sinodefence.com/strategic/spacecraft/jianbing5.asp
Ramses/Sethi SAR	Multifrequency	http://www.onera.fr/demr-en/facilities.php

Table 3.3 Selected Operational and Commercial Airborne and Spaceborne SAR Systems

are minimal and the recorded energy levels are strongly dependent on sensor-target geometry (e.g., azimuth range and topography), moisture, electrical, and structural properties of the surface. *Radar polarimetry* refers to the orientation of the electrical field relative to the wave's direction of travel. Typical radar-image products are backscatter intensity, coherence, and interferometric response.

Both imaging and nonimaging radars are used in environmental remote-sensing applications developed initially to resolve remote-sensing challenges in difficult environments; for example, sea ice detection and thickness estimation, and topographic mapping in steep terrain, for which aerial cameras and optical sensors were not optimal (e.g., perennial fog or cloud-covered regions). Synthetic aperture radar (SAR) uses the platform forward motion to artificially increase antennae length and therefore resolution; SAR sensor and platform capabilities continue to improve. Recently, the German DLR F-SAR was introduced to meet the demand for simultaneous all polarimetric capability and single-pass interferometric capability in X- and S-bands (Horn et al. 2008). In the United States, a low-frequency SAR system called BioSAR ™ is a patented aerial remote-sensing device designed for the measurement of carbon and biomass in heavily wooded areas (over 100 tons per hectare) and tropical rainforest (Imhoff et al.. 2000, Nelson et al. 2007).

Optical imagery is by far the most frequently used remote-sensing imagery in environmental studies. The principal source of radiation is solar, and the wavelength regions in which sensors are designed to form images range from the ultraviolet to the short-wave infrared (350–2500 nm). The visible, near- and short-wave infrared wavelengths detected by optical sensors are influenced strongly by surface cellular and structural properties that control reflectance, transmissivity, and absorption. Common optical sensors include many different types of cameras and film combinations (e.g., Faulkner and Morgan 2002, www.kodak.com/go/aerial web site accessed August 6, 2008). Various types of multispectral scanners, spectrographic imagers, digital frame cameras, video systems, and certain other optical instruments have been developed using two basic types of digital detectors designed to be sensitive to spectral response in visible and near-infrared wavelengths—complementary metal oxide (CMOS), and the charge-coupled device (CCD) (Aronoff 2005; Table 3.4).

Active optical remote sensing is typically accomplished with medium-power lasers illuminating a surface and recording the amount of elapsed time for a speed-of-light laser pulse reflection to return (St.-Onge et al. 2003). Wavelengths in the range of 500 to 1600 nm are used, with most Light-Detection-and-Ranging (LiDAR) operating at near-infrared wavelengths between 800 and 1000 nm. The principal data provided by LiDAR sensors are used to determine precise surface geometry. Rapid improvements in LiDAR technology have created phenomenal data rates and new data characteristics of interest;

Multispectral and Hyperspectral Instruments
• AAHIS • ADAR • ABS—airborne bispectral scanner • AHI 256 • AA3505 Bispectral Scanner System • AHS-75, AHS-160—airborne hyperspectral scanner • AMS—airborne multispectral scanner • AIRCAM™ • AISA • AISA Eagle • AISA HAWK • APEX • AVIRIS • AVIS—The airborne visible and infrared imaging spectrometer • CASI 550, CASI 1500 (Compact Airborne Spectrographic Imager) • CHRIS—compact high resolution imaging spectrometer • GER—geophysical & environmental research corporation) • HYDICE • HyperCam™ Fourier transform hyperspectral imager • HyMap • HYPERION • HySpex™ • IMSS • MAS 50 • MASTER • NovaSol 1100, 1200, and 1600 Series • PHILLS—portable hyperspectral imager for low light spectroscopy • PROBE • ProVision Technologies • ROSIS—reflective optics system imaging spectrometer) • SASI—shortwave infrared airborne spectrographic sensor • SOC-700 • SpecTIR • SpectraView 4 • Spectra Vista (former GER line of airborne scanners) • TEEMS—texaco energy and environmental multispectral imaging spectrometer: Proprietary hyperspectral imager • VASIS: Combination of CAMIS and C2VIFIS systems (above) • The VNIR-20 • The VNIR-90 spectrometer • The VNIR 100E

Source: Adapted with permission from Bolton, J. 2008. Listing of instruments available for airborne remote sensing http://fullspectralimaging.net/LIAARS.aspx 24 February 2009.

TABLE 3.4 A Selection of Commercially Available Aerial Digital Multispectral and Hyperspectral Instruments Currently Used in Environmental Remote Sensing

for example, the Optech Airborne Laser Terrain Mapper (ALTM) 3100EA has data acquisition rates of up to 100,000 pulses per second and features intensity, full waveform digitization, simultaneous first/last pulse measurement, roll compensation, and digital camera integration (http://www.optech.ca/ web site accessed 6 August 2008). Three different LiDAR instruments are used to develop structural information on the illuminated surface, including:

1. 3-D terrain, vegetation canopy information, and object heights
2. Bathymetry and atmospheric constituents (using two wavelength systems)
3. Ground-based survey-grade measurements (e.g., Optech's iFLEX is designed to collect survey-grade LiDAR data at over 100,000 measurements per second with a 360° FOV, while maintaining a Class 1 eye safety rating)

Image Resolution

There are four types of image resolution, each of which influences scale, image interpretability, and remote-sensing experimental design (Franklin 2001):

- *Spatial resolution.* Spatial resolution is the projection of the detector element through the sensor optics within the sensor instantaneous field of view (IFOV). Spatial resolution is a measure of the smallest object that can be distinguished by the sensor, but the actual spatial detail apparent in any image is a result of more complex interactions. For example, in optical imagery, spatial resolution is a function of the IFOV and the signal sampling interval, which determines the actual pixel dimensions in the resulting imagery. Contrast between features on the surface and their background can strongly influence their detectability—narrow roads, for example, often contrast sharply with surrounding vegetation, and may be visible in some images even though the spatial resolution of the sensor would suggest such features are too small to be discerned. In SAR imagery, spatial resolution is calculated in the range and azimuth directions as a function of the length of the radar pulse and the depression angle of the antennae (Luscombe et al. 1993). Again, interactions among targets and surface features may increase spatial detail—small but strong corner reflectors, for example, will often be apparent in relatively coarse spatial resolution SAR images.
- *Spectral resolution.* Spectral resolution is the number and dimension of specific wavelength intervals or bands in the electromagnetic spectrum to which a sensor is designed to be sensitive. *Multispectral sensors* typically record spectral response

data in several different, relatively wide spectral bands (perhaps one or two each in the blue, green, red, near-infrared, short-wave infrared, and thermal infrared portions of the electromagnetic spectrum). *Hyperspectral sensors* record many more individual bands (often several hundred) in very narrow wavelength intervals at much higher spectral resolution.

- *Radiometric resolution.* Radiometric resolution is the sensitivity of the detector to differences in the signal strength, or radiant flux of energy, in specific wavelengths. This resolution is analogous to photographic film speed; the faster the film speed, the lower the light conditions that can be used to create photographic images. Higher radiometric resolution allows smaller differences in spectral response patterns to be discriminated. Typically, the detector signal has a gain applied before quantization with an analog-to-digital converter. The quantization determines the number of bits of data received for each pixel, and together with the signal-to-noise ratio of the system, determines the number of levels that will be represented in the image.
- *Temporal resolution.* Temporal resolution is the frequency with which an image is acquired over a given location. Temporal resolution is a product of the mission design and sensor capability. Aerial- and field-based remote sensing are often very flexible designs for temporal resolution considerations. Some of the initial civilian satellite remote-sensing systems, such as Landsat, had low temporal resolution because sensor revisit capability was determined solely by orbital parameters. Satellites placed in higher orbits, such as the TIROS-N and NOAA AVHRR platforms, carried sensors with a much higher temporal resolution. Many current satellite systems have programmable features such that sensors can be aimed at particular surface locations from multiple orbital positions. This off-nadir viewing capability effectively increases the satellite sensor temporal resolution.

Multiple observations of the same location from different positions introduces complex view-angle effects (King 1991), but also the capability of parallax-based computations (Cooper et al. 1987) and more precise determination of the angular characteristics of spectral response (Diner et al. 1999). An important consequence of increased temporal resolution in optical remote sensing is the ability to better estimate the surface reflectance.

Trade-offs in image resolution create a complex dynamic in information content that can have dramatic consequences in remote-sensing data acquisition and experimental design (Franklin 2001). How much energy is available to be sensed, how the energy is divided into different bands, how large an area is visible in the instrument IFOV, and the signal-to-noise ratio of the system, all need to be considered. The amount of energy available for sensing is fixed within the integration time of a detector. Divide the energy into too many bands over too small an area, and the signal within each band will be weak and noisy. An increase in the number of spectral bands (eg, moving from a multispectral to a hyperspectral sensor configuration over the same wavelength region) will be accompanied by a decrease in spatial resolution, all other components of the system being equal. If spatial resolution is not decreased, radiometric resolution will be lower. To acquire more and narrower spectral bands, the sensor must view an area on the ground for a longer period of time, and therefore, the size of the area viewed has to increase to maintain radiometric resolution. If the radiometric resolution is increased, so that a smaller radiant flux can be detected, the spatial detail, the number of bands, the narrowness of the bands, or all three, must be reduced (Mather and Koch 2004). In addition, the size of the viewed area (the IFOV, and therefore the pixel dimensions), will influence the relationship between surface features and image spatial detail in a complex dynamic.

Image Scale

In mapping, scale identifies a distance equivalency defined by a ratio of map, or image, to ground distance. By geographic convention, a small scale study will suggest a large area mapped with a low level of detail. A large scale study will be interpreted to mean a small area is mapped with a high level detail. Note that this is the inverse of the meaning intended in ecology of the words "large scale" and "small scale."

Mapping or image scale is related to remote-sensing spatial resolution but is not an equivalent concept (Franklin 2001). Low spatial resolution imagery will often be suitable for a small mapping scale or large-scale ecological study (i.e., covering a large area but without much spatial detail). High spatial resolution imagery will often be appropriate when using a large mapping scale in a small scale ecological study (i.e., covering a small area but with great spatial detail). Instead of specifying mapping or image scale when choosing imagery for a particular purpose, it is often more useful to characterize the ecological phenomena of interest relative to the capability of the imagery to capture that phenomena. This would be similar to categorizing aerial photography by referring to the smallest visible feature rather than by the photo scale (Aronoff 2005). For example, Coops et al. (2007) suggested the following broad guidelines for choosing an information source when detecting, analyzing, and mapping forest disturbances with remote sensing:

- Disturbances that occur over hundreds or thousands of meters (such as wildfires); use low spatial resolution imagery (covering large areas)

- Disturbances that occur over tens or hundreds of meters (such as forest harvesting patterns); use medium spatial resolution imagery (covering medium-sized areas)
- Disturbances that occur over centimeters to meter (such as individual tree mortality caused by insects); use high spatial resolution imagery (covering small areas)

Image Acquisition

Field Instruments

Many remote-sensing projects will include a field-based (or *in situ*) remote-sensing component ranging from plot or site photography to thermometry and spectroscopy using hand-held instruments. *In situ* sensing devices have been carried or mounted on ladders, vehicles, tripods, towers, booms, trees, ships, boats, floating rafts, submersibles, or any number of other reasonably stable platforms that can hold remote-sensing devices in advantageous positions for data acquisitions in the field (Blackburn and Milton 1997). Generally, the objectives of field data acquisition are to provide a fundamental understanding of spectral response and target interactions that will apply to remote sensing at all scales (Fournier et al. 2003), to provide data to develop and test models of spectral response, and for calibration/validation purposes (Milton et al. 1995, Teillet et al. 1997). To be effective, *in situ* observations require a detailed understanding of the experimental design and a well-thought-out and carefully executed field data collection protocol.

Remote-sensing data acquisition possibilities *in situ* are virtually limitless, and therefore only a few examples using optical instruments are discussed briefly here to introduce the concept and practice. Later sections of this book provide more complete discussion of applications with such field instruments.

Optical *in situ* measurements are collected for a wide variety of purposes and with an enormous variety of light-sensing instruments. For example, those working in forest or grassland environments often need to estimate vegetation canopy structure and LAI. At least five different types of optical *in situ* instruments based on principles of diffuse light transmission and direct solar irradiance are available for this task (Fournier et al. 2003). A traditional approach is to use photographic systems (specially designed and calibrated fish-eye single-lens-reflex or hemispherical cameras) for plot or sites to characterize vegetation conditions from a position beneath the canopy (Frazer et al. 1997, Fraser et al. 2001). Active optical sensors, such as hand-held laser range-finders and tripod-based 3D imaging LiDAR, are often deployed in the field to measure canopy gap fraction (Welles and Cohen 1996), tree diameter and height (Watt and Donoghue 2005), and tree crown conditions (Asner et al. 2006).

Spectrometers and *spectroradiometers* are widely deployed field instruments designed to detect a detailed spectrum of reflected solar energy (Ustin et al. 2004). The absorbing and scattering properties of natural surfaces are defined by their chemical bonds and structure (Curran 1994, Schut 2003). Reflectance spectra are comprised of many individual absorption features. In terrestrial ecosystems, such data can be used to identify vegetation species, and estimate leaf chlorophyll, carotenoid, and antho-cyanin pigment composition and content, plant water content and dry matter residues, and other aspects of foliar, mineral, and soil chemistry. In marine applications, controlled fluorescence spectrometry experiments can identify substrate and coral extracts (Matthews et al. 1996); *in situ* spectroradiometer measurements may be related to coral species (Color Plate 3.1; Karpouzli and Malthus 2003) and to ecosystem health (Holden and LeDrew 2001, Guild et al. 2002). *Imaging spectrometers*, frequently also called hyperspectral sensors, obtain the spectral data for each pixel and enable mapping of biogeochemical features in the field and over large landscapes and waterscapes when deployed on field, aerial or satellite platforms (Schut et al. 2002, Molema et al. 2003, Ustin et al. 2004).

Aerial Sensors and Platforms

Aerial remote-sensing platforms are extremely versatile and are available for deployment in virtually any location that can be reached using air transportation—which essentially means anywhere on earth. Looking back, human ingenuity has contrived to acquire aerial images from one remote-sensing device or another using virtually anything that can fly, from pigeons to powered parachutes to passenger jets. Aerial instrument-sensor technology continues to improve, and the costs decrease, at an amazing and accelerating pace (Dare 2005a).

Again, because of the virtually limitless number of options available, only a select few are briefly reviewed here to illustrate the range and diversity of aerial remote sensing (Fig. 3.3):

- Development of balloon, blimp, and airship remote-sensing solutions has a long tradition in environmental monitoring, and many such designs have been used effectively to bridge the crucial gap in observational scales between ground observations (plots) and satellite remote-sensing observations (landscapes) in studies of animals (Flamm et al. 2008) and plants (Inoue et al. 2000, Murden and Risenhoover 2000, Stow et al. 2000). Recently, Vierling et al. (2006) developed a novel system to collect thermal, multispectral, and hyperspectral

(a) NASA ER-2 high-altitude platform

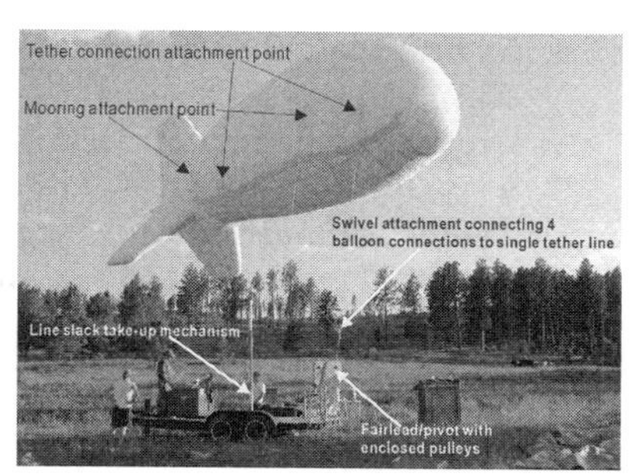

(b) SWAMI tethered balloon platform

(c) A sky arrow light aircraft platform

(d) A powered parachute platform

FIGURE 3.3 Some example aerial platforms used in environmental remote-sensing applications: (a) ER-2 *Courtesy of* NASA Dryden Flight Research Center; (b) Balloon system (*With permission from* Vierling, L., M. Fersdahl, X. Chen, and P. Zimmerman. 2006. *Remote Sensing of Environment* 103: 255-264.); (c) Ultralight aircraft, *Courtesy of* Dr. Rommel Zulueta, San Diego State University; and (d) Two-seat powered parachute, *Courtesy of* Sun Dog Powerchutes Inc. (http://sundogpowerchutes.com/index.html) and Air Sports Net (http://www.usairnet.com/powered-parachute/ [March 12, 2009].

imagery from a remotely controlled, tethered balloon. The principal advantages of the balloon were considered to be:

1. Extended flight duration
2. Highly controllable flight altitude
3. Ease of use in a wide range of environments and international locations where logistical or political constraints may preclude other options
4. Relative low cost
5. Relative ease of moving the platform [e.g., in one study by Buerkert et al. (1996) a camel was harnessed by a tow rope to guide a balloon across the Sahelian landscape]
6. Low platform vibration
7. Wireless target selection and instrument control from the ground

Other features thought to be of value were the inherent safety of the system and potential for simultaneous data collection with other nonimaging instruments (e.g., atmosphere trace gas and carbon flux).

- In the early 1990s, McCreight et al. (1994) developed and flew the Near Earth Observation System (NEOS) as part of the Oregon Transect Ecosystem Research (OTTER) project. Based on an ultralight aircraft and a removable pod to house digital frame cameras, spectrometers, and video systems, NEOS acquired spectral response patterns several times a day from different canopy layers and forest tree species to support ecosystem process model development. A typical application of the system was reported in mapping spruce budworm defoliation in a subalpine conifer forest environment as part of a multiscale remote-sensing project to determine forest damage (Franklin et al. 1995). Many other such ultralight aerial remote-sensing applications have been reported in geomorphological and agricultural mapping (Gregoire and Zeyen 1986) and in detection, counting, and mapping of animals and habitat (Grigg et al. 1997). Ultralights can vary considerably in design, from "rag and tube" open cockpit microlights to composite material, fully enclosed aircraft (Dare 2005a). The ultralight platform shares with the various kite and balloon (or blimp) systems some significant advantages when choosing flight parameters and instrument payloads (Howard and Barton 1973, Dare 2005b). However, low cost, ease of deployment, and flexibility in experimental design are among the most attractive remote-sensing dimensions.
- And finally, still dealing with air-breathing machines but at the other altitudinal extreme, the NASA ER-2 (a civilian version of the U-2 high altitude military reconnaissance plane) can obtain images from 21 km flying height while carrying over one tonne of instrument payload (Table 3.5). Originally deployed in a civilian capacity to help simulate imagery to be acquired from the first Landsat platform (called ERTS) in 1971, the ER-2 program has since provided primary mission and supplemental data in many remote-sensing applications. For example, McGill et al. (2003) described the application of the Cloud Physics LiDAR (CPL) flown on an ER-2 to measure cloud backscatter at 1.064 and 0.532 microns. These data were used to validate MODIS satellite observations across large areas in a study of linkages between atmosphere and land processes in the southern African region. Other ER-2 high altitude mapping applications have focused on algal bloom dynamics using the airborne ocean color imager (Richardson 1996), forest fires, and other disturbances in Yellowstone National Park and parts of Alberta using RC-10 metric cameras and specialized near-infrared films (Moore and Polzin 1990), and surveillance and control of arthropod vectors of disease using color infrared photography and Daedalus TM Simulator data (Washino and Wood 1994).

Selected ER-2 Instruments and Sensors
1. Multispectral Scanners • AOCI—Airborne Ocean Color Imager • AVIRIS—Airborne Visible and Infrared Imaging Spectrometer • MAMS—Multispectral Atmospheric Mapping Sensor • MAS—MODIS Airborne Simulator • TIMS—Thermal Infrared Multispectral Scanner • TMS—Thematic Mapper Simulator
2. Aerial Film Cameras • High Definition Electo-optic Camera System • VIS—Visible Imaging System
3. Other Sensors • ACATS—Airborne Chromatograph for Atmospheric Trace Species • ALIAS—Aircraft Laser Infrared Absorption Spectrometer • AMPR—Advanced Microwave Precipitation Radiometer • ATLAS—Airborne Tunable Laser Absorption Spectrometer • CLS—Cloud LIDAR System • CNC—Condensation Nucleus Counter • CPFM—Composition and Photo-dissociative Flux Measurement • EDOP—ER-2 Doppler Radar • FCAS—Focused Cavity Aerosol Spectrometer • HIS—High Resolution Interferometer Sounder • HO_x—Harvard Ozone (O_3), Carbon dioxide (CO_2), Hydroxyl Experiment • LAC—Large Area Collectors • LASE—LiDAR Atmospheric Sensing Experiment • LIP—Lightning Instrument Package • MIR—Microwave Imaging Radiometer • MMS—Meteorological Measurement System • MTP—Microwave Temperature Profiler • MTS—Mit Millimeter-wave Temperature Sounder • NASIC—NASA Aircraft Satellite Instrument Calibration • H_2O—NOAA Water Vapor Experiment • NO_y—Reactive Nitrogen Experiment • QCM/SAW—Quartz Crystal Microbalance/Surface Acoustic Wave • RAMS—Radiation Measurement System • WAS—Whole Air Sampler • WOX—Harvard Water Ozone Experiment

Source: With permission from http://www.nasa.gov/centers/dryden/research/AirSci/ER-2/sensors.html (February 24, 2009).

TABLE 3.5 Selected Characteristics of the NASA ER-2 Remote-Sensing Instrument Payload

Standard remote sensing and aerial mapping reference texts, such as Faulkner and Morgan (2002), provide many more examples of aerial remote sensing with digital frame cameras, traditional photography, videography, line scanners (including hyperspectral sensors and spectrographic imagers), synthetic aperture radar (SAR) systems, and LiDAR instruments (Color Plate 3.2). Each of these general instrument categories contain some important and interesting variations; for example, there are the three basic types of LiDAR systems already mentioned, each designed for a different remote-sensing application (direct terrain mapping, dual wavelength bathymetric and atmospheric sensing, and ground-based laser ranging). There are multispectral and panchromatic conical scanners, spin scanners, pushbroom scanners, hyperspectral scanners, and spectrographic imagers (Aronoff 2005). There are multipolarization and multifrequency aerial (and spaceborne) SAR systems (Waring et al. 1995, Space Studies Board 1998, Rees 2001). All have certain advantages or disadvantages for aerial data acquisition and analysis. And, of course, there has been an exciting proliferation of all kinds of digital and film-based aerial cameras.

Aerial remote-sensing cameras range from the inexpensive, small-format digital or film-based commercial and consumer-type products [originally termed *supplemental or miniature cameras* by Zsilinsky (1970)], to the relatively expensive, large-format, mounted multispectral digital cameras, such as the Leica Airborne Digital Sensor (ADS40) high performance digital sensor. Such cameras are capable of delivering photogrammetric accuracy and coverage as well as digital multispectral data. In 2006, the UltraCamX system large-format digital camera was introduced. Using CCD technology, the most recent camera (the UltraCamX_p) employs 7.2 micrometer pixels. The resulting image format is 14430 × 9420 pixels with a frame rate of 1 frame in 1.35 seconds (Table 3.6, Fig. 3.4; Gruber and Schneider 2007).

The UltraCamXp system is being used by Vexcel and Microsoft to create the patented *Virtual Earth Appliance* (http://www.vexcel.com/viscollaborate/veappliance/index.asp) currently containing color stereo-imagery of more than 100 US cities at 3 to 5 cm spatial resolution. Google's *3D Earth Warehouse* (http://earth.google.com/intl/en/3d.html) is another web-resource providing increased access to high spatial detail, purpose-acquired aerial remote-sensing data. *Google Ocean*, *EarthMine*, and *Streetview* are all Internet-based image services that signal an advancing wave of virtual representations of the real world based on high spatial resolution field and aerial camera sensing and simulations; all part of an emerging *metaverse* or cyberinfrastructure (Blais and Esche 2008) of human interface tasks, mobile applications, crowd- and geowiki-networking, and location-aware, ubiquitous computing spatial and remote-sensing services.

The philosophy behind these developments speaks for itself: "The entire world needs to be digitized" (Gruber 2008).

Image Product Specification	
Image format	Analogous to an aerial film image at a format of 23 cm × 15 cm, scanned at 15 μm
Image data formats	JPEG; TIFF with options for 8, 12, and 16 bits, scan-line, stripped or tiled
Image storage format in level 2	Full resolution panchromatic, separate color channels at color resolution
Color at level 3	Full resolution R, G, B, near-IR channels, planar or pixel-interleaved
Digital Camera Technical Data (Sensor Unit S-X)	
Panchromatic image size	14,450 * 9420 pixels
Panchromatic physical pixel size	7.2 μm
Input data quantity per image	435 Mega bytes
Physical format of the focal plane	104 mm × 68.4 mm
Panchromatic lens focal distance	100 mm
Lens aperture	f = 1/5.6
Angle-of-view from vertical, cross track (along track)	55° (37°)
Color (multi-spectral capability)	4 channels—RGB & NIR
Color image size	4810 * 3140 pixels
Color physical pixel size	7.2 μm
Color lens system focal distance	33 mm
Color lens aperture	f = 1/4.0
Color field of view from vertical, cross track (along track)	55° (37°)
Shutter speed options	1/500 to 1/32
Forward-motion compensation (FMC)	TDI controlled
Maximum FMC-capability	50 pixels
Pixel size on the ground (GSD) at flying height of 500 m (at 300m)	3.6 cm (2.2 cm)
Frame rate per second (minimum inter-image interval)	1 frame per 1.35 seconds
Analog-to-digital conversion at	14 bits
Radiometric resolution in each color channel	>12 bit
Physical dimensions of the camera unit	45 cm × 45 cm × 60 cm
Weight	~ 55 kg

Source: With permission from Vexcel Imaging and Microsoft Corporation, http://download.microsoft.com/download/8/5/a/85a3648b-b4f5-46d6-80e6-c0698a2ee109/ULTRACAM-Specs.pdf (March 3, 2009).

Table 3.6 Selected UltraCamXp Aerial Digital Camera Technical Specifications

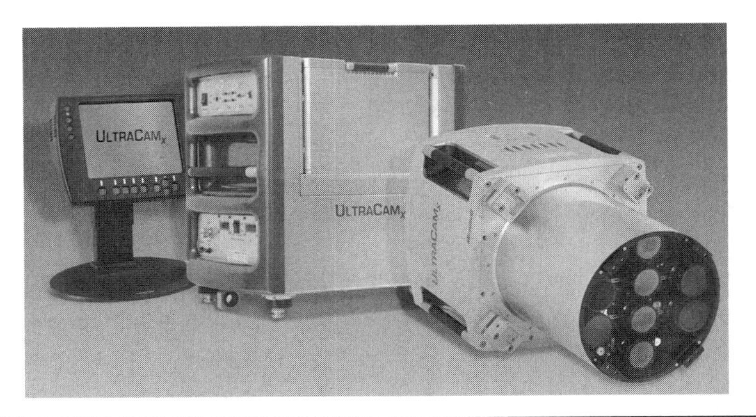

FIGURE 3.4 The UltraCamX$_p$ large-format digital camera based on CCD technology (*Courtesy of* Vexcel Imaging and Microsoft Corporation).

Current Satellite Systems

Fortunately, the whole world has been digitized—many times over, by multiple satellite remote-sensing systems. Currently, more than 150 Earth observation satellite systems are in orbit (Tatum et al. 2008). However, the spatial resolution of these satellite systems has often been less than the resolution desired by those involved in aerial remote-sensing applications, including those highlighted in the previous section. Spatial resolution continues to improve with satellite sensors, from tens of meters to less than one meter spatial resolution in recent years. This capability is still some distance from the subcentimeter capabilities of aerial and field-based systems. But one of the main reasons to use a satellite sensor, instead of an aerial sensor, is that high-spatial resolution may not always be the most important data acquisition concern, or even a necessary consideration in a given application. Instead, synoptic, large-area coverage may be more "mission-critical," or sensing capabilities not present in the available aerial systems may be desirable.

An astonishing number of improvements in satellite system functionality have occurred since the first satellites completed Earth's orbit more than 50 years ago (Tatum et al. 2008): *Sputnik 1* (October 1957), *Sputnik 2* (November 1957), *Explorer 1* (January 1958) and *Vanguard 1* (March 1958). An impressive array of satellite and sensor options is currently available (*see* Apps. III and IV). This situation will increase in complexity in future as remote sensing becomes an even more diverse and substantial international enterprise (Dousse and Wheeler 2008). One coordinating program, called GEOSS (Global Earth Observation System of Systems), is an international effort supported by over 100 governments and participating organizations to coordinate investments and strategies for earth observations in nine environmental themes, including biodiversity. This "system of systems" links existing and planned remote-sensing systems, promotes common technical standards, and offers access to imagery and image analysis software. A voluntary registration of remote-sensing assets, a best practices wiki, and a data portal are all available or under development (http://www.earthobservations.org/geoss.s html web site accessed 6 October 2008).

In May 2006, a marketing study ("The Market for Civil and Commercial Remote-Sensing Satellites" by Forecast International) projected deliveries of more than 125 satellites worth $16.3 billion over the next 10 years, with data suppliers, and commercial launch services, competing fiercely for market share. Most of the new satellites were expected to be relatively low earth orbiting remote-sensing systems, with both civilian and security (intelligence and defence) capabilities. A few examples of primarily environmental remote-sensing systems that were highlighted in the Forecast International study and have since advanced in development include:

- The United Arab Emirates and the Gulf Cooperation Council (GCC) has commissioned a South Korean-built remote-sensing satellite to be named DubaiSat-1. The system is designed with panchromatic and multispectral sensors for use in urban and infrastructure applications, natural disaster monitoring, and research. At the time of writing, DubaiSat was "space-ready" with the launch expected to occur in 2009.
- India's Polar Satellite Launch Vehicle C9 placed Cartosat-2A and IMS-1 in low Earth orbit on schedule in April 2008. The launch was notable for another reason; the payload was comprised of a record 10 satellites. In addition to the two Indian remote-sensing systems, eight research *nano-satellites* were deployed for Japan, Germany, Canada, the Netherlands, and Denmark. The India Space Research Organization was projected by Forecast International to be one of the top unit producers with a supply of 14 new satellites in the coming decade.
- Canada's Radarsat-2 next-generation mission was successfully launched on a Soyuz vehicle from the Baikonur Cosmodrome on December 14, 2007. The first images were available in January 2008. Radarsat-2 carries the world's first multi-polarization synthetic aperture radar satellite sensor acquiring multilook 3 m spatial resolution imagery in programmable configurations.

The competing, and sometimes complementary, demands of the civilian and security application communities have always been a major theme in remote sensing (Sattinger 1966, Burroughs 1986, Morley 1992). In 1783, Benjamin Franklin, American Ambassador to France, observed the first Montgolfier balloons in flight over Paris and predicted their use in "conveying Intelligence" about "an Enemy's

Army" (Hall 1992). The postwar V-2 rocket-based photography in New Mexico, and other pioneering military developments, stimulated the intelligence community to consider possible applications for space satellites. As early as 1955 the US Air Force developed plans for a *Strategic Satellite System*. Some observers have suggested that design and operational discussions around the Landsat program involved competing military and civilian perspectives; apparently, the National Security Council originally recommended that Landsat be classified (Hall 1992).

In Table 3.7, a selection of currently operating satellites with applications in ecological studies and environmental monitoring are listed; *see* Apps. III and IV for a more complete listing of past, current, and selected future satellites and sensors.

Satellite Platform	Sensor	Resolution: Spatial (m)	Resolution: Spectral (nm)	Extent (km²)
NOAA 17	AVHRR	1100	500–1250	2940
SPOT 4	VEGETATION HRVIR	1000 10–20	430–1750	2250 60
Terra, Aqua	MODIS ASTER	250 15	366– 530–1165	2330 60
Landsat	ETM+ TM	30 30	Variable	185 185
IRS Resourcesat-1	AwiFS LISS IV	23.5 5.6	520–1700	141
EO-1	Hyperion	30	433–2350	37
Orbview-2	SeaWiFS	1000	402–885	
Quickbird-2	BGIS 2000	0.6–2.44	450–900	16.5
IKONOS	OSA	1–4	450–850	13.8
Worldview-1	WV60	0.5	PAN	
GeoEye 1	GHIRIS	0.41–1.64	PAN	
ALOS	Prism AVNIR-2 PALSAR	Variable	Variable	
TerraSAR-X	SAR	Variable	X-band	
RADARSAT	SAR	Variable	C-band	

TABLE 3.7 A Selection of Currently Operating Satellites with Applications in Ecological Studies and Environmental Monitoring (*see* Apps. III and IV for more complete listings of past, current, and future satellites and sensors)

Image Analysis

Image Radiometry

Remote-sensing images are affected by multiple sensor-dependent and scene-dependent radiometric factors that occur during the image formation process. All commercially available remote-sensing image-processing systems, and increasingly geographical information systems, have multiple tools to help users deal with image radiometric evaluation and corrections (Barnsley 1999, Peddle et al. 2003b, Jensen 2007). However, some of these effects are simply too complex, or have yet to be adequately quantified, to be dealt with effectively using commercially available models or empirical and standard processing routines. For example, adjacency effects of optical reflectance measurements remain relatively poorly characterized (Jianwen et al. 2006), but are known to be important in wide angle remote sensing over large areas with high contrast features (Fig. 3.5; Richter et al. 2006a) and in mountainous areas (Woodham 1989). The influence of sensor-dependent radiometric terms, such as sensor integration time, variable point-spread-function, selected view-angle variations, and problems in image quality originating in signal-to-noise ratio or variable sensor performance will require corrections that are sensor-specific (Aronoff 2005). Common radiometric equations for image development are included in Table 3.8.

Radiometric corrections resulting from data acquisition or mission design are typically handled before data delivery by commercial

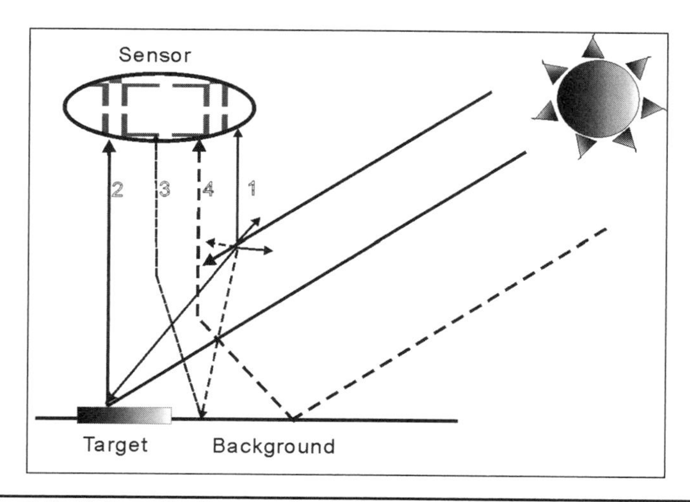

Figure 3.5 Schematic diagram of radiation components showing contributions from the atmosphere: (1) the target; (2) and two adjacency effects (3 and 4). (*Modified from* Richter, R., M. Bachmann, W. Dorigo, and A. Mueller. 2006a. *IEEE Geoscience Remote Sensing Letters* 3: 565–569.)

Image Radiometry	Equation	Variables
At-sensor reflectance	$\rho = \frac{\Pi d^2 L_s}{E_0 \cos\theta_z}$	ρ = apparent reflectance d = normalized Earth/Sun distance L_s = at-sensor radiance (W m^{-2} μm^{-1} sr^{-1}) E_0 = irradiance (W m^{-2} μm^{-1}) θ_z = solar zenith angle
SAR backscatter	$P_s = \frac{P_t G_t}{4\pi R^2}$	P_s = power density of the scatterer P_t = power at the transmitter G_t = antenna gain R = distance from the antenna
Brightness temperature	$F_R = \varepsilon \sigma T_K^4$	F_R = radiant flux ε = emissivity $\sigma = 5.67 \times 10^{-12}$ W cm^{-2} K^{-4} (constant) T_K = kinetic temperature

TABLE 3.8 Selected Radiometric Equations for Remote-Sensing Imagery

data providers. For example, imagery acquired from a multispectral scanner may show differences in radiometric response as a result of the solar illumination angle—a complicating factor is the sensor viewing position relative to the solar position (e.g., in aerial photography this phenomena may be more familiar as the "hotspot" reflection). Empirical corrections for these and other sensor radiometric calibration problems can be applied with mission-specific information on the sensor-solar geometry and acquisition time (Peddle et al. 2003b). But the geometry of the sensor-solar-target relationship will introduce significant radiometric factors that are still more difficult to manage. For example, topography introduces complex terms of radiation transfer, including adjacency effects and non-Lambertian reflectance properties. Therefore, the geometry of spectral response patterns will depend partially on the position of the target, the position of the illumination source, the position from which the observations are made, and the presence and influence of surrounding features.

Topography will have both radiometric and geometric implications. Geometric pixel effects for some remotely sensed imagery may be predicted by incidence and view-angle characteristics during acquisition (Barnsley 1984). The radiometric *topographic effect* is defined as the variation in spectral response from inclined surfaces, compared with the response from a horizontal surface, as a function of the orientation of the surface relative to the radiation source and the sensor position (Holben and Justice 1980). Corrections for the radiometric topographic effect range from a simple cosine correction based on terrain slope and solar zenith angle, to the application of complex geometrical optical models that consider shadowing effects and estimation of the bidirectional reflectance distribution function (Richter 1997, Peddle et al. 2003b). An essential requirement in corrections for the topographic effect is the use of high-quality digital elevation model (DEM) data. However, vegetation and land cover will create additional radiative transfer interactions. For example, the main illumination differences between trees growing on slopes and on flat surfaces is in the amount of sunlit tree crowns and shadows that are visible to the sensor, rather than the differences in illumination predicted by the underlying slope (Gu and Gillespie 1998). In SAR imagery, topography will induce slope foreshortening and can dramatically influence image usability (Van Zyl 1993).

Cumulatively, these geometric effects can create significant image analysis challenges; however, spectral response at a particular wavelength for a given feature of interest will depend not only on the sensor characteristics, the viewing angle, the illumination geometry, and the topography, all of which may vary significantly within a single remote-sensing image and almost certainly will vary from image to image over space and time, but also on the *atmospheric conditions* present at the time of image acquisition.

In many portions of the electromagnetic spectrum, the atmosphere has a significant impact on the signal sensed by satellite or aerial sensors due to the presence of clouds (Esche et al. 2002). Cloud contamination within large pixels may be detected with a filtering approach applied to high temporal resolution image profiles (Cihlar and Howarth 1994, Bradley et al. 2007). Cloud shadows in such data can be detected with "texture- or wavelet-thresholds" and data replacement procedures implemented (e.g., Simpson and Stitt 1998, Berber and Gaber 2004). Traditionally, in higher spatial resolution imagery, clouds and cloud shadows are removed manually and masks applied, with data "filled-in" from alternate sources; for example, available SAR imagery, clear imagery of a different date, or from areas of the existing image with similar surface conditions (Vani et al. 2006). In this latter instance, automated procedures to remove clouds and cloud shadows have improved with the development of *fast-fragment image completion* techniques (e.g., Chen et al. 2005). The main steps in the process (also sometimes called image approximation) include:

- Detection of clouds and cloud shadows (usually with a threshold or shape tool)
- Generation of a confidence map and level-set
- Search for source fragments (in current image or through alternate image sources)
- Compositing of fragments (the overall image approximation is updated after each fragment composition)
- As fragments are added, the mean confidence of the image converges to one, completing the image (Chen et al. 2005)

Radiative scattering and absorption by gases and aerosols are important atmospheric effects (Richter et al. 2006b, Coops et al. 2007). The analysis objectives and the influence of the atmospheric component on surface reflectance or emittance properties must be understood in order to choose appropriate image correction methods (Song et al. 2001). Recommended approaches for scientific remote sensing involve absolute correction based on physical radiative transfer principles following direct measurement of atmospheric properties at the time of image acquisition (Peddle et al. 2003b). The source of information on water vapor and temperature can be obtained from local radiosonde data or climate stations, and is necessary to determine atmospheric constituents, which are then used to parameterize the radiative transfer equation to identify relative contributions of atmosphere and the surface to the signal observed at the sensor.

This requirement of detailed atmospheric data at the time of image acquisition is a constraint in many applications. Therefore, empirical and model-based options have been developed to approximate or estimate the required data for the radiative transfer equation (Richter 1990, Song et al. 2001, Peddle et al. 2003b). One option is to develop a dynamic model driven by analyzed meteorological data; a second option is to consult an online source of radiosonde data or a sunphotometer network that may have acquired the necessary atmospheric information near the time and location of the image; a third option is to acquire simultaneous surface observations of targets and atmosphere using *in situ* instruments to calibrate the remote-sensing imagery. However, if the necessary atmospheric observations are not available, most commercially available image processing systems support much less-demanding image atmospheric corrections based on a standard atmosphere, a variation of the pseudo-invariant spectral properties approach, or an image normalization procedure. Each of these is briefly summarized in the remainder of this section.

A *standard atmosphere approach* assumes fixed standard atmosphere values or uses estimates of the necessary values acquired over time and in similar conditions (Richter 1990). This involves selection of atmospheric conditions thought likely to have been in place while the image was acquired. The analyst may examine features in the image to determine their similarity to features viewed through different types of atmospheric conditions; when a match is found, the optical depth and assumed constitution of that type of atmosphere is used to develop the approximate corrections to apply to all image spectral response patterns (Richter 1997).

Sometimes it is possible to estimate the required atmospheric parameters from the image data themselves (also known as the *dark-object, dark-target approach* (Chavez 1988, Campbell and Ran 1993), or *pseudo-invariant spectral properties approach* (Wu et al. 2005). In this procedure, some terrain features are assumed to have been imaged with near-constant spectral response or a pseudo-invariant spectral response. Deep, dark lakes, or images of dark asphalt/rooftops are thought to be essentially invariant—in other words, they appear always to have very low, close to zero, radiance under almost any imaging conditions (Teillet and Fedosejevs 1995). The analyst checks visible band spectral response over these dark targets, and adjusts the observed values to more closely match the expected very low spectral response for such features. The difference between the observed dark target spectral response and the expected dark target spectral response is attributed to the additive effects of the intervening atmosphere.

A final option is to apply *image normalization*. The imagery of interest is adjusted statistically to match the radiometric properties of another image or to a common radiometric base (Peddle et al. 2003b). Multiple image analysis tools, such as histogram matching (perhaps called color balancing) or relative radiometric normalization (RNN), are typically available to support image normalization. This approach has been effective in reducing solar zenith angle differences and uniform haze corrections in multiple image data sets acquired from the same sensor system (Hall et al. 1991, McDermid 2005).

Image Geometry

Geometric image processing is designed to determine spatial or locational accuracy and thematic positioning of remotely sensed data. Locational errors arise because each remotely sensed image will contain spatial distortions that are a function of the acquisition system and external factors, such as the distortion caused by orbital perturbations (Toutin 2004) or topographic relief (Barnsley 1984, Coops et al. 2007). Corrections for many geometric effects are applied by data providers long before the interpretation of images or the implementation of analysis procedures selected by image analysts. However, some geometric image processing associated with mapping applications, such as matching images to other spatial data and creating image mosaics, may be required as part of the applications' image analysis and interpretation steps.

Orthorectification may be implemented if suitable digital elevation model data are available or can be derived from the imagery. This geometric procedure is designed to remove geometric distortions caused by relief, which may introduce complex topographic shifting of pixels that also influence both geometric and radiometric characteristics of the surface features (Cheng et al. 2000). A typical sequence of steps to bring an image to a known projection (or orthoprojection) with map coordinates was outlined by Steiner (1974):

1. Identify all possible geometrical errors; for example, scale, rotation, and translational changes, view angle, or panoramic distortion, earth curvature, and topographic relief displacement (Hayes and Cracknell 1987).

2. Determine an appropriate form of transformation to be applied; available choices will often include different types and orders of polynomial transformations, depending on the type of error identified, and parametric functions (Toutin 2004).
3. Locate corresponding ground control points (GCPs) in the image and a reference system; typically, the analyst interactively chooses control locations, such as road intersections or distinctive water features, that are visible in the uncorrected imagery and for which known locations are available, perhaps from survey field work, a reference basemap, or a previously corrected image (Mather and Koch 2004, Sertel et al. 2007).
4. Formulate a mathematical transformation for the image based on the ground control locations in the image coordinate system and a selected map projection; this decision may become quite complex depending on the nature of the image acquisition, the environmental conditions, and the desired spatial precision in the output remote-sensing products (Fogel and Tinney 1996).
5. Implement the transformation, and subsequently resample the image, to derive image values for each location; the choice of resampling routines might consider desired image or map output quality, and possible impacts of resampling on subsequent planned analysis procedures (Hyde and Vesper 1983).

These steps rely on accurate and reliable identification of GCPs, a time-consuming and sometimes difficult task depending on the image quality, data display, and analyst experience. Software to automatically generate a geometric correction without ground-control has been developed in satellite remote sensing with precise spacecraft position and altitudinal information (Cheng and Chaapel 2006, Cheng 2007). In other situations, user approval of automatically generated GCPs may be necessary, but the automated and semiautomated registration procedures are now widely employed in remote-sensing imagery, Geographic Information System (GIS), and photogrammetric data processing (Ehlers 1997, Yukiko et al. 2000, Wong and Clausi 2007). *Feature-based geometric corrections,* in which distinct entities such as roads and drainage networks are automatically extracted and used to match images acquired at different times and to map referencing, have also been developed based on advances in image correlation and extraction tools (Dai and Khorram 1999, Chen et al. 2003a,b).

Image Classification

The objective of image classification is to create generalizations of image spectral response patterns that then map onto the landscape as a series of land cover or other land-attribute classes. A suite of image processing support functions are needed to support image classification (including sampling, display, and map output). The basic steps are threefold:

1. Development of a classification system comprising of individual and hierarchical classes; for maximum effectiveness, the list of classes should be mutually exclusive and exhaustive, and defined by a logical description of class attributes using surface characteristics useful for the final product (often a map).
2. The implementation of a decision rule to place each image pixel or image object into a class; multiple choices of classifiers exist, ranging from well-known statistical classifiers, such as maximum likelihood decision rules, to more complex neural networks and evidential reasoning algorithms. Selection of the optimal classifier decision rule may require a detailed understanding of the classification problem and data characteristics.
3. The evaluation of the success of the decision rule using an accuracy assessment technique; typically, accuracy will be based on an independent sample of classes derived from field work or aerial photointerpretation, in a careful sample design to consider spatial variability. Different types of accuracy measures may be of interest (e.g., map accuracy overall and weighted by class, individual class errors of omission and commission).

Three standard approaches to image classification include supervised classification, unsupervised classification, and modified approaches (Swain 1978). In *supervised classification*, the process is guided by a human interpreter, usually through the provision of training areas that represent spectral response patterns in known land cover or attribute classes. *Unsupervised classifiers* rely on the statistical properties of the spectral response patterns to reveal the underlying class structure, which is then labeled by the image analyst in a later step leading to map production. Modified approaches might involve some adjustments to either the supervised or the unsupervised process, perhaps by implementing an unsupervised clustering of training areas prior to their selection and use in a supervised classifier.

Increasing classification accuracy has been a major remote-sensing research focus for decades. Some now-well established techniques include:

1. The use of *multitemporal imagery* (Zhu and Tateishi 2006)
2. Incorporation of *ancillary data* (also termed auxiliary or colateral data) such as geomorphometric variables extracted from digital elevation models (Hutchinson 1982)

3. *Fuzzy classification* and the use of proportional or soft class membership (Foody 1999)
4. Image *texture analysis* (Puissant et al. 2005)
5. Image *context classification* procedures, which operate spatially as well as spectrally in an attempt to provide information to the classifier about surrounding features (Gurney 1981)

This latter approach has developed into powerful image understanding systems that incorporate spatial reasoning procedures (Guindon 2000). Increasingly complex image reasoning may use rule-based systems, which are a flexible option to incorporate GIS data into image classification. For example, in agricultural areas, a lack of radiometric resolution can reduce classification accuracy of crop types; incorporation of crop history (rotations), context (altitude, soil type), and structure (field size and shape) as ancillary data in a series of classifier rules or contextual reasoning steps can increase agricultural land-cover classification accuracy significantly (Raclot et al. 2005).

Image *segmentation* is a process to group pixels into homogeneous objects that are then used in place of pixels in classification (Kettig and Landgrebe 1976). This approach has benefited from the recent widespread availability and increasing use of high spatial resolution remote-sensing imagery in many applications, and increasingly well-designed new software (Keratmitsoglou et al. 2006, Blaschke et al. 2008). The most powerful object-based image analysis (OBIA) and classification approaches use a large number of characteristics of the segmented objects to classify imagery, including shape and texture (Lee and Warner 2006).

Vegetation Indices

Image transformations are among the most powerful ways to extract information from imagery, and of the many available transformations, vegetation indices have been most frequently adopted. Relying on observations in different portions of the electromagnetic spectrum, a simple computation, such as a ratio, will combine observational data to produce a single, one-dimensional index related to a surface target or feature of interest. For example, the *normalized difference vegetation index* (NDVI) is calculated with observations in the near-infrared and red bands. NDVI is related to vegetation properties, such as LAI, at least to the extent that LAI is a function of the absorbed-photosynthetically active-radiation (APAR). Foliage reflects less energy in the red band because most of these wavelengths are absorbed by photosynthetic pigments, whereas much of the near-infrared energy is reflected by foliage (Gausman 1977, Tucker 1979). NDVI enhances these differences in wavelength interaction with vegetation (and soils) in a single index value with a known range.

A change in a vegetation index, such as NDVI, over time is used to construct seasonal profiles of spectral response related to phenology and land-cover change (Coops et al. 2007, Hird and McDermid 2009). Over longer time scales, such data have revealed relationships between major disturbances (e.g., fires) and the global carbon budget (Potter et al. 2003). Considerable effort continues to be focused on identifying different change indices, and testing the best methods of combining indices to perform optimal change detection in different applications and environmental conditions (Le Hegarat-Mascle et al. 2006).

Linear transformation using principal components analysis of Landsat data has generated a number of other indices that can emphasize structures in the spectral response patterns which arise as a result of particular biophysical characteristics of vegetation and soils. The *Tasseled Cap Transformation*, for example, was derived to represent individual components of brightness/greenness/wetness in Landsat Thematic Mapper data, following an initial formulation for Landsat multispectral scanner (MSS) data in agricultural areas (Crist and Cicone 1984). Brightness is a positive linear combination of all six reflective TM bands, and responds primarily to changes in features that reflect strongly in all bands (such as soil reflectance). Greenness contrasts the visible bands with two infrared bands and is strongly influenced by vegetation amount. The wetness index is dominated by the contrast between the visible and near-infrared compared to the short-wave infrared bands, and may be more strongly related to vegetation structure, vegetation moisture, and total water content (Cohen et al. 1995, Collins and Woodcock 1996).

Data Fusion and Synergy

Image transformations are also useful in merging different data sets and creating data *synergy*—a situation in which the combined interaction of two or more data sets is greater than the sum of their individual contributions. Such an effect can be observed when combining panchromatic and multispectral imagery of different spatial resolutions (Wunderle et al. 2007), or when combining imagery from different regions of the electromagnetic spectrum (Tzeng and Chen 2005). A significant challenge has been to provide analysts with measures to control the spectral and spatial quality of the fused image; high spectral quality in data fusion often implies low spatial quality, and vice versa (Lillo-Saavedra and Gonzalo 2006).

Data-fusion techniques have emerged as key visualization tools as well as providing improvements in classification accuracy, image sharpening, missing data substitution, change detection, and geometric correction (Solberg 1999). Fusion procedures are generally categorized by their input/output characteristics, whether they operate on the original image data, or on derivative products (such as maps), and the level of processing applied. An efficient and accurate fusion process depends on an understanding of the characteristics of the imagery and GIS data available (Zhu and Tateishi 2006). When integrating or fusing multitemporal imagery, an essential problem is the determination of *transition probabilities*, which represent consistency between images acquired at different times. Analyst expertise is typically required, based on field experience or work with imagery and aerial photography showing similar land-cover and -change patterns.

Advanced fusion techniques incorporate contextual information from surrounding pixel or object neighborhoods. Sensor-dependent data fusion may employ a specific image formation model during the fusion process. Sensor independent methods are available that operate separately on individual data sets (Madhavan et al. 2006). For instance, region boundaries produced by texture segmentation of SAR images can be merged with classification results produced from a supervised classification of multispectral scanner imagery. The sensor independent fusion model works to create a fusion image that is independent of the origin of the data.

Change Detection

Using digital image analysis, change detection procedures can be automated. The objective is to identify changes in the imagery that are a result of changes in the surface features of interest, rather than the characteristics of the imagery themselves (e.g., differences in atmospheric or illumination conditions). Selection of the optimal change detection procedure requires matching the changes that are of interest to the characteristics of the data and the methods (Gong and Xu 2003); for example, to detect relatively rapid changes to the environment, such as forest harvesting patterns or fire damage, a high temporal resolution and high spatial resolution data set may be required (Rogan and Yool 2001). Visual image interpretation or a classification approach may be appropriate. Detection of changes in forest vegetation regrowth, on the other hand, which tend to occur more slowly, may be more effective using vegetation indices derived from low temporal resolution, relatively coarse spatial resolution imagery (Rogen and Chen 2004). In this situation, a classification approach may not have sufficient sensitivity and, instead, a continuous variable *threshold model* based on the relationship between a vegetation index and forest structure or age may be more accurate (Cohen et al. 1995). No single approach is optimal and applicable in all cases (Lu et al. 2004), but instead, there is a high degree of complementarity among the various change detection methods (Coppin et al. 2004).

Pixel-to-pixel change detection approaches rely on precise geometric processing and application of analytical techniques such as classification comparison, trend analysis, and image differencing (Singh 1989). Post-classification change comparison is perhaps the most common method to extract thematic land-cover change information (Prenzel and Treitz 2006). Essentially, each image in a time-series is classified separately and then overlain to derive differences. Improvements in the overlay process are implemented by appropriate use of iterative thresholds and masks of change/no-change developed with vegetation indices or other spectral analysis techniques.

Specific area-to-area comparisons—made possible through object-based image analysis, for example—will require still more complex data handling (Walter 2004, Blaschke et al. 2008). The spectral response within each object identified by an object-based classifier can be compared together with object shape identifiers such as homogeneity, perhaps defined using an analyst-specified threshold. However, a change in the landscape may only be detected over time if a significant part of the object is affected.

Geomatics Solutions: Remote Sensing, GIS, GPS, and Spatial Models

GIS Integration

A common theme visible almost since the beginning of the development of remote sensing has been the suggestion that greater integration with other geospatial technologies, notably GIS, is both desirable and necessary (Estes 1985). Early developers thought that the enormous potential of remote sensing and GIS could only be realized through an integrated approach, in which the technologies complement each other (Townshend 1981, Goodchild 1992, Ehlers et al. 1997, Estes and Star 1997). This insight has proven prescient. Today's remote-sensing systems are fully integrated and intimately involved in virtually every aspect of GIS, spanning the technologies and applications of photogrammetry, cartography, web services and development environments, and spatial modeling.

GIS integration has emerged as a powerful and flexible approach to management and analysis of spatial data of all types, including remote-sensing information. Updating GIS databases with remote sensing is now routine in many applications, and many mapping applications previously unachievable, or only inadequately developed, are now possible with an integrated remote sensing and GIS approach. For example, an integrated approach has made possible new insights into the environmental correlates of disease and disease-vectors in the health sciences (Albert et al. 2000), the linkage between pattern and process in analysis of forest disturbances (Wulder and Franklin 2007), and in dynamic monitoring of diverse components of coastal environments (Richardson and LeDrew 2006). In all of these applications, and many others, remote-sensing image analysis will require access to other data layers available only within a GIS environment; DEMs, for example, or existing and historical cultural, vegetation, and soils maps.

A recent trend has been to open source GIS and remote-sensing systems and services. One outcome of this trend is the increased *interoperability* of many of the largest commercially available systems with more standardized data formats, metadata, and processing links (Reichardt 2004). Commercial developers of GIS and remote-sensing systems recognized early the benefits of greater integration, including reduced time spent on data conversion and a productive flattening of the learning curve.

There are many providers of GIS and remote-sensing software systems. Commercial systems are not reviewed here, but a quick visit to the *GeoWorld Magazine* web site (http://www.geoplace.com/ME2/Default.asp) reveals dozens of product reviews, thorough comparisons, and features describing applications' development. Some developments are influencing the entire spectrum of spatial sciences; for example, the growing awareness and adoption of web-based services has had a pervasive and powerful impact on GIS and remote-sensing systems.

Other developments are focused on documenting improvements in specific applications across a broad spectrum of spatial and temporal scales (Blais and Esche 2008). In natural disaster operations, for example, current field, GIS and remote-sensing data challenges include managing software support systems for pre- and post-event information collection, integration, analysis, and distribution (*an information supply chain*) (Leidner 2007). Of critical importance are the identification of automated coordinate (using GPS) and remote-sensing data collection protocols, and determination of essential data that must be available to first-responders (e.g., the presence and location of dangerous substances). Multiple agencies are often involved in a complex infrastructure and rapidly changing disaster-situation; therefore, information sharing responsibilities, standards to facilitate collaboration and data understanding, and update and accessibility procedures that support the decision making are required (Li et al. 2007). The maturation of geospatial systems and technologies working together in an integrated fashion have provided the means to manage disaster situations far better than in the past, leading Leidner (2007) to predict that "the chance can be greatly reduced that anyone will die, suffer injury, or be left behind because critical information wasn't available when or where it was needed."

Another trend has been increased availability of some inexpensive (or even free) GIS software systems that require only modest commitment to training, maintenance, and support. Strand et al. (2007) provide a listing of such systems and document a range of options of value in biodiversity monitoring applications. To cite only one example here, version 3.3 of the Integrated Land and Water Information System (ILWIS) was developed and released in 2005 by the ITC (the Netherlands), and is now available as a free download recommended by the World Institute for Conservation and Environment:

> ...ILWIS comprises a complete package of image processing, spatial analysis, and digital mapping...full online help, extensive tutorials, 25 case studies of various disciplines...combines raster (image analysis), vector and thematic data operations in one comprehensive software programme on the desktop. ILWIS delivers a wide range of possibilities including import/export (e.g., open and export [*.shp] shape-files very easily), extremely user-friendly digitizing, editing, analysis and display of data as well as production of quality maps... analyze stereo photographs from your monitor with a stereoscope mounted on your screen... and handles very powerful problems, including mathematical watershed models.
>
> *(http://www.gis4biologists.info/gis_software_options.htm web site accessed 12 August 2008)*

Effective July 2007, ILWIS software is freely available as open source software (binaries and source code) under the 52°N initiative (GPL license). This software version is called *ILWIS 3.4 Open* (http://www.itc.nl/ilwis/default.asp web site accessed 12 August 2008).

Remote sensing and GIS data availability and web presence, continue to expand (Ball 2008). The Open Geospatial Consortium (OGC) Web Mapping Service (WMS) specification makes it easy to publish imagery on a WMS server, and several other solutions now allow ready import to Google Earth and Microsoft Virtual Earth. The free MapCruncher tool will help users embed customized information within Virtual Earth, import many different types of underused maps (e.g., hiking and bike routes), register maps and other image data, and share and discover MapCruncher layers.

Location Sensing Technology

Global Navigation Satellite Systems (GNSSs), together with other location sensing technologies and position-aware devices, are thought to comprise an essential feature of a coming era of "ubiquitous-computing, ubiquitous-positioning" applications (Meng et al. 2007). The constellation of 24 operational NAVSTAR satellites originally developed by the US Department of Defence as a military navigation system has revolutionized civilian applications in wide range of fields. This American GPS system has been joined by a Russian system (GLONASS, undergoing significant revitalization to full operational capability in 2009), a European network (Galileo, planned to be fully operational by 2013), and a Chinese system (COMPASS, also planned to be fully operational by 2013) providing continuous positioning and timing information under any weather conditions anywhere in the world and in any open environment (El-Rabbany 2006).

These global systems are likely to be increasingly augmented with regional systems, such as the Singapore Satellite Positioning Reference Network (SiReNT), a differential GPS system that the Singapore Land Authority put in place in 2006 to support surveying work, but with mature applications in homeland security, transportation, and emergency services. Another regional system is comprised of three Japanese satellites positioned strategically in high orbits over Tokyo. These augmented satellites typically read signals from all of the major systems and add regional signals to achieve higher precision within obstructing environments (such as urban building canyons) where currently weak satellite signals received by mobile GPS receivers are significantly attenuated. These developments will eliminate current poor positioning solutions and difficult situations in which no position fixes are acquired at all.

Other location-sensing technologies, such as ground-based pseudolites, ultra-wideband (UWB), and radio-frequency identification (RFID) have developed over the years to address the challenges of location sensing, with each approach solving a slightly different problem or supporting different applications (Meng et al. 2007). Many issues of design, reliability, and data quality have been addressed, and an entire literature has quickly developed to service the demanding engineering and technological aspects of the new systems (Bossler et al. 2002, Kumar and Moore 2002).

Applications developments have also expanded quickly, although few predicted the full revolutionary impact of location sensing potentially being used in virtually every home and business within a decade (Lachapelle 1991, Kennedy 2002, Lachapelle 2004). A quick glance through various industry websites or trade magazines such as *GPS World* (http://www.gpsworld.com/web site accessed 13 August, 2008) reveals a tremendous enthusiasm in location sensing and GPS-inspired innovation in surveying and construction, marine, avionics and transportation systems, utilities, communications and location-based services (smartphones and wireless technologies), and personal navigation support. Increasingly, major construction and engineering industries, for example, are aware that spatial data issues are critical and can affect competitiveness, costs, and planning (Caldas et al. 2006, Bansal 2007, Song and Caldas 2008). More than 40 high-precision GNSS applications were identified by Schrock (2006) in industry sectors of machine-guidance, research, social order (emergency services and compliance), traditional spatial applications (e.g., surveying and mapping), and transportation. High-precision GNSS use not only the satellite constellations, but networks of many thousands of individual earth-sensors, or Continuously Operating Reference Stations (CORS).

A major impact of advances in location sensing in environmental studies has been in the ability to carefully track animals and infer location-specific behavior (Gillespie 2001). The choice of an appropriate set of tracking devices is a major decision (Ren et al. 2008), and depends to a certain extent on the species, known propensities such as foraging or predation characteristics, and the type of environment. In the case of GPS, satellite reception determines location accuracy; certain animal behavior and environmental conditions (such as water depths and dense vegetation canopies) may significantly attenuate GPS signals resulting in differential acquisition rates across the land- or water-scape. Without correction, these influences can lead to possible misinterpretation of the data (Frair et al. 2004, Haines et al. 2006). The type of animal behavior of interest will suggest different data requirements; for example, measurements to calculate species home range or day range will likely require two different sets of considerations for the GPS telemetry time interval and the number of days that must be sampled to obtain reliable estimates (Ren et al. 2008).

Miniaturized GPS data loggers offer the possibility of unprecedented detail about animal movements compared to direct geolocation (radio or satellite) tracking (Adrados et al. 2008). Such devices have proven to be particularly useful in marine environments with species that spend some proportion of time at the surface (Ryan et al. 2004). A diversity of associated devices—to measure a wide range of parameters from meal size to heart rate, and to record relevant environmental data, such as temperature and water depth—can be combined with the GPS unit. This approach has been termed *bio-logging* in which analysts attach devices to animals to measure parameters that form the basis for the network of variables that influence the life histories of individuals (Ropert-Coudert and Wilson 2005). In one terrestrial application, Hunter et al. (2005) designed a combined GPS/digital camera collar to map grizzly bear movement and infer behavior such as sleeping and feeding from photographs taken obliquely. An important consideration in such designs is the trade-off in battery lifespan and sampling rate.

GPS is a one-way ranging communication technology. But other applications of the 20 cm wavelength GPS signals have emerged that were not fully appreciated during the initial development stage of the technology (El-Rabbany 2006). For example, the scattering of GPS signals from surface features is a new source for geodetic, tectonic, and ocean altimetry measurements (Gleason 2007). The signals can be used to form a bistatic radar technique, with separated transmitter (a GPS satellite) and receiver (on a host platform). This configuration has been used to sense sea winds and waves, and may have potential to assess soil moisture.

Spatial Models

Spatial modeling is a generic term used to characterize a very broad area of geospatial tools that are designed to work with the types of spatial data available in field, GIS, remote sensing, and GPS environmental studies. Spatial models differ from aspatial models, which do not incorporate the location of landscape conditions or change, and instead consider the landscape as a single unit (Huettmann et al. 2005). Spatial models may come in the form of an interactive software package to develop understanding of observations and relationships, and to help predict future conditions or anticipate possible alternate outcomes.

Models begin with a conceptual stage and statement of assumptions leading to a mathematical formulation (Shugart 1998). Models involve a process of abstraction through a series of carefully designed stages; classification schemes for different types of models typically consider the components of the system that are of interest, perhaps with an emphasis on form, pattern, and description, or on underlying mechanisms, and certain processing methods. Barnsley (2007) distinguished different types of models based on:

1. The extent to which the model is derived from observations (empirical models) or theory (theoretical models)
2. The degree to which random events and effects play a major role in the model (deterministic vs stochastic or probabilistic models)
3. How the model deals with environmental processes that can be considered to operate in a discrete or continuous manner with respect to time and space (discrete vs continuous models)

Models can also differ on operating characteristics (such as desired model outcomes, prediction vs inversion objectives) and handling of model parameters. Many spatial models will make use of complex spatial statistics, and are focused on form and pattern. An important distinction is made between static models, in which the system being modeled does not change, and dynamic models, in which the system being modeled changes over time. *Spatially distributed dynamic modeling* (SDDM) provides an umbrella term for the environmental models or submodels in which the underlying processes change over time and space (Atkinson et al. 2005).

An enormous variety of spatial models and spatial modeling techniques have been used in biodiversity and wildlife studies (Skidmore 2002, Gardner and Urban 2007). Typically, simple modeling is accomplished in a point pattern analysis, or a regression model; more complex modeling may use agent-based or rule-based approaches. Rather than focus on ever more detailed ways to categorize models, which in practice often blur into one another or coexist together in multimodal and hierarchical processing environments, a few model classes are cited here for illustrative purposes. In later chapters of this book, many more examples and case studies highlighting the power and limitations of different types and versions of spatial models are presented.

Species and Habitat Models

Animal distribution and habitat may be modeled using species occurrence, or expected distributions, and statistical relationships developed between these data and habitat attributes and environmental spatial data. This class of models is sometimes referred to as *element distribution modeling* (EDM) (Beauvais et al. 2006). A feature of the element-distribution modeling approach is the use of the EDM to quantify habitat and niche limits for a species by correlating environmental factors with or without occurrence data and detailed knowledge of habitat preferences. Typical EDM outputs are suitability maps, range or niche maps, predictions of species distribution, and habitat quality assessments. An EDM spatial model for a given species almost always involves a unique combination of data quality, quantity, study area (extent and grain), history, biology, intended model use, and other factors, including model algorithms, methods of mapping, and evaluative procedures (Beauvais et al. 2006). Two recent examples:

- Tweet et al. (2007) developed a spatial EDM of northern bobwhite (*Colinus virginianus*) population distribution for conservation planning within the West Gulf Coastal Plain Bird Conservation Region in the southcentral United States. Northern bobwhite spatial distribution was predicted using regression procedures based on landscape characteristics, such as the amount of grassland edge derived from 1992 National Land Cover Data. Estimated bobwhite abundance closely approximated a limited sample of field observations of bird detections. Subsequent projections of population change and sustainability provided guidance for targeting habitat conservation and rehabilitation efforts for restoration activity.
- A more complex EDM example was provided by Gottschalk et al. (2007) in their assessment of the potential distribution of the Caucasian black grouse *(Tetrao mlokosiewiczi)* in Turkey. The approach was to model suitable habitat using satellite imagery, a digital elevation model, and landscape metrics (such as distances to roads and forest patches). Resource selection was derived by means of a generalized linear model analysis; observations of occurrence of the grouse were used to develop a model of the resources the grouse needed to survive and reproduce. Input data layers in the empirically

defined spatial regression model of population estimates included alternative assumptions on population density and habitat separation. Comparisons to field observations of presence/absence revealed differences that were thought to be biologically reasonable. The resulting potential distribution map was used to select priority areas for conservation and to specify additional survey locations.

Animal Movement Models

Animal movement, utilization distribution, and habitat selection are intertwined but may be explained using spatial models that consider movement or change across landscapes based on habitat features, individual animal observations, and GPS telemetry data. In one example, Christ et al. (2008) developed a random walk model of brown bear movement and habitat selection in Alaska; this approach simulated a velocity field to describe the strength of attraction toward (or aversion from) different types of forest habitat and other landscape features, such as streams. Model results confirmed earlier observations of strongly positive forest habitat and stream effects, and avoidance of some land-cover types (nonforest). This type of model introduces significant capability in home range estimation, and can be used in hierarchical multiple-animal modeling to allow for the estimation of resource selection at both a population and an individual level.

A movement spatial model to predict and visualize the East African wildebeest (*Connochaetes taurinus*) migration-route patterns in variable climate conditions in the Serengeti-Mara ecosystem of East Africa was developed using DEM data and NDVI extracted from MODIS satellite imagery by Musiega et al. (2006). The migrating wildebeests' response to food resource distribution and terrain complexity apparently impacts their movement characteristics, such as speed, direction, and turning frequency. The routing model selected was a least-cost-pathway approach determined by terrain relief and grass recovery patterns depicted in the NDVI data. The model included a route attractivity index to simulate year-to-year variation in rainfall and forage quantity and quality. Such analysis may be useful in predicting migration routes in drier years.

Process Models

A wide variety of process models have been developed to simulate landscape processes. For example, Walsh et al. (2002) examined the linkages between people and the environment in northeastern Ecuadorian Amazon by combining longitudinal household survey, satellite remote-sensing imagery acquired at multiple times and scales, field images and sketch maps, GPS coordinates for features, such as roads, household farms and built structures, and landscape metrics (Fig. 3.6). A multilevel cellular automata model (Wolfram 1984), a subclass of agent-based or rule-based process models, was used. The

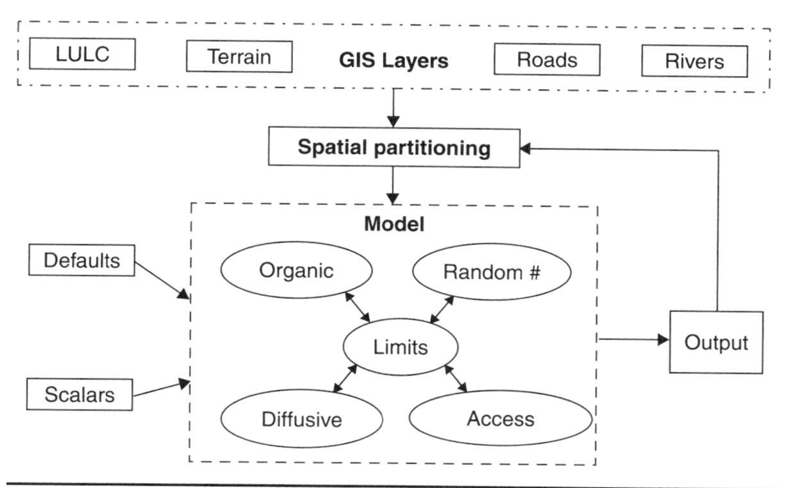

FIGURE 3.6 General schematic representation of the cellular automata model used to simulate land cover and land-use dynamics in Ecuadorian Amazon. (*With permission from* Walsh, S. J., R. E. Bilsborrow, S. J. McGregor, B. G. Frizzelle, J. P. Messina, W. K. T. Pan, K. A. Crews-Mery, G. N. Taff, and F. Baquero. 2002. In Fox, J., et al., eds. *People and the Environment: Approaches for Linking Household and Community Surveys to Remote Sensing and GIS*. Springer, Dordrecht. 91–130.)

model consisted of a regular grid of cells in one of a number of possible states, which were then updated in discrete time steps according to a local interaction rule, usually specified as a transition function or growth rule that addressed possible neighborhood configurations and processes (such as road building). Linear, nonlinear, and hierarchical relationships of land cover and land use emulated biological patterns for each cell based on terrain, soil quality, vegetation, hydrology, and the infrastructure context of the cell (e.g., proximity to roads, markets, schools, time since settlement etc). Expected land cover and land-use patterns in future time periods were explored and considered relative to perturbations in the base variables and stochastic elements interpreted relative to environmental policy.

Mechanistic ecosystem process models generalize biogeochemical and hydrological cycles across lifeforms and climate (Hanson et al. 2004). Typically, these models require climate and environmental data (air temperature, radiation, precipitation, humidity, atmospheric CO_2, soil texture, and depth) to calculate hourly, daily, and annual carbon dynamics, water, and nutrient cycles. The processes responsible for these dynamics (e.g., photosynthesis, respiration, litterfall, decomposition, allocation) are measured in the field, simulated, or estimated from remote sensing (Coops and White 2003). Examples of ecosystem process models incorporating remotely sensed data include:

1. Land-use change detection model
2. Biomass estimation model
3. A model simulating soil erosion and carbon sequestration
4. Fire fuel-loading model
5. Gross and net primary production models (Zheng et al. 2006, Nightingale et al. 2007)

Such models typically require accurate maps of vegetation and land-cover conditions, and a number of other parameters, such as initial soil moisture and texture. These estimates may be obtained independently from remotely sensed data, to drive the model, or are used in model validation and comparisons.

CHAPTER 4

Remote Sensing of Ecosystem Process and Structure

For several days I have walked the forest trails in utter frustration as I realized how very little of the forest I could actually see...How desperately I wanted information concerning the topside of the forest and what a terribly narrow swath I was seeing and sensing.

—*D. M. Gates, 1967*

Land Cover and Vegetation Structure

Land Cover

Detection and identification of land cover or other attributes are accomplished through analysis of spectral response patterns using visual image interpretation techniques (Colwell 1966, Oswald 1976, Story et al. 1976, Howard 1991) or computer pattern recognition procedures, such as digital-image classification (Swain 1978, Robinove 1981, Pettinger 1982). The key decisions that can strongly influence remote-sensing land-cover classification and mapping success include the data quality, the choice of land-cover classification scheme and level of detail, optimal decision-rule configurations (manual and digital), and the use of appropriate methods to evaluate success and utility of resulting map products (Buiten and Clevers 1993). The choice of land-cover classification scheme and level of detail are critical; for example, land cover may be defined according to physiognomic criteria (e.g., hardwood forest vs conifer forest), or structural criteria (e.g., species composition, such as pine-dominated forest vs spruce-dominated forest). Deciding on a list of classes to map will influence options in the selection of the best data for classification, which in turn will be highly dependent on the purpose, the analysis, and the available remote-sensing sources.

Visual image interpretation capability is an extremely valuable skill set for use with digital displays of remote-sensing data, including increasingly available high spatial resolution optical and SAR aerial and satellite imagery, and of course, photographic output (Campbell 2005). A visual image analysis is based on the interpretation of seven *image elements:* tone, texture, shape, size, shadow, pattern, and site association. These elements are diagnostic or distinctive of features and objects visible in a remote-sensing computer display or in quality photographic products (Hall RJ 2003). The correct recognition and use of these elements of visual analysis of imagery are essential for construction of a dichotomous or selection key, a check-list-based interpretation key, or an analogue (sometimes called a catalogue) model (Avery 1968, Townshend 1981). The skill required to develop and use manually interpreted information in image analysis may be currently under-appreciated in some applications (Sader and Vermillion 2000, Lillesand et al. 2004). One of the most important features of visual image interpretation is related to the extraction of detailed environmental information not easily obtained using digital classification methods (such as the inference of microhabitat conditions). In other words, typically, visual and digital image analyses are not mutually exclusive, but instead, are complementary (*see* http://www.ccrs.nrcan .gc.ca/resource/tutor/fundam/chapter4/01_e.php web site accessed 10 April 2009).

Earlier, a comparison between analog and digital methods of image interpretation may have focused on the different levels of accuracy that could be obtained for the same set of general land cover or habitat classes of interest, or the different sets of skills required to manage a photointerpretation project compared to a computer-based analysis of digital imagery (Ryerson 1989, Buiten and Clevers 1993). However, recently, in certain image analysis tasks, such as land-cover classification, accuracy levels have become much more comparable between the two different approaches. The selection of analog or digital methods might depend on other factors, such as cost. For example, Nichol and Wong (2008) reported that manual interpretation of aerial photographs for broad habitat and land-cover classes in Hong Kong was 95 percent accurate overall; a similar level of accuracy was achieved with digital image classification of high spatial detail Ikonos satellite imagery, at approximately one-third the cost.

Typically, in a visual classification of land cover, a hierarchical approach is implemented in which the image elements are used to delineate and identify objects and units of homogeneous cover, sometimes called *photomorphic units* (Philipson 1997). In a visual interpretation of a standard metric black and white aerial photograph, image elements of a light, uniform tone, smooth texture in a rectangular or square shape could strongly suggest an agricultural field. In the photomorphic approach, this field will be outlined and separated from surrounding areas, and labeled by the analyst. Surrounding areas then, in turn, are interpreted in context of the identified field and appropriate image elements. The analyst will interpret features in increasing detail and consider each outlined area with a manual or computer-assisted knowledge-based labeling process. Some analysts have commented on the similarity in logic and approach between the visual analysis of image elements in the delineation and classification of photomorphic units, and current digital methods of object-based segmentation and image classification (Herold et al. 2003).

Hudson (1991) outlined the typical steps in an aerial photointerpretation selection key process applied to forest mapping, and provided an example of species recognition in the Dominican Republic. West India pine (*Pinus occidentalis*) trees were first identified in the available large-mapping-scale black and white metric aerial stereo-photography by their uniquely shaped crowns (narrowly rounded, typically asymmetrical, occasionally flat and spreading), which appeared as a light gray tone, with an irregular, deeply serrated shape. Darker-toned broad-leaved trees appeared with a more rounded, smoother crown shape. The pine stands were more roughly textured, with a less uniform pattern than the broad-leaved stands, which were smoothly textured but with larger, though less frequent, canopy gaps. Stand structure, including crown closure, could be readily estimated following this visual classification into species-dominated forest stands.

An example of visual interpretation applied to digital imagery was provided recently by Campbell (2005). Digital orthophotos produced from color infrared aerial photography of a portion of the Eastern Rivers National Wildlife Refuge Complex on the Chesapeake coastline of Virginia were interpreted for vegetation patterns grouped into land-cover types, such as wetlands, different forest types (represented by species-dominance), and agricultural patterns. The open water could be clearly discriminated by distinctive dark blue colors and patterns. The land-cover classes of interest were associated with tonal and textural elements visible in the imagery; darker red colors with smooth texture were more often closely associated with dense vegetation, such as forests; lighter shades of red and pink suggested less dense vegetation (such as grasslands); and various shades of blue and lighter colors in particular shapes and sizes could be readily interpreted as exposed surfaces, agricultural crops, and soils. Bird sightings were then readily related to the interpreted land-cover classes and the positions of edges between cover classes.

Well-designed digital image enhancements are a critical image processing step that can vastly increase the visual discrimination and interpretational value of remote-sensing imagery for a particular feature or application of interest (Chen 2007, Jensen 2007). In mapping forest, shrub and herb communities associated with Glacier National Park avalanche paths, for example, Walsh et al. (2004) developed a digital transformation of multispectral Ikonos images to assist in visual delineation of morphological zones. The transformation was based on the principal components analysis algorithm, which produced brightness/greenness/wetness dimensions similar to those originally derived from Landsat imagery using the Tasseled Cap Transformation. Considerable expertise in *both* image interpretation and understanding the dynamic geomorphological and ecological processes of interest are essential in interpreting avalanche and other mass wasting features on the transformed imagery; such features, because they are relatively well-defined on some types of aerial and satellite imagery, with a characteristic shape and texture, are also an ideal candidate for object-based digital image analysis methods (Barlow and Franklin 2007).

The logic of computer multispectral classification and mapping of land cover is based on the digital use of remote-sensing data as surrogates for the land attributes of interest (Robinove 1981). The term *spectral signature* is sometimes used to describe the link between the spectral response pattern and land cover, but this term has the unfortunate effect of suggesting a degree of precision that is unlikely to occur in many situations. Typically, statistical methods are needed to manage variability in the spectral response pattern (Pettinger 1982).

Digital classification of land cover is one of the best developed uses of remote-sensing spectral response patterns and has benefitted from a concerted research effort to improve and refine methods to achieve high accuracy, flexibility, reproducibility, robustness, and consistency in results (Hutchinson 1982, Cihlar et al. 1998). A general goal has been to create objective methods of digital classification in which results are not overly dependent on analyst's decisions. Optimal decision-rules in digital classification are selected from a wide range of alternatives. The *maximum likelihood decision rule*, based on estimates of class mean and covariance, is one of the most frequently used digital classifiers in remote sensing (Franklin et al. 2003). Class estimates are obtained from training areas in a supervised approach, or through an iterative data clustering procedure.

A recent Landsat mapping study of 1974–2001 forest land cover in central Siberia illustrates the essential steps (Bergen et al. 2008):

1. Unsupervised classification using the ISODATA (Tou and Gonzalez 1974) algorthim was applied to develop clusters for classes of water, forest, disturbance, bare soil, wetland, agriculture, and urban.
2. Polygons (objects) of multiple pixels exhibiting similar spectral properties were grown from seed pixels located in training areas.
3. Class separability was evaluated through use of histograms, ellipse plots, and by calculating matrices of interclass distances.
4. Iterative supervised classification using the maximum likelihood decision rule separated out the forest land cover into four classes (conifer, deciduous, mixed, and young).

5. Other land-cover and -change classes such as burn, cut, insect damage, regeneration, and some agriculture land-cover classes were obtained by comparing image classifications at three different time periods and automatically relabeling objects.
6. Accuracy assessment was performed using independent samples of pixels from selected areas yielding an 81 percent overall land-cover classification accuracy.

In comparisons with maximum likelihood classifiers, fuzzy classifiers and decision rules based on *evidential reasoning* often perform as well or better (Peddle 1995), particularly in situations where the spectral response patterns do not meet assumptions about variable distributions (e.g., normality) and covariance structure (e.g., independence and multicollinearity) required by parametric classifiers such as maximum likelihood (Franklin et al. 2003). Fuzzy classification rules are desirable because of their ability to soften the decision for each pixel or object in the image; instead of only a single class assignment, each pixel or object would be assigned to each of the classes, but with different degrees of membership (Foody 1999). In still more demanding remote-sensing classification situations, such as those covering regional, continental, or global areas over multiple time periods, and when integrating diverse types of GIS data, even more powerful classification approaches have been implemented (Rogan and Miller 2007). Examples have included the use of generalized linear models (GLMs), nonparametric machine learning algorithms (MLAs), and decision trees (Friedl et al. 1999).

Species and Species Assemblages

Detection and identification of vegetation species and species assemblages by remote sensing is often closely related to land-cover classification and may be accomplished by classifiers and other image processing techniques aimed at a specific level in a hierarchical land-cover classification scheme (Boyd and Danson 2005). The success and desirability of remote-sensing applications of species mapping and identification of communities will vary by ecoregion. For example, in tropical ecosystems, the ability to accurately map tree species "will facilitate ecosystem characterization, tree demographic studies, mapping endangered or endemic species, identifying important food sources for wildlife, and quantifying carbon pools and carbon sequestration rates" (Castro-Esau et al. 2006). Such species-level discrimination, then, is a key contribution of remote sensing in ecology (Bradley and Fleishman 2008).

Unfortunately, the literature on vegetation species separation using remote sensing primarily in the optical, thermal infrared, and microwave portions of the electromagnetic spectrum is confusing, even contradictory, prompting Nagendra (2001) to urge "more studies like Gong et al. (1997) that would assist in developing general guidelines on the bands most suited to species discrimination." Gong et al. (1997) had used *in situ* hyperspectral radiometer data acquired above sunlit and shaded sides of canopies in identification of six California conifer tree species. The use of an artificial neural network classification algorithm with these data showed that:

- Smoothed reflectance and first derivative spectra of sunlit samples (alone) resulted in an overall accuracy of greater than 91 percent.
- The effects of site background and illumination conditions on tree species spectra were significant.
- The discriminating power of visible bands was higher than that of near-infrared bands.
- Higher classification accuracies were obtained in the blue to green or the red-edge spectral region as compared with four other spectral regions.
- A smaller set of selected bands generated more accurate identification than all spectral bands.

Leaf-, pixel-, and crown-scale analyses were conducted with seven species of emergent tropical rain forest tree species by Clark et al. (2005). *In situ* spectroradiometer and HYDICE aerial push broom indium-antimonide hyperspectral sensor data in 161 bands (437–2434 nm) were analyzed; optimal regions of the spectrum for tree species discrimination were found to vary with scale. Accuracy decreased when moving from leaf-scale observations *in situ* to canopy level observations in the aerial imagery. However, classification accuracies approached 88 percent overall using a maximum likelihood classifier with pixel-scale data, and increased to over 90 percent when classifying crowns separated out as image objects before classification. Near-infrared bands were important across all scales, but data acquired in the visible and shortwave infrared bands were more important at the pixel and crown scales of analysis.

High spatial resolution imagery is typically acquired at lower spectral resolution than hyperspectral imagery, but with appropriate band selection and image processing, such data are very effective in vegetation species classification and identification of species assemblages. In forest cutovers in northern Ontario, for example, distinct spectral properties of branches, stems, leaves, and reproductive structures were used with multispectral aerial imagery to identify accurately many species of shrub, conifer, and hardwood vegetation required to estimate conifer regeneration competition (Pouliot et al. 2006). A key to the success of this application was the use of imagery acquired at a time when the conifer species were at maximum visibility and distinguishable from deciduous vegetation (fall season leaf-off conditions) (Pouliot and King 2005).

Automated tree crown delineation is an effective processing step prior to implementation of species recognition and classification (Gougeon 1995). Although the combined effects of illumination angle, view angle, and tree geometry, will introduce variations, tree crowns are usually brighter at the apex with darker neighboring pixel values along the crown edge, in shadows, and between crowns (Gerylo et al. 1998, Culvenor 2003). Local radiometric maxima (the brightest point) and local minima may be used to isolate or delineate the crown (Wulder et al. 2000; Fig. 4.1) using a variety of image processing algorithms or a template matching procedure (similar to a traditional aerial photo-interpretation catalogue).

In structurally diverse mixed species forests in Queensland, a tree crown algorithm was developed for 1 m spatial resolution multispectral aerial imagery (Bunting and Lucas 2006). The objective was to map and classify tree crowns that were expansive (e.g., eucalypts with a large branching habit), or partially overtopped, small and open (e.g., subcanopy pine). In another approach, in a heterogeneous broadleaf woodland in the United Kingdom, LiDAR data were acquired together with the multispectral high spatial resolution imagery and processed in an integrated method (Hill and Thomson 2005). An unsupervised classification of 3D image objects generated by image segmentation was successful in separation and identification of individual tree crowns. A similar experimental design was used to separate tree crowns from background based on a LiDAR canopy height map and NDVI- and texture-based digital image classification (Koukoulas and Blackburn 2005). Accuracies for beech, oak, and birch trees were 81, 91, and 86 percent correct, respectively.

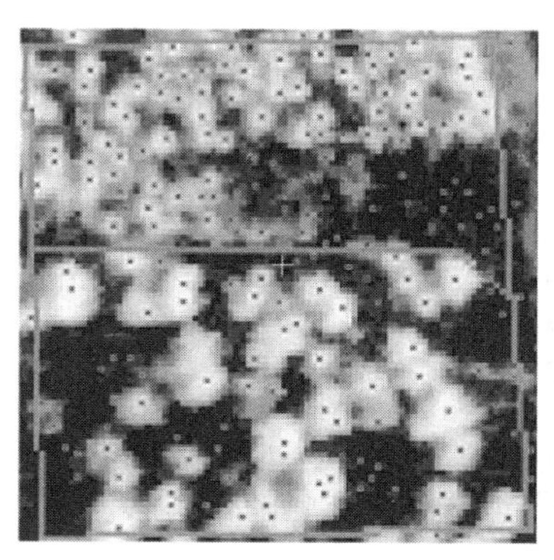

(a) Red with 3 × 3 LM filter (correct 0.67, commission 0.25, omission 0.33)

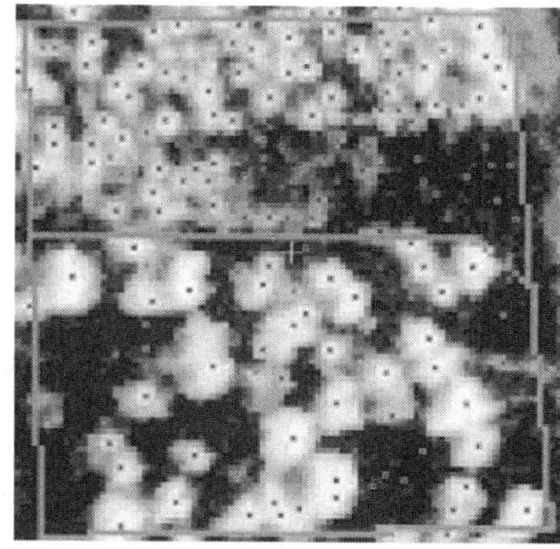

(b) PAN with SB LM filter (correct 0.62, commission 0.11, omission 0.38)

Figure 4.1 Use of local maxima to identify tree crowns in high spatial resolution imagery. (*With permission from* Wulder et al. 2000. *Remote Sensing of Environment* 73:103–114.)

In a rare study of LiDAR intensity returns, Moffiet et al. (2005) attempted tree species separation in subtropical woodland in Queensland, Australia. One species of interest, Poplar Box (*Eucalyptus populnea*) exhibited larger gaps in crown foliage when compared to White Cypress pine (*Callitris glaucophylla*), which displayed a relatively dense needle structure. This foliage type provided less opportunity for a strong second reflection either from within the crown or from the ground. This difference in *vegetation permeability* to LiDAR signals was useful in species classification. The LiDAR intensity variations were also sensitive to differences in ground texture and enabled an unsealed road only partially visible beneath the canopy to be differentiated from surrounding soils.

Microwave remote sensing—and active microwave remote sensing in particular—is useful in ecological applications requiring an understanding of vertical profile, dielectric properties, and vegetation geometric structure (Waring et al. 1995). Multipolarization, multifrequency, and multiresolution SAR data supplements two-dimensional remote sensing with information in the vertical dimension, and like LiDAR, it has the potential to generate high-accuracy vertical profiles over small areas that are often useful as discriminatory attributes. Mapping broad vegetation cover conditions, such as wetlands (Taft et al. 2003), forest types (Hoekman 2007), or agricultural crop types (Cloutis et al. 1996, Hoekman and Vissers 2003), are reasonably well-established SAR applications using classification techniques similar to those employed to classify multispectral optical imagery. Detection and mapping of land-cover change and deforestation patterns is more frequently attempted than vegetation species classification (Theil et al. 2006), although some reasonable species classification results have been achieved in relatively simple forest conditions based primarily on species differences in structure (Franklin 2001). SAR fusion with optical imagery is often a desirable approach to achieve data synergy (Truehaft et al. 2004).

Lower spatial resolution remote-sensing data, obtained by SPOT HRV, Landsat TM and ETM+, ASTER (advanced spaceborne thermal emission and reflection radiometer) and other similar sensors, have been used in some vegetation species classification projects, with mixed results. For example, Marçal et al. (2005) classified nine Portugese Forest Inventory classes in the Vale do Sousa region using a selection of spectral bands derived from orthorectified and segmented ASTER imagery. Of interest were eucalyptus, pine, and mixed forest classes. Comparisons of fuzzy classifier, support vector machine (SVM) classifier, the k-nearest neighbor (KNN) algorithm, and a logistic discriminant classifier (LD), revealed accuracies of 44.6, 70.5, 60.9, and 72.2 percent, respectively, over all classes tested with 277 independent validation sites. The main reason given for the relatively poor accuracy obtained with some classification techniques was the observed spectral similarity between the selected species mapping classes; some classifiers were found to handle the spectral similarities better than others.

Techniques such as *spectral mixture analysis* have increased the accuracy of vegetation species mapping from relatively low spatial resolution data sets. One study designed to classify bald cypress (*Taxodium distichum*) and Tupelo gum (*Nyssa aquatica*) trees in Thematic Mapper imagery in South Carolina provided estimates of the relative amount of these trees found within each pixel in 10 percent increments (Huguenin et al. 1997). The approach accounts for subpixel variations introduced by background features, such as soil and understory, and then estimates the fractional cover of each feature contributing to the spectral response pattern (Asner et al. 2003). The assumption is that pixel spectral response is the sum of the contributions of each contributing feature, weighted by their fractional presence within each pixel (Color Plate 4.1). Highest accuracies using spectral mixture analysis have been obtained when detailed *endmember spectra*, acquired by *in situ* measurement, but perhaps also estimated from the imagery themselves in "pure" training areas, are available to develop the mixture model of contributing features and weights (Fig. 4.2).

A review of remote sensing in mapping coral reef cover highlighted the complicating effects on spectral mixture analysis of the water column (Knudby et al. 2007). One approach is to develop a spectral library; this is a database of light measured after reflection through different depths of a water column with differing optical properties (Dierssen et al. 2003). Uncertainty in fractional estimation may still exist, even if all endmember spectra are available in the

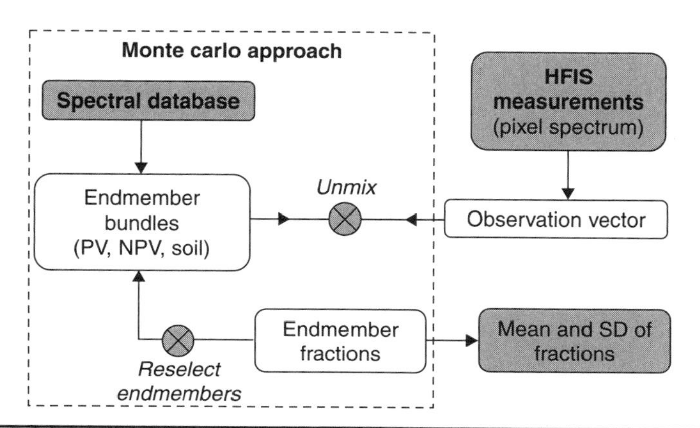

Figure 4.2 Schematic diagram of the spectral mixture analysis model used to map rangeland vegetation and exposed soil in Northern Chihuahua, USA from ER-2 AVIRIS data. A sequence of the resulting fractional cover change images are shown in Color Plate 4.1. (*With permission from* Asner, G. P., and K. B. Heidebrecht. 2005. *Global Change Biology* 11:182–194.)

spectral library or are measured in the field, because of the similarity of the spectral response of different contributing features (such as substrate types influenced by bottom depth, water column optical properties, input conditions such as sky irradiance).

Spectral mixture analysis was used to identify species assemblages and dominant-species classification of a large dune ecosystem in Kenfig National Nature Reserve in the United Kingdom, the site of at least two rare plant species of interest (Lucas et al. 2002). The analysis was based on multispectral high spatial resolution aerial imagery and a fuzzy classifier. The mixture model identified areas with different proportions of sand, vegetation, and shade, and the success of this step depended strongly on the spectral separability of these features and the quality of endmember spectra developed from the imagery using principal components analysis. The final map portrayed the floristic composition of vegetation communities at a subpixel scale, the high degree of spatial heterogeneity in the environment, areas of degraded vegetation, and all shadowed terrain and water bodies. A similar approach has been recommended for the use of remote sensing in invasive alien species discrimination, examples of which are discussed in later sections of this book.

Understory Plants and the Forest Floor

Direct detection of vegetation understory species is rarely possible with satellite remote sensing, and usually only possible with high spatial resolution satellite or aerial remote sensing in certain favorable circumstances where the canopy is relatively open or contains large gaps and few shadows. With high spatial resolution satellite data and aerial remote sensing, understory species and canopy layering may be more readily detected in stands with larger canopy gaps (Wilson and Ference 2001). Carefully separating spectral response in these gaps from shadowed areas and other features requires detailed image analysis, often with a high degree of analyst interaction. LiDAR data may be a more promising option to map three-dimensional vegetation structure and understory (Vierling et al. 2008); for example, in one study, a full waveform LiDAR configuration was used to construct images of canopy layers in order to quantify foliage height diversity and identify significant understory growth (Clawges et al. 2008). In another study of aerial LiDAR intensity data, in Finland, Korpela (2008) reported the detection of reindeer lichen (*Cladina* sp.) understory within homogeneous conifer forest stands. Approximately 75 percent classification accuracy was obtained.

A few rare low spatial resolution satellite remote-sensing studies have been conducted in situations where the understory (or lack of one) does influence the relatively low spatial resolution spectral response enough so that direct mapping can be attempted (Spanner et al. 1990, Stenback and Congalton 1990, Ghitter et al. 1995, Iiames et al. 2008). Good results have been reported with supervised classification of multitemporal Landsat data in relatively simple, open forest structures (particularly valuable to employ leaf-on, leaf-off image acquisitions over deciduous forests with a conifer understory) (Hall et al. 2000). Predictive models of understory conditions may be derived, alternatively, from modeling of remote-sensing spectral response patterns and field plot data (Borry et al. 1993), or a combination of geological, soils, and DEM data (Chastain and Townsend 2007). This latter approach was attempted in spatially extensive and ecologically important conifer understory communities dominated by rosebay rhododendron (*Rhododendron maximum*) and mountain laurel (*Kalmia latifolia*) in Appalachian forests. Landsat ETM imagery and DEM topographic data yielded an overall classification accuracy of 87.1 percent with a maximum likelihood decision rule (Chastain and Townsend 2007).

Other habitat features on the forest floor are difficult targets for a remote-sensing approach. Some success has been reported with:

1. Manual interpretation of high quality and high spatial resolution aerial photography
2. The implementation of predictive modeling approaches
3. The development of empirical relationships to selected obvious remote-sensing parameters (e.g., tree size, canopy structure)

All of these ideas have been implemented to detect small, seasonal wetland features, associated with seeps and *vernal pools*. Such features are known to be of critical habitat significance and at risk in many woodland environments (Calhoun and deMaynadier 2004).

A logistic regression procedure based on independent variables such as surficial geology, land use and land cover, soils, topography, and surface hydrology was partially successful in locating vernal pools in Massachusetts (Grant 2005). Earlier work had outlined the procedures for accurately and reliably mapping such features using standard aerial photointerpretation techniques (Stone 1992). Lathrop et al. (2005) then applied this visual interpretation approach to 1 m spatial resolution leaf-off color digital orthophotography in a GIS environment. Vernal pool habitat features with a minimum size of approximately 0.02 ha were reasonably accurate; errors of commission were relatively small (12 percent) and were readily attributed to the interpretation protocol, which emphasized detection of potential habitat as well as distinct vernal pool features. However, large errors of omission were thought likely in detection of such surface features beneath forest canopies using any remote-sensing approach (Van Meter et al. 2008); a combined method, in which visual image interpretation results could be considered within a predictive model sampling framework, based on a digital remote-sensing model, was recommended to ensure higher accuracy and a more representative sample of existing vernal pool habitats was obtained.

Another forest floor habitat attribute of wide interest is the specific configuration or amount of *downed woody debris*. Interpretation of woody debris in forests has only occasionally been attempted using remote-sensing data alone, but one promising strategy has been to combine suitable remote-sensing data with predictive modeling following the development of field-based empirical relationships (Ståhl et al. 2001). High spatial detail remote-sensing imagery, for example, may be used to select areas of interest or to guide the location of transects to areas with differing amounts of woody debris based on remotely sensed stand structure characteristics. An exception has been in the assessment of downed trees and forest damage (e.g., caused by hurricanes), which have been frequently obtained by interpretation or classification of multitemporal aerial and satellite remote-sensing imagery (e.g., Womble 2005, Shedd et al. 2006).

Invasive Alien Species

An invasive alien species (IAS) is defined as a non-native species whose introduction causes or is likely to cause harm to the economy, environment, or human health (Schnase et al. 2002). Traditional methods of IAS detection often include grid-based floristic surveys, plot census, and herbarium collections (for plants), rapid immunoassays for fungal parasites, and field trapping and museum collections (for insects). Remote sensing is considered the only feasible mapping approach for IAS assessment over large areas (Mack 2005).

The characteristics of a particular invasive alien species will influence the choice of a remote-sensing approach to detection and mapping (Madden 2004). In addition to predictive modeling methods based on GIS and remotely-sensed data layers, currently two direct remote-sensing approaches are commonly implemented to detect and map IAS:

- Detection of *unique IAS phenological and biochemical attributes*; this approach is described in the following sections as part of the larger presentation on remote sensing of *foliar chemistry and stress (health)*
- Analysis of *structural properties* and mapping of spatial patterns of IAS infestation

Structural characteristics and mapping of spatial patterns of IAS can be accomplished with a variety of sensors, but accuracy will be strongly dependent on the IAS having a particular influence in the environment or structure that may be detected. Death of native vegetation species and declining plant growth are two obvious examples of IAS impacts—exemplified in insect defoliation and vegetation damage studies, such as the mapping of Gypsy moth (*Lymantria dispar*) forest damage in New England (Dottavio and Williams 1983), tree mortality caused by Dutch elm disease (*Ceratocystis ulmi*) in Saskatchewan (Wilson et al. 1998), or detection of aquatic invasive plants which alter water quality and color in the southeastern United States (Madden et al. 1999). These applications are highly successful because the impact of the IAS is to create a unique structure to the spectral response compared to areas without the

IAS. Distinct color patterns associated with IAS flowering may be another unique structural identifier apparent in appropriate remote-sensing imagery, although other structural characteristics caused by IAS (such as height differences) may not be as dramatic. However, with appropriate remote-sensing instruments and processing techniques (such as LiDAR for height differences), such structural changes induced by IAS may be readily detected (Underwood and Ustin 2007).

Gillieson et al. (2006) presented a case study of root-rot fungus (*Phytophthora cinnamomi)* impacts in Queensland rainforests based on the detection and mapping of canopy decline and mortality with color aerial photography and multispectral videography. Such infestations are responsible for significant forest decline and dieback in an increasing number of Australian ecosystems (McComb and Hardy 2003). A high spatial detail approach based on vegetation indices, such as NDVI, was thought necessary in order to ensure a high degree of accuracy in the detection process. Dieback patches were classified accurately in simple and complex notophyll vine forest, several mesophyll forest types, and closed forest with eucalypts. Also presented was a case study of 41 weed species, including the aggressive weed Pond apple (*Anonna glabra*), in the World Heritage site of Bridle Creek, near Cairns, Australia (*see* Fig. 4.3). Three types of data were used:

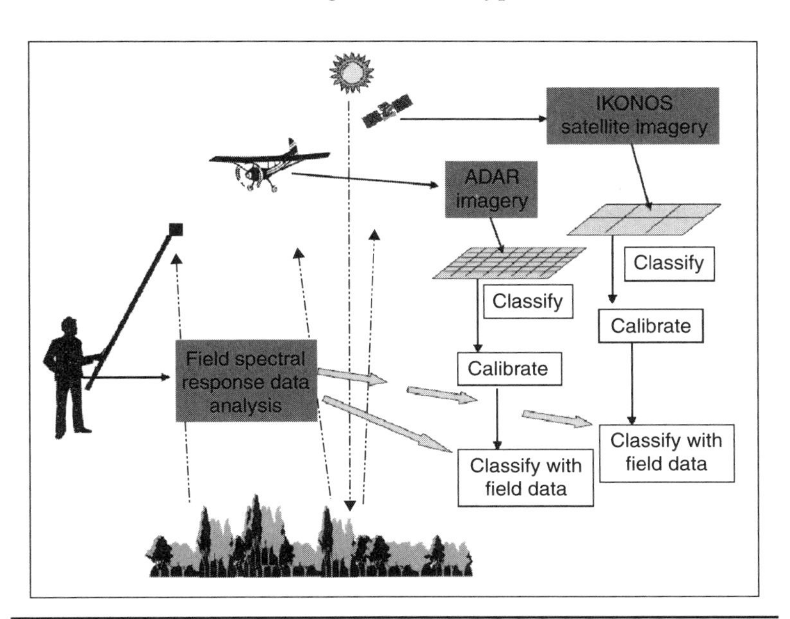

FIGURE 4.3 Integrated research methodology used for weed assessment and scaling up from field observations to satellite image data. (*With permission from* Gillieson, D. S., T. J. Lawson, and L. Searle. 2006. *Applications of High Resolution Remote Sensing in Rainforest Ecology and Management.* Cooperative Research Centre for Tropical Rainforest Ecology and Management, James Cook University, Cairns.)

- Percentage cover and species identification in a stratified sampling method to obtain field plot estimates of vegetation composition
- Field vertical photography and measures of spectra from a tripod-mounted and hand-held spectroradiometer
- Imagery from the IKONOS satellite sensor at approximately 4 m spatial resolution and the airborne data acquisition and registration (ADAR) digital camera system at approximately 1 m spatial resolution.

Statistical separability of species was possible with the *in situ* spectral response data, but variability of spectral response acquired by the aerial and satellite sensors within some weed species samples was much higher than the apparent differences between species (Gillieson et al. 2006). Additional work was recommended on remote sensing with multispectral imagery at spatial resolutions better than 2 m, detailed attention to the proper calibration of the aerial sensor data, and implementation of fuzzy classification techniques to provide softer classification decisions to accommodate the apparent variability of weed species and natural vegetation spectral response patterns.

Successful *in situ* hyperspectral remote sensing of tropical dry forest trees infested with lianas has been reported in Central America (Sánchez-Azofeifa and Castro-Esau 2005). Crown cross-sectional reflectance measurements in five different trees with varying degrees of liana growth were obtained within one hour under clear sky conditions. Species-level discrimination of lianas was not attempted; instead, derivative spectra and vegetation indices were related to the amount of liana infestation. Higher reflectance was observed at 550 nm in liana-infested trees, possibly because liana leaves have a reduced chlorophyll concentration compared to host tree leaves (Castro-Esau et al. 2004). However, liana infestation also served to lower spectral separability among tree crowns, which may reduce tree species discrimination capability by remote sensing. One possible explanation was that:

> ...lianas tend to grow randomly in the canopy and their abundance is dynamic. Species of lianas with different leaf structures, water content, pigment concentration, and abundance in the canopy are not the exception but the rule. Thus, the total mixed response received by a sensor represents a combination of several different biophysical and biochemical properties that are dynamic and nonlinear in nature, producing a mixture that causes confusion...the observations presented...could be used to start developing new tools for mapping the extent of liana coverage using high spatial and spectral resolution data sets...Variables such as leaf longevity, leaf area index, canopy architecture, illumination conditions, leaf structural properties (e.g., leaf thickness), pigment

concentrations, canopy water content, and the combined impact of these variables needs to be studied further to gain an understanding of their impact on remote-sensing observations.

(Sánchez-Azofeifa and Castro-Esau 2005: p. 2106)

Invasive alien species dynamics provide an ecological forecasting test case that will require reliable, accurate, and timely remote sensing and field information at unprecedented temporal and spatial resolutions (Clark et al. 2001). An initial list of data needs will overlap with the needs of those interested in managing for other biodiversity indicators, and likely would include (Schnase et al. 2002):

- Ecosystem biophysical structure—for example, biomass, vertical structure, ocean particulates, pigment florescence, trace gas fluxes, near surface atmospheric carbon dynamics, stream chemistry
- Ecosystem functional capacity and physiological state—for example, pigment concentrations, live biomass, biomass turnover rates, photosynthetic and respiratory capacity
- Biological population mapping, including species distributions, communities, functional-type mixtures

Vertical Complexity, Canopy Gaps, and Structure

Successful estimation of vegetation structure (e.g., vertical complexity, canopy gaps and other structural variables such as LAI, age, height, biomass), has been reported in numerous environmental settings using a wide selection of remote-sensing instruments and experimental designs. Early success with visual aerial photointerpretation was quickly translated into operational procedures and investment in multipurpose vegetation inventory databases (Kreig 1970, Hudson 1991, Gillis and Leckie 1993, Hall 2003). As digital remote-sensing approaches to mapping vegetation and ecosystem structure have gained wide acceptance, a vast literature has developed comprising of specific studies and experiments.

In practical applications of computer-based vegetation structure estimation, five broad approaches are frequently implemented and are discussed briefly in this section (Franklin 2001):

- *Texture analysis procedures* applied to high spatial resolution imagery
- Physically-based *radiative transfer models* applied to SAR imagery (backscatter) and low spatial resolution optical imagery (reflectance) to predict vegetation structure based on radiation interactions

- Continuous variable estimation, and *vegetation indices*, applied to low spatial resolution optical imagery
- *LiDAR data analysis* (intensity returns and vegetation hits or laser echoes per unit area)
- Remote-sensing *data synergy*, in which two or more sensors acquire data and complementary information extraction methods are applied to obtain unique information contained in each image

Texture Analysis of High Spatial Resolution Imagery

The interplay between different features contributing to pixel spectral response in high spatial detail imagery, such as the dominant species and canopy conditions, understory, shadows, and exposed soil or background, are the key to developing appropriate methods for the extraction of information on vegetation structure from high spatial resolution imagery. Image texture analysis may be based on spatial autocorrelation functions such as semivariograms (St.-Onge and Cavayas 1995), frequency domain (or Fourier transformation power spectrum techniques), structural measures, or measures statistically derived from *spatial co-occurrence* matrices (Haralick et al. 1973, Haralick 1979). The statistical texture approach—usually based on first-order measures such as window-, filter- or kernel-variance, and second-order measures, perhaps derived from spatial co-occurrence or wavelets—has gained wide acceptance, and is usually included as an image processing technique in most commercial image analysis systems.

Modeling and mapping forest structural complexity using texture analysis of 20 cm spatial resolution aerial multispectral imagery was described by Pasher and King (2009a,b) for a region of temperate hardwood forest in Quebec (*see* Color Plate 4.2). Forest structural complexity was defined as "the abundance and variability of different horizontal and vertical structural attributes present in the forest." Because more complex forest structure should provide for greater biodiversity as a direct result of the greater number and variety of habitat niche available, structural complexity was considered a potentially useful biodiversity indicator (*see also* McElhinny et al. 2005). The imagery was processed for brightness corrections and georeferenced; a range of spectral, spatial, and object based variables were extracted from the imagery at each of 50 field plots. The texture variables included spatial co-occurrence texture measures as well as semivariance range and sill. Object-based information, to reflect characteristics of tree crown shapes and sizes as well as shadow shapes, sizes, and patterns, was extracted following a semiautomated segmentation routine.

The final model related five variables representing different spectral and spatial aspects of the imagery specifically measuring spectral response and variations within canopy gaps and tree crowns, as well as specific topographic features, to field-observations of structural complexity in the 50 field plots. This model was applied across the entire study area to predict and map structural complexity, and field validated at previously unvisited locations that were classified into broad categories of structural complexity. Predicted areas of high-and low-complexity were found to be 88.2 and 81.3 percent accurate, respectively (Pasher and King 2009a,b).

Two of the challenges when using a texture image analysis approach are:

1. Managing the very high data volumes associated with texture
2. Choosing a selection of highly discriminating texture measures

Kayitakire et al. (2006) encountered these two issues in an application of oak (*Quercus* spp.), beech (*Fagus sylvatica*), and Norway spruce (*Picea abies*) forest structure analysis using texture measures obtained from IKONOS-2 panchromatic (0.4 m spatial resolution) imagery in the Hautes-Fagnes region of eastern Belgium:

- An early assessment of texture measures was conducted with available digital aerial photography to determine the most applicable measures for the forest types and age of the stands; three spatial co-occurrence variables were selected (variance, contrast, correlation).
- Different window sizes for the texture measures were tested in the four main texture directions; three window sizes were chosen to represent different scales of information (5 × 5 pixels, 15 × 15 pixels, and 25 × 25 pixels).
- The final data set was comprised of 168 texture measures, four independent spectral variables, and five dependent forest variables (age, height, basal area, density, circumference); all were tested for normality and some transformations implemented (e.g., forest age was log transformed).
- For each forest variable, the texture measure with the highest absolute correlation coefficient was selected (simple linear model of the form $y = b_0 + b_{1x}$, where y is a forest variable, x is a texture measure, and b_0 and b_1 are the coefficients of the model).
- Results included coefficients of determination higher than 0.7 for all five forest variables predicted by at least one of the spatial co-occurrence texture measures, with relative prediction errors within or very close to the usual sampling inventory errors for four variables; height was the most accurately predicted forest variable, and basal area was the least accurate; different texture window sizes (scales) were useful in predicting different forest variables; no particular trend in the residuals was observed.

Models and Texture Analysis of SAR Imagery

Estimating structure with multipolarization aerial C-band SAR imagery in relatively pure stands of boreal spruce and pine was attempted by

Wilson (1996) using an empirical texture image analysis approach. Compared to typical results using high spatial resolution optical imagery, lower coefficients of determination were obtained, but some relationships were strong enough to develop predictive models; for example, standard errors were less than 15 percent of observed tree heights. The study confirmed that texture in SAR imagery is a consequence of differential backscattering of the radar signals from the tree crowns, branches, boles, and ground features (the "double-bounce").

A physically based radar backscattering model can be used to clarify these differential backscattering interactions; essentially, such a model represents the intensity with which microwave radiation interacts with vegetation (Dobson 2000). In forests, the backscattering model will simulate canopy-trunk-ground interactions, and structural variables of interest can then be derived by inversion and comparison to acquired remote-sensing imagery. Multiple layers of canopy can be introduced in the model with various combinations of trunk size, branching characteristics, and foliage possible to more closely approximate conditions in actual forests of different tree species, age, and understory conditions.

Liang et al. (2005) applied this type of model in Queensland White Cypress Pine and mixed eucalypt forests while acquiring multifrequency, multipolarization, and multiresolution NASA AIRSAR data with an output pixel size of approximately 10 × 10 m. A comparison between simulated and actual backscatter was undertaken in order to explain the observed SAR signals based on forest structure. In the majority of cases, the modeled backscatter was within the dynamic range of the AIRSAR data. The upper canopy was found to contribute the greatest scattering of C-band signals; there, the backscatter was a result of radiation interactions with the branches and foliage. At L- and P-bands, the double-bounce scattering primarily from large tree trunks was more noticeable. L-band variation was explained primarily by large branch structures; P-band interactions were dominated by the interactions between the trunks, large branches, and ground surface. Cross-polarization of the C- and L-band radar wavelengths resulted in more accurate predictions of forest structure.

Using Low Spatial Resolution Optical Imagery

Early studies established the potential of empirically derived forest structure, and other vegetation community estimates, from relatively low spatial resolution imagery such as that obtained from the Landsat and SPOT sensors (Walsh 1980, Franklin 1986, Walsh 1987, Danson 1987, De Wulf et al. 1990, Cohen and Spies 1992). If the canopy is relatively open, the spectral response patterns can be correlated with structural attributes, such as individual tree or shrub size and shape, crown conditions, density and gap conditions, understory characteristics, and possibly attributes of the forest floor, including soil moisture and organic matter conditions. Larger tree crowns absorb more red light, but reflect more strongly in the infrared

(Franklin 1986). Decreasing visible reflectance (darker image tones) will be associated with increasing canopy development. As the vegetation grows and matures, the areas between individuals (such as tree crowns) are no longer visible, and the shadows cast will deepen. As the cover approaches full closure, the correlation between optical spectral response and structure approaches zero. The problem is a fundamental one; the sensor now detects spectral response only from the top of the canopy (Holmgren and Thurresson 1998). If the canopy is closed, structural conditions in the stand may continue to evolve, but the optical spectral response will be uncorrelated with that development (Cohen et al. 1995).

Two approaches to develop structural information in relatively open canopy conditions from low spatial resolution optical satellite imagery are based on: (i) the development of empirical relationships to key structural variables, such as crown closure, and (ii) vegetation reflectance modeling. For example:

- Empirical relationships were derived between forest stand height, crown diameter, and crown closure and Landsat TM wetness index in 38 seral, late-seral, mature and old-growth cedar, hemlock, Engelmann spruce (*Picea engelmannii*) and subalpine fir (*Abies lasiocarpa*) stands in British Columbia (Hansen et al. 2001). A structural complexity index, constructed from field observations, was used to map stands that displayed altitudinal zonation and spatially variable stand development conditions (Fig. 4.4).

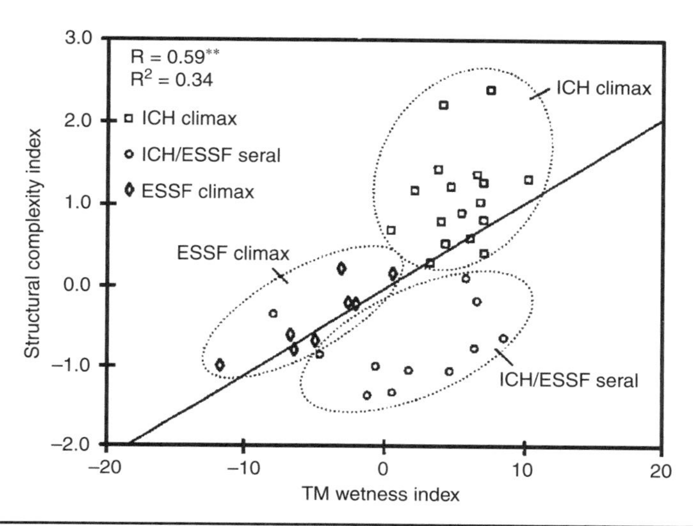

FIGURE 4.4 Altitudinal zonation and spatially variable stand development conditions expressed using a structural complexity index in mature conifer forest communities in British Columbia. (*With permission from* Hansen, M., et al. 2001. *Remote Sensing of Environment* 77:50–65, and Franklin, S. E. 2001. *Remote Sensing for Sustainable Forest Management.* Lewis Publishers (CRC Press), Boca Raton.)

The wetness index had been shown in earlier work to be strongly related to vegetation structure in forests (Cohen and Spies 1992) and agricultural areas (Crist et al. 1986). Generally, wetness index values are known to increase as a forest stand or crop matures, until maximum cover is achieved, then decrease as shadows and components other than foliage begin to influence the spectral response.

- An early example of optical canopy reflectance modeling was described by Oliver and Smith (1973) in a shortgrass prairie vegetation study to determine the optical properties of Blue grama (*Bouteloua gracilis*). Stochastic models incorporating estimates for incoming light flux, plant geometry, and intrinsic optical properties were compared to a deterministic model based on the Kubelka-Munk theory for diffuse reflectance. Direct solar and diffuse sky irradiance was measured (0.4–1.05 μm) *in situ* with a field spectroradiometer. Canopy geometry was measured with photography subsequently converted to digital format with a color scanner. The model predictions compared favorably with the field and laboratory measurements, and successfully quantified the non-Lambertian character of the Blue grama canopy.

Canopy reflectance models come in many forms and degrees of complexity (Asner et al. 2003). The geometric-optical (GO) model is an important class of canopy reflectance model that simulates the effects of illumination geometry and the structure of vegetation; GO model dynamics reveal spectral sensitivities that may be displayed in remotely sensed data (Li and Strahler 1985). Typical modeling requirements include estimates or assumptions of the amount, size, and angular distributions of leaf and non-photosynthetic contributors to spectral response, crown parameters such as height and shape, and optical properties (including pigment concentrations). Nilson and Ross (1997) described four subcomponents of one geometric-optical model simulating the optical properties of needles, shoots, tree crowns, and forest communities. Such models are related to spectral mixture analysis and can be used in *multiple forward-mode* to determine the dependence of pixel values on contributing structural factors (Peddle et al. 2003a). In forward mode, the model predicts a range of spectral response patterns that a particular configuration of vegetation structure could generate; then, the analyst matches an observed spectral response to the most likely configuration in the model output.

Structural Interpretation of LiDAR Data

Three dimensional LiDAR remote sensing and modeling of vegetation structure is a promising and fast growing application based on the ability of laser altimeters to penetrate vegetation canopies through to ground level, resolve individual tree crowns, identify layers of canopy, and provide discrete return or full waveform profiles for analysis (St.-Onge et al. 2003, Clawges et al. 2008). Initial processing of LiDAR signals is based on precise sensor geometry (from GPS and Inertial Navigation Systems control) and interpolation methods to yield DEM- and image-products (such as a canopy height model) from the laser pulse returns. Some LiDAR data providers consider interpolation methods proprietary but, increasingly, commercial image analysis systems are including several options for this task in new releases. One of the first challenges in using LiDAR data is to develop an understanding of the data characteristics and to relate these to the attributes of interest in the application (Goodwin 2006).

A typical analysis approach is to employ regression methods to select a LiDAR predictor variable (e.g., height percentiles) to relate to ground-based measures of structural information, such as tree or stand height (Morsdorf et al. 2006). For example, Bater (2008) investigated the capacity of LiDAR to estimate the distribution of living and dead trees in forest field plots. Each stem (DBH > 10 cm) was assigned to a wildlife tree (WT) class and a suite of LiDAR-derived predictor variables extracted to predict the cumulative proportions of stems within the WT classes. The coefficient of variation of the LiDAR height data was the best predictor variable, and standard errors were less than 5 percent. The derived relationships allowed for the prediction of the proportion of stems within WT classes across the landscape.

Most LiDAR height measurements are simply the difference in height between the average or maxima canopy laser hits and the ground returns, but different sampling approaches must be considered. An example of three different sampling approaches to compare forest information obtained from LiDAR and ground observations is shown in Color Plate 4.3 (Turner 2007):

- Applying a straightforward overlay of the pixel values from a canopy height surface using the 0.1 ha circular field plots
- Application of a local-maxima search filter across the canopy height surface to locate dominant crown peaks and extract statistics within each plot
- Use of a canopy segmentation algorithm to delineate dominant crowns and extract crown-derived statistics for each plot

In these eucalypt forests in Australia (Turner 2007), and in uneven aged conifer stands in North America (St.-Onge et al. 2003), LiDAR estimates of height regressed on field measurements produced a predictive equation with a strong coefficient of determination (> 0.90). Some observers have commented that it is likely that LiDAR-derived heights exceed the accuracy with which height can be estimated on the ground or by traditional stereophotogrammetry (St.-Onge et al. 2004).

However, the influences of LiDAR platform altitude, viewing angle, wavelength, and pulse density have not been fully explored and may introduce significant uncertainty (Goodwin et al. 2006). The fusion of complementary LiDAR data with aerial optical imagery to produce an accurate map with information about the location, height, and species of each tree is a promising example of remote-sensing synergy (Coops et al. 2004, Koukoulas and Blackburn 2005).

Exploring remote-sensing synergy: Remotely sensed data synergy is an important theme in remote sensing. The operating assumption is that complementary information may be obtained from multisensor or multiresolution data in a number of ways; one of the more obvious approaches is to arrange simultaneous image acquisition of the same area with different sensor packages.

Such remote-sensing synergy was explored in North Carolina by Nelson et al. (2007) in mapping loblolly pine (*Pinus tadea*) plantation above-ground woody biomass using the low frequency (80–120 MHz) VHF BioSAR™ system and a profiling LiDAR called the Portable Airborne Laser System (PALS). Both technologies had demonstrated independently a capacity to sense forest structure using different portions of the electromagnetic spectrum and with different image processing approaches. The best SAR model explained 81.8 percent of the stem green biomass variability, with a regression standard error of approximately 57 t/ha. The best LiDAR model explained 93.3 percent of the biomass variation with significantly lower standard error (approximately 34 t/ha). However, little sensor synergy in this environment was observed between the SAR and the LiDAR, possibly because this forest environment afforded little opportunity for the two sensors to detect different aspects of the forest structure. Increased SAR/LiDAR synergy can be expected in areas that provided increased opportunity for each sensor system to contribute complementary information, for example, in more topographically and species-complex sites, with a greater range of biomass and growth forms.

Another forest structure study by Hyde et al. (2006) using multisensor remote-sensing data observed synergy with active and passive optical data (e.g., between LiDAR and ETM+, and between LiDAR and QuickBird imagery). Again, low data synergy occurred when using the simultaneously-acquired LiDAR and SAR data, because both of these sensors responded in similar ways to the sampled forest structure. Low data synergy was also observed between the two passive optical sensors (ETM+ and QuickBird) when processed for similar vegetation indices. These examples highlight a significant design characteristic when attempting to create data synergy: *The objective should be to obtain different information, or different scales of information, in the original remote-sensing image acquisitions or during image processing.*

To create data synergy during an image analysis procedure, the goal is to process the available imagery for different types (or scales) of information (Ahearn 1988). For example, to extract forest structure from relatively low spatial resolution ETM+ imagery, a vegetation index approach may be suitable. But for high spatial resolution QuickBird imagery, a window-based statistical texture analysis would likely be more effective (Franklin et al. 2001a,b, Keramitsoglou et al. 2006). Together, high spatial resolution image texture and low spatial resolution vegetation index information could provide greater complementary information and might be more valuable than processing both images for only one type of information (e.g., two vegetation indices at different spatial resolutions).

In aerial multispectral imagery of different spatial resolutions acquired at different altitudes over aspen stands in Alberta, processing the higher spatial detail imagery for different kinds of texture measures revealed data synergy that was useful for discriminating multilayer forest stands (Moskal and Franklin 2002) and in aspen defoliation mapping (Moskal and Franklin 2004). Different texture image processing strategies were needed to explore fully the information content of the different spatial resolutions of optical imagery. This remote-sensing synergy originated in the multiple scales of forest information contained in the multiresolution imagery. Thus, an important practical consideration in studies of remote-sensing synergy, and in comparing information content from imagery acquired with different characteristics, such as spatial resolution, or from different sensors, is to ensure that the optimal processing or information extraction strategy is applied to each of the disparate data sets. Designing optimal methods for individual image characteristics will likely provide superior results when compared to carrying over one image processing technique from one scale of image analysis to another.

Dead and Damaged Trees

Vegetation damage assessment is accomplished by remote sensing of color and moisture changes (e.g., chlorosis, dehydration), and the detection of loss of plant chlorophyll, turgidity, foliage, or other growing organs (Hoque et al. 1992). Such effects are caused by many influences, including the activities of organisms, fire, or other biophysical processes (such as pollution) (Franklin 2001). Inventory of the microhabitat attributes of dead and damaged trees, such as cavities and geometric features is one of the most difficult remote-sensing applications, and has not yet been fully explored (Gibbons and Lindenmayer 2002). In only a few instances have comprehensive efforts been implemented to determine the effectiveness of remote sensing. For example, the number of snags or standing/downed trees was estimated from plot or transect samples, in a hierarchical design by Canci et al. (2007).

When combined with identification of dead, dying trees, and snags on high spatial detail imagery (Color Plate 4.4; Canci et al. 2007), this approach may result in a comprehensive survey of individual dead and damaged trees (*see* Bütler et al. 2004).

Cavity sampling methods in the field are complicated by inadequate access or visibility of cavities in standing trees (McComb and Lindenmayer 1999). One series of studies of bat selection of roosting habitat and tree cavities used thermal infrared sensing to discriminate cavity characteristics; tree species and conditions (e.g., bark thickness, water content) influenced internal microclimates that helped predict bat occupancy (Kunz et al. 2003). Catena et al. (1990) described a field-based thermal infrared sensing system for detecting internal cavities in trees and individual animal presence/absence in nests; subsequent experiments have compared such imagery with optical infrared data acquired using nondamaging infrared insertion devices and other sensing extensions (Bucur 2003).

Boonstra et al. (1994) found that red squirrels (*Tamiasciurus hudsonicus*) could be detected using one early thermal imaging system, although active nests were less accurately detected, possibly as a result of variations in the insulating property of nests or fur. Correlation approaches between field-observations of cavities and remote-sensing spectral response patterns are a more common approach; such models may be developed using statistical relations (e.g., suppression mortality model based on live tree inventory). Direct remote sensing of tree cavities from aerial or satellite systems has been rarely attempted.

Chemical Constituents of Vegetation, Soils, and Water

Foliar Chemistry and Stress (Health)

The interaction of electromagnetic radiation with plant leaves is determined by their chemical and physical properties (Schut 2003). From 400 to 700 nm, various pigments such as chlorophyll, xanthofyll, and carotene dominate leaf reflectance; in the near-infrared region, from 700 to 1300 nm, water absorption strongly influences reflectance. Some nutrient conditions are related to chlorophyll content (N, Mg, Zn, Fe, Mn, P, S); other plant constituents such as starch, lignin, cellulose, and sugar affect leaf reflectance. Chemical absorption regions occur at more than 40 specific wavelengths between 430 and 2350 nm (Curran 1989, Ferwerda and Skidmore 2007). Remote sensing of foliar chemistry is an exploration of these chemical absorption regions of the electromagnetic spectrum based on the assessment of foliage harvested and examined in laboratory settings and in field, aerial, and satellite instrument data. The sampling of spectra from leaves, trees, canopy, and vegetation communities to assess chemical stress and growth factors necessarily encounters the many challenges of environmental remote sensing, including IFOV/pixel size considerations, atmospheric influences, and selection of appropriate methods of analysis.

To illustrate the approach, hyperspectral imagery acquired by the AVIRIS sensor on-board an ER-2 research flight over Vancouver Island forests were used to estimate foliage chemistry of Western hemlock (*Tsuga heterophylla*) (Niemann and Goodenough 2003):

1. Foliage samples were acquired by helicopter from tree crowns, placed on ice and transported immediately to a wet chemistry analytical lab for extraction of chlorophyll *a* and *b*, carbon, nitrogen, carotenoids, and a full suite of inorganic constituents.
2. AVIRIS data were processed for instrument radiometry and atmospheric corrections were based on radiosonde data; after orthocorrection using available 1 m orthophotos of the study site, the hyperspectral imagery were reprojected with 20 meter pixels.
3. Spectral mixture analysis separated canopy reflectance and background effects based on field measurements of endmember spectra; a local maxima approach (Wulder et al. 2000) was used to identify pixels that corresponded with field-sampled hemlock tree crowns, which were also visible in the orthophotography used in the geometric correction.
4. Correlograms were used to display the relationship between reflectance, pigments, C, N, and some of the sampled inorganics; multiple regression was used to control for crown closure.

This analysis confirmed many of the strong absorption features located in earlier studies (Table 4.1; *see* Curran 1989) but also found considerable unexplained variability. Water content and multiple pigments absorption of energy in the same spectral regions caused confusion in spectral analysis (Blackburn 1998). For the Vancouver Island

Chemical Constituent	Wavelengths (nm)
Chlorophyll *a*	707
Chlorophyll *b*	1010–1100
N	667, 1415, 1507, 2035, 2055, 2198, 2301
C	1935
Carotenoids	541

Source: With permission from Niemann, O., and D. G. Goodenough. 2003. In Wulder, M. A., and S. E. Franklin, eds. *Remote Sensing of Forest Environments: Concepts and Case Studies.* Kluwer Academic Press, Boston. 447–467.

Table 4.1 Wavelength Regions for Prediction of Foliar Chemistry of Western Hemlock Using AVIRIS Hyperspectral Imagery

data, the relationships between reflectance and chlorophyll *a* and *b*, C, and N were weak, partly as a result of the selected remote-sensing experimental design (low spatial resolution but large area mapped).

Improving the availability and performance of spectral mixture analysis methods to separate canopy and understory reflectance, and components of photosynthetic vegetation, nonphotosynthetic vegetation and exposed soil are significant research challenges for hyperspectral data and other optical imagery (Asner and Heidebrecht 2005, Davidson et al. 2008). Significant progress is needed, for example, in order to better estimate chemistry and plant constituents to the degrees of accuracy and reliability required by ecosystem process models. A research effort to derive foliar N may be of particular importance; foliar N is related to a variety of ecological and biogeochemical processes, ranging from the spread of invasive species, to the ecosystem effects of insect defoliation events, to patterns of N cycling in forest soils (Martin et al. 2008).

Extracting different spectral variables can improve characterization of the relationships of interest; for example, the relationship between foliar N and chlorophyll is curvilinear, and reflectance decreases asymptotically with increasing chlorophyll (Wood et al. 1992, Kantety et al. 1996). One approach is to calculate the chlorophyll absorption width (CAW) rather than use reflectance measured in a specific image wavelength. In a controlled field experiment with a commercially available mixture of four grass cultivars (*Lolium perenne*) and a tripod-mounted spectrometer and artificial illumination, CAW was observed to change as N supply changed, and CAW was subsequently strongly correlated with dry matter yield ($R^2 = 0.95$) (Schut 2003). *Derivative spectra* are often used to reduce the influence of mixed pixel reflectance (i.e., background effects such as soil contributions), but other influences (e.g., leaf angle distribution, canopy geometry) may require more complex data handling.

Stress and reduced health may be a result of nutrient and moisture deficiencies, soil contamination, fungal and insect activity, and a variety of abiotic and biotic conditions that may be detected in remotely sensed data. The symptoms of plant stress, such as chlorosis of foliage, stunted growth, loss of organ function, and defoliation, are readily identified with *in situ* data and in optical imagery at appropriate resolutions (Hall et al. 2007). Early detection is possible by examining nonvisible wavelengths. For example, in identification of trees in British Columbia attacked by spruce bark beetles (*Dendroctonus rufipennis*), and while foliage was still green, visual interpretation of color infrared aerial photography was as effective as field surveys in identifying spruce beetle attack (Murtha and Cozens 1985):

> ...loss of greenness, the dryness, and the lack of spectral variation among the foliage age classes produces light-toned, highly reflective, mono-hued tree crowns which seem to "glow like a halo" on CIR photos
>
> *(Murtha 1985: p. 99)*

Generally, the imagery of stressed plants (compared to healthy plants) will display a blue shift and increased differences in red and near-infrared reflectances, known as the "red-edge" (Reid 1987, Essery and Morse 1992). Red-edge indices use reflectance at wavelengths 680–740 nm, where reflectance is highly sensitive to strong absorption by chlorophyll in the red region, and strong reflectance in the NIR region, due to scattering in the leaf mesophyll and the absence of absorption by pigments (Curran et al. 1990). Increased absorption by chlorophyll creates a deeper and broader absorption feature. Increased canopy moisture levels tend to reduce near-infrared reflectance, although this may be confounded by structural complexity. These differential red and near-infrared responses to changes in stress, and changes in pigments and moisture, can, in turn, change the shape of the red-edge.

These spectral features have been derived from satellite sensor data in large-area studies of forest decline (Ardö 1998) and from aerial sensor data to map a soil metal concentration gradient (Olthof and King 2000). Spectral characteristics derived from red-edge observations and spectral mixture analysis of four-band aerial digital multi-spectral camera imagery were associated with the influences of three stress agents (fungal pathogen, soil nutrient deficiency, and aphid infestation) and subsequent damage to *Pinus radiata* tree crowns in a New South Wales plantation (Sims et al. 2007):

1. Field data on nitrogen soil deficiency, crown transparency, leader discoloration were collected in near-coincident sampling protocols that focused on assessment of health in areas known to be stressed by insects, soil nutrient deficiencies, or fungal infestation.
2. Aerial imagery was acquired from the digital multi-spectral camera (DMSC) II system comprised of four individual 1024 × 1024 CCD arrays capable of acquiring data at 12 bit quantization and fitted with independent narrow 10 nm bandwidth filters positioned at 680, 720, 740, and 850 nm to quantify a simple red/infrared ratio and red-edge reflectances.
3. Calibration of the imagery and atmospheric correction was implemented with targets placed within the image acquisition area and observed by the aerial sensor and field spectroradiometer.
4. Each tree crown was manually delineated in the resulting imagery and spectral values, such as the red-edge and fractional estimates of foliage and background obtained by spectral mixture analysis, were correlated in a decision tree model with the field observations. These data were used to develop a classification rule for "healthy" and "unhealthy" crowns.

The best discriminator of crown discoloration caused by the fungal pathogen was the upper slope of the red-edge; the classification accuracy exceeded 83 percent. Poor crown development and higher crown transparency associated with soil nitrogen deficiency and aphid infestation were best mapped using the sunlit canopy fraction from the spectral mixture analysis; accuracies were lower than in mapping crown discoloration, and more variable across the site.

Food Resource Availability and Quality

A large number of remote-sensing studies have mapped habitat availability using image classification techniques and vegetation indices, and inferred or otherwise interpreted food availability and quality (Théau et al. 2004). Methods vary widely, and include, for example:

- *Aerial videography image classification.* Lake et al. (2006) report greater than 91 percent classification accuracy in aerial videography mapping of Arctic vegetation classes, including areas containing an important food source used by broods of Emperor geese (*Chen canagica*). Interpretation of food quality in a fine-scale, microhabitat analysis was possible by understanding different communities or species assemblages of grasses known to be more or less nutritious (*C. subspathacea* vs *C. ramenskii*). These communities were then categorized in the image classification.
- *Landsat image classification.* Earlier work to map geese forage classes with Landsat imagery could separate nonvegetation and vegetation cover types, but classification of macrohabitat vegetation classes used by lesser snow geese (*Anser caerulescens caerulescens*) for food resources was less successful because of the relatively lower spectral- and spatial-resolution of the Landsat imagery compared to aerial videography (Jano et al. 1998). However, a similar experimental design with Landsat imagery to identify Arctic lichen-dominated reindeer feeding areas was satisfactory (accuracies of 84 percent for summer, and 88 percent for winter pasture) (Colpaert et al. 2003).
- *Sea-Viewing Wide Field-of-view Sensor (SeaWIFS) photosynthetically active radiation (PAR) data to calculate net primary production.* Littaye et al. (2004) determined the relationship between summer aggregation of a certain species of whales and satellite-derived environment conditions in the northwestern Mediterranean Sea. WIFS PAR data are false color representations of ocean pigment concentrations (i.e., sea surface color), and are used to estimate phytoplankton, the concentration of biomass on the ocean surface, and net primary production when coupled with sea surface temperature measurements

and a light-photosynthesis model. The results described local food availability for ocean mammals and some fish and shellfish species. Evidence was interpreted to mean that whale movements were dependent on food availability rather than environmental conditions (Moses et al. 2004).

- *Terra Satellite MODIS NDVI seasonal differences in habitat quality.* Ito et al. (2006) used MODIS NDVI to track migration patterns and show seasonal differences in habitat quality associated with Mongolian gazelles (*Procapra gutturosa) on* the Mongolian steppe. Seasonal migration corresponded to shifts in the normalized difference vegetation index (NDVI) in their habitat. The mean NDVI values of their annual, summer, and winter ranges were calculated and shown to differ with seasonal migrations of gazelles. The results showed that the NDVI is a good indicator of gazelle habitat, but the NDVI alone did not explain the patterns of seasonal migration of gazelles. More work was recommended to evaluate the effectiveness and limitations of the Terra MODIS NDVI as an indicator of habitat quality at this scale.
- *Hyperspectral remote sensing.* Field and aerial spectrometers have been deployed to provide information on food availability and quality through estimates of biophysical conditions and chemical plant constituents (Van der Heijden et al. 2007). For example, Schut (2003) and Schut et al. (2006) reported work in grass communities based on digital camera (grown cover, chlorophyll content, leaf area, clover content, and N stress) and imaging spectroscopy (quantification of nutrient and drought stress, estimation of feeding value and mineral content).
- *Remote-sensing synergy.* An operational grassland and crop assessment system was developed to combine the high spatial resolution and high spectral resolution sensors in a GPS-equipped mobile vehicle with artificial illumination to reduce the dependence of the system on sunlight and the time of day (Molema et al. 2003). Starks et al. (2002) reviewed recent progress in bench-top near-infrared spectroscopy for North American range and grassland communities, and then assessed forage quality (nitrogen, neutral detergent fibre, acid detergent fibre) in the field using a hand-held hyperspectral instrument for real-time assessment of forage quality and mapping of the nutritional landscape. Subsequently, better management of pastures and supplements, and support for harvesting decisions, was thought possible. An earlier dimension to this work was the estimation of soil water content in these grassland environments with microwave brightness temperature data acquired from the L-band passive microwave electronically scanned thinned array radiometer (ESTAR) (Starks and

Jackson 2002). An empirical model incorporating soil texture and dielectric properties was used to calibrate the final imagery, which were then compared favorably to field observations and model results based on climate data.

- *LiDAR remote-sensing data.* Hill et al. (2008) determined the structural influences on habitat selection for two woodland birds, the Great Tit (*Parus major*) and Blue Tit (*Parus caeruleus*). The approach was to identify, in the field, the key components determining habitat quality for reproductive success; then, with an integrated multispectral digital camera, airborne LiDAR and ground-based LiDAR system, the necessary information on underlying terrain, canopy height, density, gap fraction, surface roughness, and the understory shrub layer, were combined with plant species and bird census data (nestbox and territory mapping). The final map output was a habitat quality map that related species distributions, abundances, and patterns of persistence to three-dimensional habitat structure and composition.

Soil Processes and Conditions

Remote sensing of soil processes and conditions has been successfully accomplished using two main approaches: (i) direct detection by thermal, passive microwave, and active microwave remote sensing, and (ii) modeling approaches relying on imagery and DEM variables to predict soil information (e.g., slope and aspect configurations leading to soil variability) (Band et al. 1991, McDonnell 1996). A significant research effort with these data and other types of imagery has revealed a number of approaches of potential interest in biodiversity and habitat applications. A selection of the key remote-sensing soils variables developed in such applications is based on the following methods:

Soil Moisture

Aerial thermal (Gauthier and Tabbagh 1994) and microwave imagery (Jackson 1993) have long been used to generate estimates of soil moisture and other soil properties (such as soil texture and erosion) from remote-sensing temperature data. Observed microwave emission, for example, depends on both soil moisture and soil temperature. Dry soils are a poor electrical conductor, but conductivity increases predictably with increasing moisture content; therefore, wet soils, which are usually darker tones in optical imagery, typically appear brighter in SAR and passive microwave imagery (Aronoff 2005).

Soil Organic Carbon

Yadav and Malanson (2007) reviewed progress in soil organic matter research in recent decades and noted that remote sensing has proven particularly effective in estimation of soil organic matter stocks in certain types of topsoil. A number of sensing approaches have been employed. For example, Gomez et al. (2008) compared predictions of soil organic carbon using visible and near-infrared reflectance hyper-spectral field and Hyperion satellite data in the Narrabri region, New South Wales (NSW), Australia. Empirical models based on partial least-squares regression showed that predictions of soil carbon using the Hyperion spectra were similar to, but less accurate, than those of the field spectra data resampled to similar spectral resolution. The most important estimation differences appeared to be a result of the different spatial resolutions of the field and satellite-based data sets.

Soil Nutrient Status

Early work in prediction of soil organic matter and the relationship to soil texture (e.g., Krishnan et al. 1980) using particular optical wave-lengths has been extended to incorporate soil nutrient estimation in a procedure related to vegetation chemistry analysis (e.g., Ferwerda and Skidmore 2007). Ecosystem process models require available nitrogen (net N mineralized and N inputs) to estimate allocation and internal dynamics of the leaf/root/stem carbon allocation fraction; spatial estimates of leaf nitrogen concentration may be generated through a relationship with LAI and soil conditions inferred through leaf litterfall and decomposition rates (Lucas et al. 2000).

Soil Temperature

Thermal satellite imagery—from MODIS sensors, for example—have been used to produce standardized land surface temperature (LST) products, from which soil temperature and soil heat flux can be derived (Huang et al. 2008). In the microwave portion of the electromagnetic spectrum, Jones et al. (2007) retrieved surface soil temperature infor-mation from the EOS Advanced Microwave Scanning Radiometer for seven boreal forest and Arctic tundra biophysical monitoring sites across Alaska and Northern Canada. Their approach required a mul-tiple-band iterative radiative transfer process-based method to pro-duce dynamic vegetation and snow cover correction quantities. The seasonal pattern of microwave emission, and relative accuracy of the soil temperature retrievals, were influenced strongly by landscape properties, including the presence of open water, vegetation type and seasonal phenology, snow cover, and freeze-thaw transitions. An over-all root-mean-square error of less than 4 K was obtained in summer thawed conditions, with a larger error occurring in winter during periods of dynamic snow cover and freeze-thaw state.

Water Quality

Water quality and water habitat assessment are key environmental con-ditions that are increasingly amenable to remote-sensing approaches. The improvements in remote sensing of rivers, for example, has been reviewed recently by Marcus and Fonstad (2008), who concluded that it is now possible to generate accurate and continuous aerial remote-sensing maps of in-stream habitats, depths, algae, wood, stream power, and other features, if the water is clear and the aerial view is unobstructed. They predicted that the availability of such maps will transform river science and management, and recommended a synergistic combination of LiDAR, SAR, thermal, and multispectral and hyperspectral optical data in operational remote-sensing stream habitat applications. A confirmation of the value of this remote-sensing approach to water quality and habitats was provided by Gilvear et al. (2008) in an application to lamprey habitat mapping in Scotland's River Tay based on Airborne Thematic Mapper imagery and color aerial photography.

Currently, the retrieval of concentrations of different water quality factors, such as estimates of turbidity and chlorophyll *a*, from remote-sensing measurements is most accurate when based on the analysis of hyperspectral or multispectral optical data (Hellweger et al. 2004, Voutilainen et al. 2006). Following the implementation of appropriate image radiometry and geometry corrections, two general approaches are available based on: (i) empirical relationships, or (ii) modeling techniques.

Empirical approaches involve predicting the water quality concentrations of interest using a regression equation in which the independent variable is typically selected as a ratio of measurements of two appropriately positioned spectral bands. For example, a total of 38 parameters including total suspended solids, chlorophyll *a*, dissolved organic matter, diffuse attenuation coefficient, and calcite were estimated using the MODIS sensor onboard the Terra and Aqua satellites by Hellweger et al. (2004). A typical remote-sensing/field sample design was used in which samples obtained in the field were related to spectral observations acquired near-simultaneously. Some of the challenges of this approach are related to spatial resolution (for example, image pixel size compared to field data collection over a much smaller area), sampling water quality at the surface versus at depth, and temporal resolution. It is also the case that, with many regression models, the results are highly site and sensor specific, without a robust capability of extension to new areas and data.

Bio-optical reflectance models have continued to mature and are a proven methodological research tool; typically, a semiempirical or mathematical approach, in which the relation between concentrations of water quality constituents are described, and a model inversion, are used to explain observed spectral response patterns over water. The mathematical approach is to link three model components: a flow field, a convection-diffusion model, and an observation model (Voutilainen et al. 2006). The computations can become quite complex; for example, it may be difficult to specify the boundary conditions, model the volume and boundary sources, and model the effect of wind on the flow field.

Semiempirical bio-optical model approaches are readily implemented based on the pioneering ocean optical properties development work of Gordon et al. (1975). The inputs to the simple model are the concentrations of water quality factors of interest, linked to total absorption and backscattering coefficients through a series of empirical relationships for the water column that allow spectral variations to be estimated. A continuation of these ideas has been implemented in a bio-optical model for use with Compact Airborne Spectrographic Imager (CASI) data (Ammenberg et al. 2002):

1. The model considers the inherent optical properties of water and optically active substances that are known to influence these properties. A large 25 year data set of measurements of the concentration of chlorophyll *a* and phaeophytine *a* (Chl), suspended particulate inorganic material (SPIM), and the absorption coefficient of colored dissolved organic matter (CDOM) was used for a site in Lake Malaren, Sweden.
2. The model receives as input the optically active substances and outputs a reflectance spectrum just above the water surface to simulate absorption (at 420 nm) for comparison to the CASI observed reflectance patterns, which are then used to prepare maps of variability in Chl, SPIM, and CDOM.
3. Validation was accomplished by near-coincident (within three hours) continuous field measurements of fluorescence and beam attenuation and water analysis results from nine water samples.
4. Model predictions and CASI reflectance observations of concentrations of the optically active substances were in close agreement with field measurements within a reasonable level of accuracy ($R^2 > 0.73$ for Chl, 0.82 for SPIM; for CDOM there appeared to be good agreement, but too few samples for statistical comparison).

Some of the challenges in this approach are related to coordinating the timing of the field and aerial remote-sensing observations, constraining the area over which the field sampling and remote sensing is conducted, and degree to which the bio-optical model inversion is dependent on local conditions:

> There is of course a risk in our method, to tune the model too well to the studied lake, and thereby get the same kind of site-specific algorithms (as regression-based approaches). The model will be used on more remote sensing scenes and on more lakes in the near future, to further investigate its potential.
>
> *(Ammenberg et al. 2002: p 1636)*

A similar approach was described with Hyperion satellite data in a study of water quality in Lake Garda, Italy's largest lake (Giardino et al. 2007). Water quality was assessed by:

1. Defining a bio-optical model with specific inherent optical properties of the lake
2. Converting the Hyperion at-sensor radiances into subsurface irradiance reflectances using the MODTRAN-based atmospheric correction model
3. Adopting a bio-optical model inversion technique to retrieve chlorophyll *a* and tripton concentrations from two Hyperion bands

The correlation coefficient between *in situ* water samples and Hyperion-derived concentrations was approximately 0.77 for chlorophyll *a* and 0.75 for tripton. One problem encountered was an atmospheric adjacency effect beyond 700 nm in the Hyperion data that was not adequately modeled in the atmospheric correction. However, this approach provided the ability to assess water quality independently of *in situ* water quality data, and was thought ready to support on-going management applications.

Ice and Snow Conditions

Mapping snow and ice is a requirement for certain habitat and wildlife management applications. Many of the snow and ice remote-sensing applications originally identified by Meier (1980), in an early review of remote-sensing potential, have now been made possible on a routine basis using satellite or aerial imagery; for example:

- Forecasts of snowmelt runoff and reservoir operation
- Soil and irrigation water supply management
- Flood warning and flood potential mapping
- Changes in climate, freezing and breakup of rivers and lakes, monitored over large areas
- Seasonal snow and sea ice cover mapping for navigation and climate modeling

Rees (2005) presented a thorough review of remote sensing of cryosphere processes and highlighted a number of approaches that may be of interest to those studying habitats and biodiversity in frozen environments. Mapping sea ice extent, for instance, can be accomplished with optical sensors because sea ice reflects a high proportion of incoming solar radiation compared to open water and vegetation. A snow cover on sea ice can further increase the shortwave albedo and may in addition provide a thermal contrast detectable with thermal and microwave sensors. Algorithms to retrieve aerosol and water/snow/ice properties, including snow grain size and surface temperature, from ADEOS-II Global Imager (GLI) data were evaluated by Stamnes et al. (2007). The general methods include cloud masking, surface cover discrimination, and multilinear regression with radiosonde data and radiative transfer models.

Microwave radiometry detection of a thermal gradient in such environments is readily achieved by comparing observations in more than one band (or by polarization differences). Polynyas are areas of thin ice or open water in polar regions that are often identified as important macrohabitat features—identification and monitoring can become a complex endeavor relying on multiple sources of imagery and models (e.g., Marcus and Burns 1995). Maps of sea ice thickness and open water from ASMR imagery, for example, were used by Kern et al. (2007) with weather data and a modeling technique to create a time series of polynya extent and dynamics in Antarctica. Vincent et al. (2001) used a series of historical records, aerial photographs, and satellite imagery (e.g., Radarsat-1) to document the loss of ice cover or an increase in specific ice-water habitat features (e.g., meltwater lakes) known to be important to macro-invertebrates, cyanobacteria, microscopic animals, and algal communities, in the Canadian High Arctic.

Ecosystem Processes

Gross and Net Primary Productivity

Remote sensing in estimation of gross primary production and net primary production is a powerful multiscale application. The complexity of ecosystem process models varies greatly with the purpose of the model and the scale of the principal outputs—estimates of conditions (e.g., standing biomass) or processes (e.g., photosynthesis) over biomes to landscapes to individual plants.

Generally, to obtain NPP, *landscape ecosystem process models* require assumptions or initial estimates of leaf area index, available water capacity of soil, and gridded daily meteorological variables (shortwave radiation, maximum and minimum temperatures, humidity, and precipitation) (Lucas and Curran 1999). A flowchart showing the major components of the ecosystem process model called BGC++ is shown in Fig. 4.5. BGC++ model development was based on previous model understanding gained in a wide variety of terrestrial ecosystem model simulations and applications (Running 1990, Running and Hunt 1993, Hunt et al. 1996). This class of process model was dubbed the *integrative-model* (Waring and Running 1999) because of the strong connection between remote sensing, ecosystem process dynamics, and mechanistic model simulations. Other examples of this class of

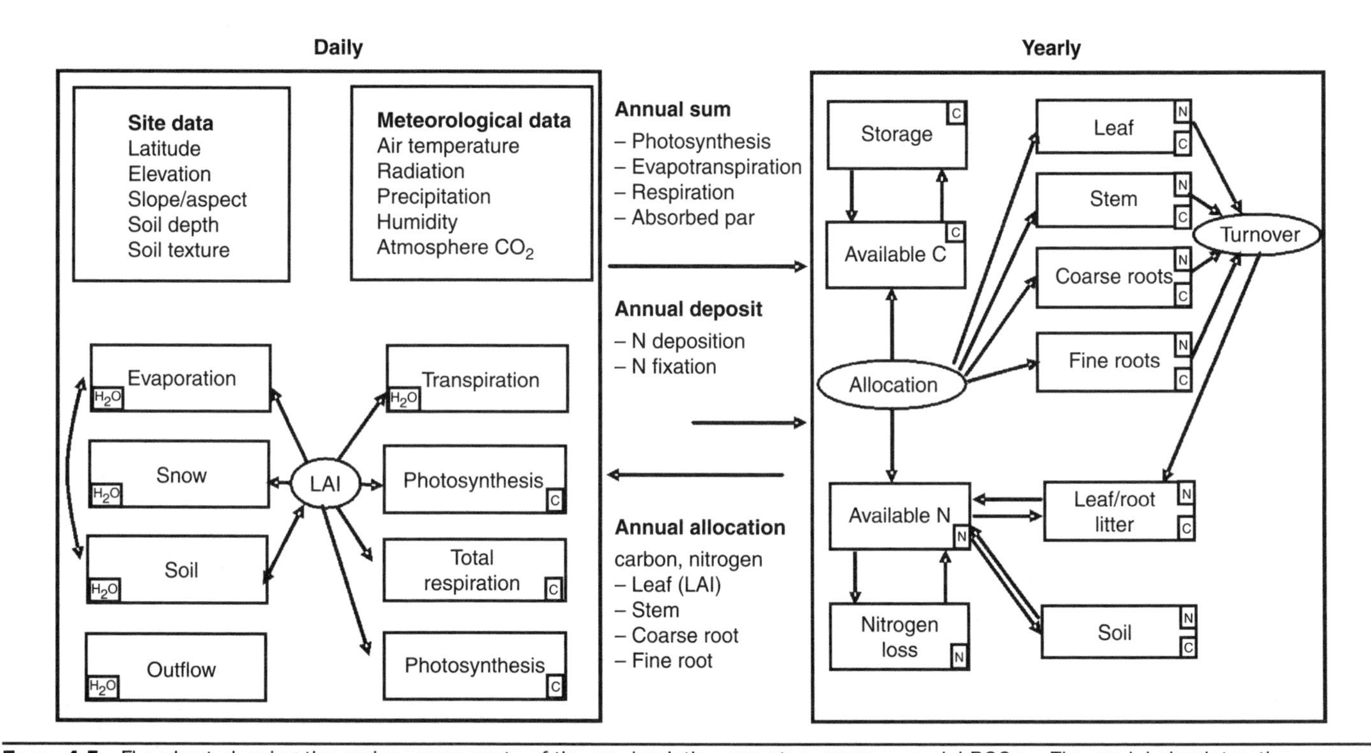

FIGURE 4.5 Flowchart showing the major components of the mechanistic ecosystem process model BGC++. The model simulates the biogeochemical cycles of carbon, water, and nitrogen. The boxes show current amounts and the arrows between the boxes show the fluxes. LAI controls the daily fluxes, and allocation and turnover control the annual fluxes. (*With permission from* Franklin et al. 1997. *International Journal of Remote Sensing* 18:3459–3471.)

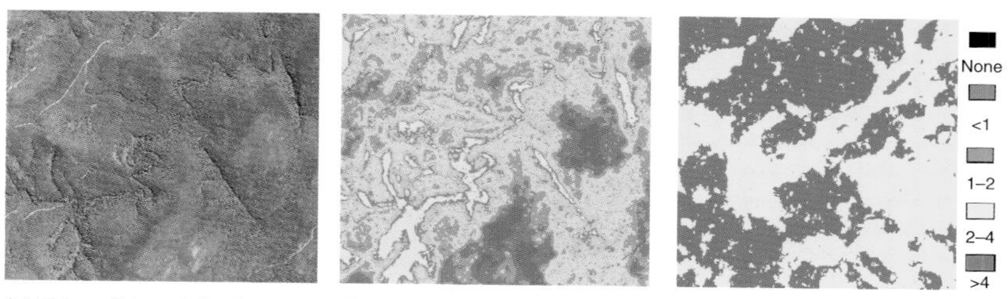

(a) High spatial resolution imagery predicting *Brown Antechinus* individuals per 100 trap nights

(b) Ground and QuickBird imagery of features and animals at the Bronx Zoo

Color Plate 1.1 High spatial resolution satellite imagery applications: (a) predicting abundance of *Brown Antechinus* (*Courtesy of Dr. Nicholas Coops, University of British Columbia*); (b) ground and QuickBird imagery of features and animals (*Courtesy of Rob Gutro, NASA*) (http://www.nasa.gov/vision/earth/lookingatearth/elephants_space.html).

(a) *In situ* spectroradiometer taking radiometric measurements in a sea grass area

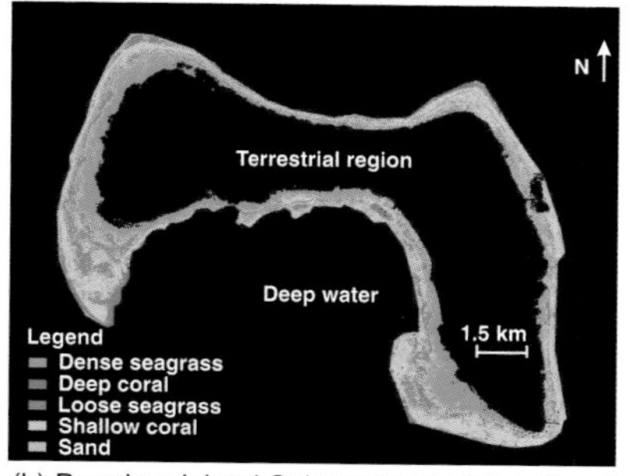

(b) Bunaken Island Submerged Habitat Classification Based on IKONOS imagery

Color Plate 3.1 Marine remote sensing using multispectral imagery and field observations: this example illustrates seagrass and coral species habitat mapping in Indonesia (*Courtesy of Dr. Ellsworth LeDrew, University of Waterloo*).

Color Plate 3.2 Examples of field, aerial, and satellite remote sensing digital image products for forest management, habitat and biodiversity monitoring (*Courtesy of Dr. Ron Hall, Canadian Forest Service*).

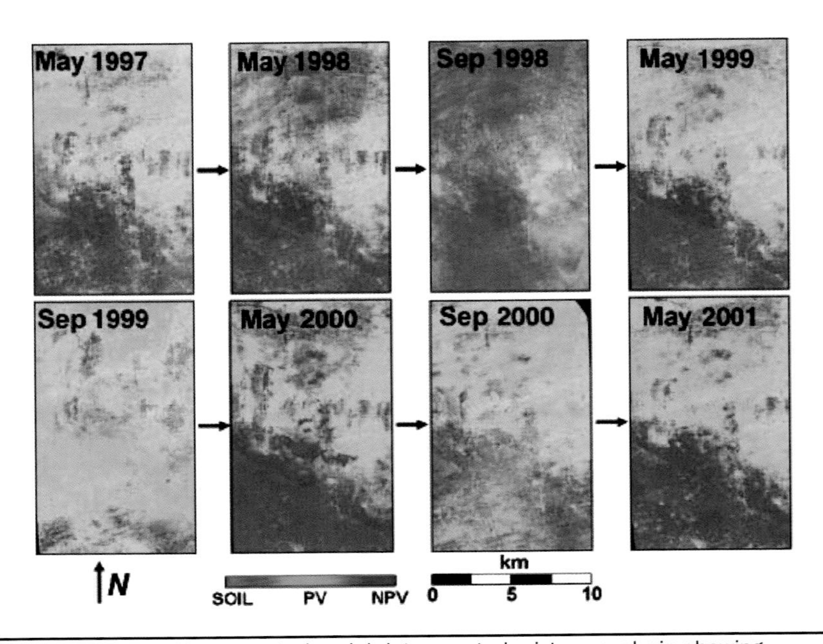

Color Plate 4.1 Hyperspectral aerial data spectral mixture analysis showing fractional cover change in rangeland vegetation (green), nonphotosynthetic vegetation (blue), and soils (red) at Jornada Experimental Station, Northern Chihuahua, USA, over time. (*With permission from Asner, G. P., and K. B. Heidebrecht.* 2005. Global Change Biology *11: 182–194.*)

(a) Field photo showing low structural complexity

(b) A low complexity area at full CIR image resolution (as outlined in C and D)

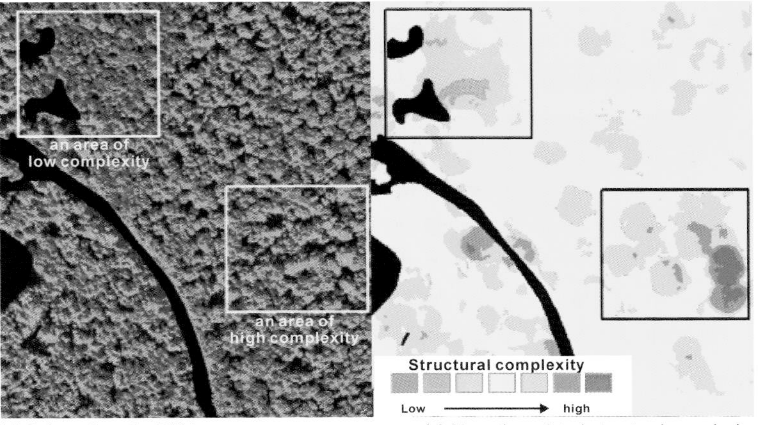

(c) Subsection of a CIR image

(d) Map of predicted structural complexity corresponding to the image area in C

(e) Field photo showing high structural complexity

(f) A high complexity area at full CIR image resolution (as outlined in C and D)

Color Plate 4.2 Examples of forest structural complexity modeling and mapping using high spatial resolution multispectral aerial imagery and topographic data (*Courtesy of Dr. Doug King and Jon Pasher, Carleton University*).

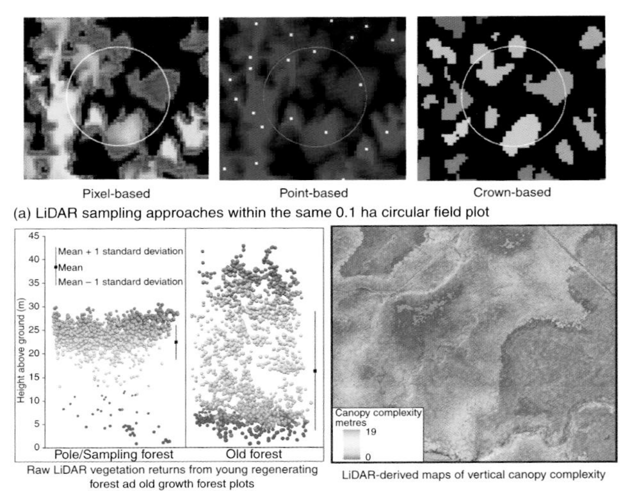

Color Plate 4.3 LiDAR sampling approaches and application example: (a) Sampling LiDAR aerial data. (*From Turner, R. 2007. An overview of airborne LiDAR* applications in New South Wales state forests. *Proceedings of ANZIF Conference*. Coffs Harbour. http://www.forestry.org.au/pdf/pdf-public/conference2007/papers/Turner%20Russell%20Lidar.pdf accessed 18 August 2008. *With permission*.); (b) aerial LiDAR vegetation height and complexity estimation (*Courtesy of Dr. Nicholas Coops, University of British Columbia*).

Color Plate 4.4 High spatial resolution aerial imagery in detection of individual tree mortality. (*With permission from Canci, M., J. Wallace, E. Mattiske, and A. Malcolm. 2007. Evaluation of airborne multispectral imagery for monitoring the effect of falling groundwater levels on native vegetation*. International Archives of the Photogrammetry, Remote Sensing and Spatial Sciences *34: abs*).

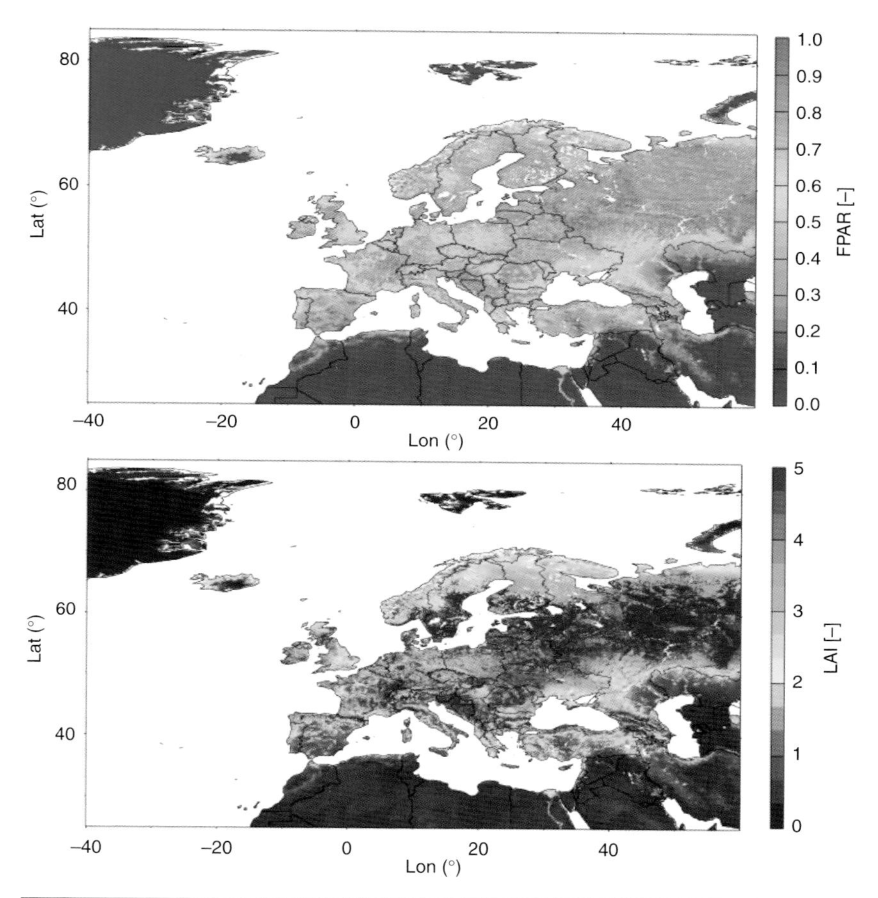

Color Plate 4.5 Plant phenology and biophysical attributes mapped across Europe using a 20-year AVHRR land-surface parameter dataset. (*With permission from Stöcklii, R., and P.L. Vidale. 2004*. International Journal of Remote Sensing *25: 3303–3330*).

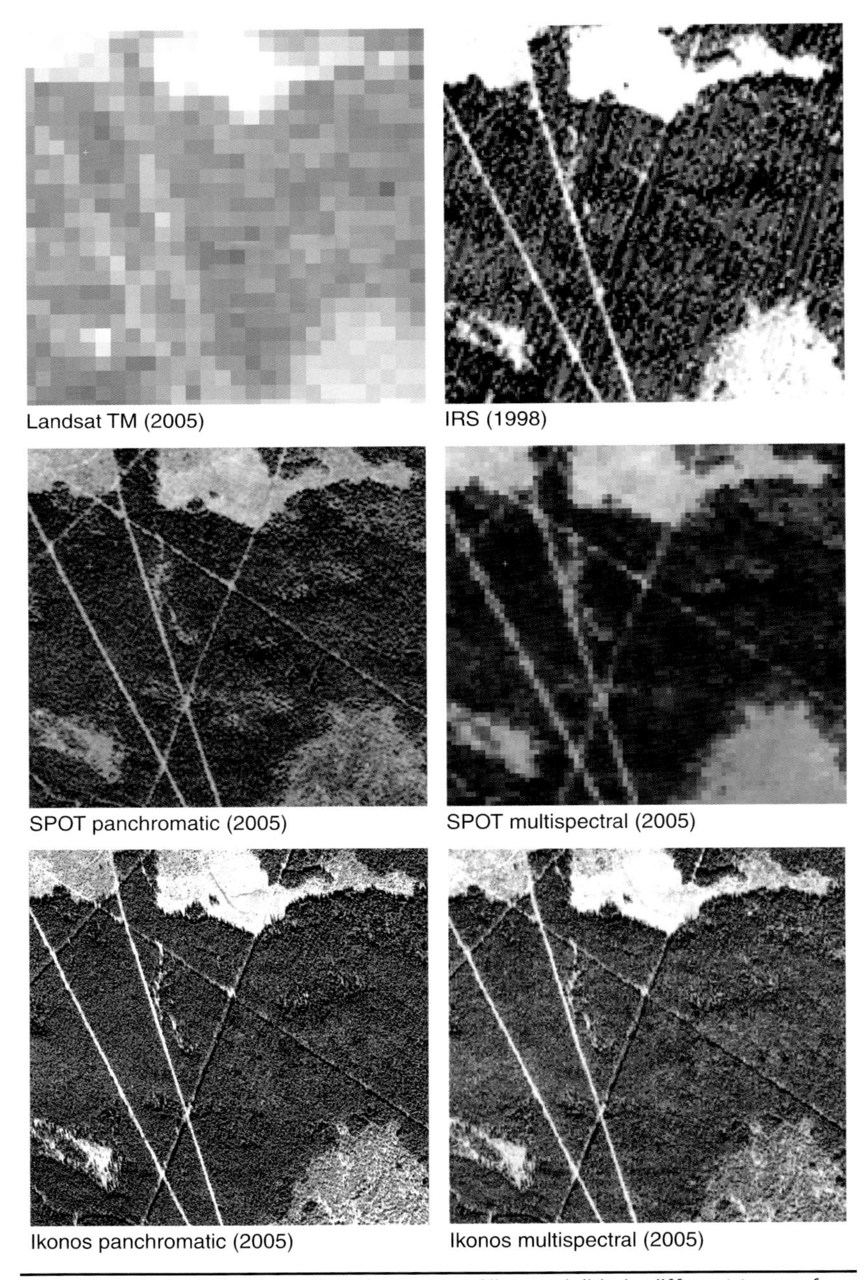

Color Plate 5.1 Seismic cutlines in western Alberta visible in different types of satellite imagery (Landsat, IRS, SPOT, Ikonos). Area shown is approximately 0.8 km^2. (*Courtesy of Dr. Yuhong He, University of Toronto Mississauga*).

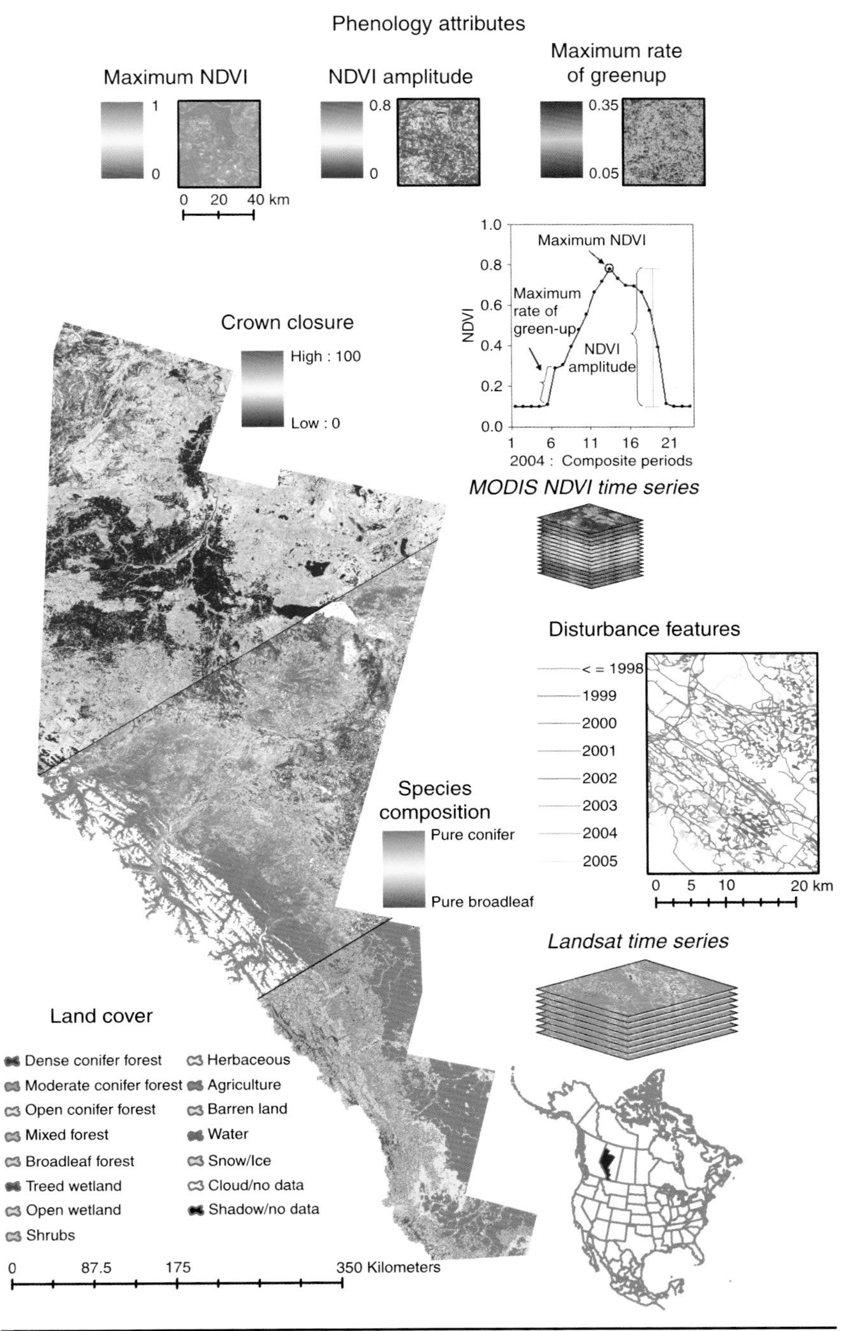

COLOR PLATE 5.2 Grizzly bear habitat mapping and modelling image products in western Canada (*Courtesy of Dr. Greg McDermid, University of Calgary*).

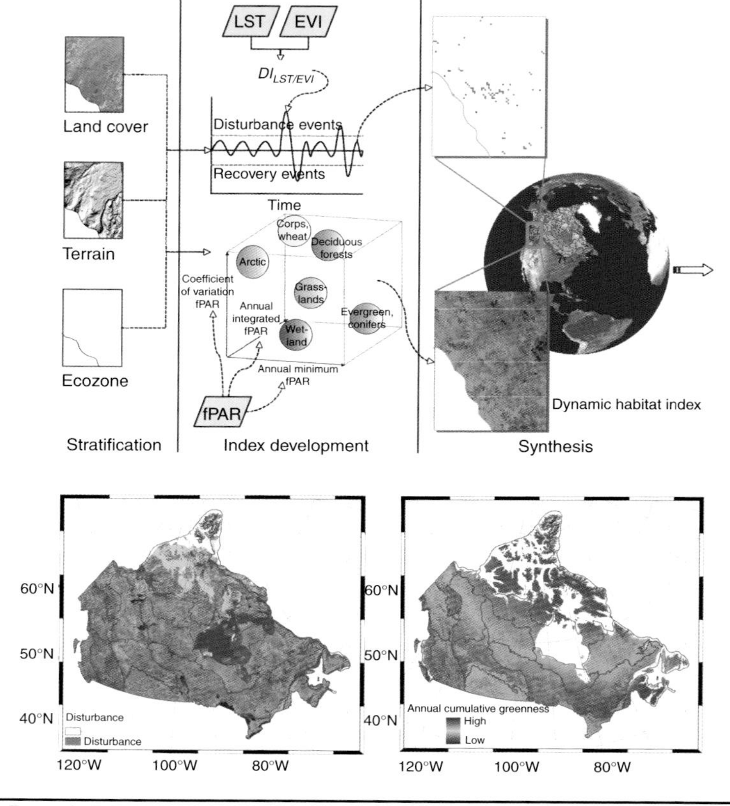

Color Plate 6.1 *Biospace* Canadian biodiversity monitoring program image products (*Courtesy of Dr Michael Wulder, Canadian Forest Service*).

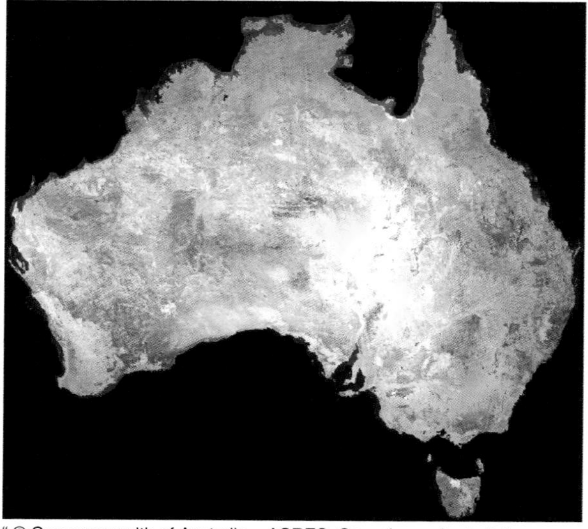

"© Commonwealth of Australia – ACRES, Geoscience Australia"

Color Plate 6.2 Landsat 7 mosaic of Australia comprised of 369 individual satellite images acquired in 1999 and 2000 for the National Carbon Accounting System.

model include the boreal ecosystem productivity simulator (BEPS) (Liu et al. 1997) and physiological processes predicting growth from satellites (3-PGS) (Coops 1999).

These models simulate the landscape- or biome-scale dynamics in the biogeochemical cycles of carbon, water, and nitrogen. Typically, LAI controls the daily fluxes, and allocation and turnover rates control the annual fluxes. LAI may be assumed to be a flat stack of leaves (Sellers 1985) or may be stratified or otherwise structurally represented throughout a volume of photosynthesizing surfaces (Chen et al. 1999). Soil water content is modeled to generate the soil water balance with inputs of rainfall, snowmelt, canopy interception, evapotranspiration, and overflow. Stomatal conductance (leaf and canopy level) is calculated as a function of radiation, air temperature, vapor pressure deficit, and leaf water potential and N concentration. Daily photosynthesis is a function of conductance constrained by LAI and daylength; respiration estimates or assumptions are required for each vegetation type and biomass class, and may be based on temperature data or assumptions about below- and above-ground standing biomass. Growth respiration (often assumed to be a constant fraction of gross photosynthesis) is subtracted to yield NPP estimates.

At least three intersection points occur between this type of ecosystem process model and remote sensing (Coops and White 2003):

1. The use of remotely sensed data in model initiation or parameterization; depending on the model and remote-sensing system available, examples might include model requirements for initial conditions of vegetation type, leaf area index, leaf mass, leaf N concentration, photosynthetic capacity, standing biomass, litterfall, maximum canopy stomatal conductance, soil temperature, type and fertility (Waring and Running 1999).
2. The incorporation of remotely sensed data within modeling, as driving variables of forest condition, function, or the biophysical environment; examples here might include estimates of structure, photosynthetically active radiation on a daily time step to "fine-tune" model calculations that may not be sensitive to the effects of soil drought, and derivatives of multiple-sensor observations such as surface resistance (Coops and White 2003).
3. The use of remote sensing in model validation; examples have included validating ecosystem process model predictions of forest growth and biomass with SAR data (Kasischke and Christensen 1990), validation of modeled forest succession patterns using Landsat (Hall et al. 1991, Coops and White 2003), and validation of energy and water balance predictions with thermal imagery (Silberstein et al. 1999).

At the landscape scale, Zheng et al. (2006) found it difficult to separate modeling from remote sensing because ecosystem process models have incorporated remotely sensed information into virtually all aspects of necessary inputs. Looked at another way, landscape-level applications of remote sensing (e.g., classification methods, change detection analysis) that include modeling logic are simply another instance of a powerful technological synergy.

Global scale estimation of primary production will typically require multispectral imagery from satellite sensors at relatively low spatial resolutions (Turner et al. 2003). The MODIS sensor has been providing continuous coverage at spatial resolutions of 250 m, 500 m, and 1 km from the Terra and Aqua satellite platforms, since 1999 and 2002, respectively. A range of MODIS land data products, including vegetation indices such as NDVI, land surface albedo, land surface temperature/emissivity, and more comprehensive measures of vegetation production, vegetation duration, and vegetation cover (Running et al. 2004), are available online with a considerable data management and distribution support network in place (http://modis.gsfc.nasa.gov/).

The NASA MODIS land algorithm user's guide (Heinsch et al. 2003) is recommended to those beginning to work with these data products. The proper use of the MODIS land data sets designed to represent primary production, GPP and NPP, requires considerable investment in understanding product development. The basis of the MODIS GPP and NPP algorithms is the NDVI, which together with information on land cover, climate and vegetation characteristics is used to derive measures of above-ground gross and net primary production. At least one study relating these MODIS measures to bird species richness has shown that the new MODIS productivity measures improve on NDVI, especially in areas with large variation in vegetation cover and density (Phillips et al. 2008).

Phenology

Understanding changes in vegetation phenology is important to wildlife and biodiversity management, as alterations in the timing of phenological events are among the first responses at plant and ecosystem levels to climate change (Badeck et al. 2004), and may also have implications for understanding and modeling plant chemistry and productivity (White et al. 1997). On global scales, satellite remote sensing now offers the possibility of routine mapping of vegetation phenology.

The NOAA/NASA AVHRR Pathfinder data set (James and Kalluri 1994) uses polar orbiting satellites to collect multiple daily observations over large areas. European plant phenology and climate as seen in the NASA/NOAA AVHRR pathfinder NDVI and land-surface parameter data set are shown in Color Plate 4.5 (Stöcklii and Vidale, 2004). The data are corrected for Rayleigh scattering, ozone absorption, instrument degradation, but not corrected for aerosols or viewing geometry effects.

A fine temporal resolution in vegetation phenology can be useful for global-scale climate modeling and mesoscale weather simulations. Increased knowledge of the temporal variability of vegetation is a key requirement to quantify carbon pools and exchanges on local and global scales (Sellers et al. 1994).

At the landscape-level, vegetation phenology may be an important contributor in understanding some ecosystem process and patterns. MODIS NDVI-based phenology products were used in one regional study of grizzly bear habitat and resource selection (Table 4.2; Stenhouse and Graham 2008 http://www.fmf.ab.ca/GB/GB_report9.pdf web site accessed 20 August 2008). Prior to development of these phenology products, the MODIS time-series requires correction of noise introduced by atmospheric and view angle variations, snow and ice, and cloud contamination (Beck et al. 2007, Hird and McDermid 2009). Cloud contamination, dust, and zenith angle effects are important concerns in large-area remote-sensing imagery (Beck et al. 2007).

Validation of large-area products such as phenology maps can be a major challenge, requiring significant data collection over highly diverse terrain for comparison to image products themselves comprised of composite time-series inputs (currently 1-, 8-, 10- or 16-day intervals). Ideally, algorithm testing and validation occurs at different steps of the data processing stream (Cihlar et al. 2003), for example:

1. Geometric and compositing algorithms are checked to ensure the spatial integrity of the original satellite imagery are preserved.
2. Different cloud screening and corrective measures are compared for effectiveness.
3. Comparative studies of BRDF and view-angle modeling approaches result in appropriate data normalization characteristics.
4. Higher-level seasonal and phenology products are compared to field plots or higher spatial resolution remote-sensing data (Iverson et al. 1989), and across geographic areas and scales of interest.

Typical map user strategies are to rely on correlation coefficients and root-mean-square-error estimates between field observations and well-understood remotely sensed phenology metrics, such as onset of spring and end of autumn, length of growing season, or timing of maximum greenup. However, validation of national or global scale image products is a quality control and accuracy assessment commitment to ensure that the product's usefulness for a specific purpose may be established (Cihlar et al. 1997).

Metric	Description	Methods of Estimation
Start of growing season (SOS) and end of growing season (EOS)	Timing and level of photosynthesis at the start and end of measurable photosynthesis	Thresholds, inflection points, curve derivatives, or Fourier-based methods
Length of growing season (LGS)	The duration of measurable photosynthesis	Number of composite periods or days between SOS and EOS
Mid-Season Timing and level of photosynthesis at the middle of the growing season	Point midway between time at which 90% of the seasonal amplitude is reached and time at which 10% of senescence has occurred	Integrated NDVI (INDVI)
Overall productivity	biomass produced during the growing season	Area under the NDVI curve between SOS and EOS
Rate of Green-Up	Speed at which spring green-up occurs	Slope of line between SOS and maximum NDVI
Green-Up Fraction	Portion of growing season spent in green-up	Number of days/composite periods between SOS and maximum NDVI over the total LGS
Start of Senescence	Timing and level of photosynthesis at the start of rapid decrease in photosynthesis	Inflection point method
Rate of Senescence	Speed at which autumnal senescence occurs	Slope of line between maximum NDVI and EOS

Source: With permission from Hird J. N. and G. J. McDermid. 2009. *Remote Sensing of Environment* 113:248–258.

TABLE 4.2 Example Vegetation Phenology Metrics Based on MODIS Image NDVI Time Series Data

Fragmentation

Spatial heterogeneity creates patterns in the landscape. Quantifying this pattern and relating changes in pattern, or structure, to ecological processes are of abiding interest in biodiversity and wildlife management (Linke et al. 2007). Fragmentation processes are one interpretation of changes in spatial heterogeneity over time that appear to have significant influences in many biodiversity and wildlife management situations. Early examples of landscape quantification and fragmentation analysis using remote sensing often focused on the landscape scale expressed most consistently in Landsat image classifications, and aerial photointerpretation-based vegetation inventory databases (Riitters et al. 1995, Wickham et al. 1997). Today, these and other options exist to quantify landscape structure. Digital high spatial resolution imagery has only infrequently been employed in the past for this application because of high data volumes and questions related to ecological scale.

Some interesting methodological challenges must be considered when approaching questions of spatial heterogeneity and fragmentation with remote sensing. For example:

1. The *generation of accurate and repeatable landscape maps* for use in landscape quantification; many studies have noted the importance of scale, minimum mapping unit, resolution, extent, and classification scheme on landscape metrics (Gergel 2007), and others have considered the issue of comparing landscape metrics derived from maps produced with different geospatial technologies and methods (Benson and MacKenzie 1995, Frohn 1998, Franklin et al. 2002a, Frohn and Hao 2006).
2. The *extraction of meaningful landscape metrics* across a wide range of dimensions including measures of area, edge, shape, richness, juxtaposition, and connectivity (McGarigal and Marks 1995; McGarigal et al. 2000); recent studies have shown that metric redundancy can be managed by focusing on parsimonious metric gradients (Linke and Franklin 2005) and general behavioral groupings of metrics (Neel et al. 2004).
3. The relation of changes in metrics to changes in a biological process of interest (wildlife behavior, for instance); perhaps the most important challenge in this topic is to answer the question: *What is the ecological meaning of the patterns detected for the process of interest?* In addition to the growing literature comprised of mostly empirical studies, there is emerging consensus that theoretical advances are necessary to understand the broader context in which landscape patterns reside (Fahrig 2003, Fortin and Dale 2005, Linke et al. 2005).

Gergel (2007) suggested a few directions for research that may be helpful in advancing capability to measure and monitor fragmentation with remote sensing. An important issue is to determine which landscapes are outside or approaching the limits of their natural range of variability—Which landscapes are more heterogeneous than others? How did this heterogeneity arise? What portion of the heterogeneity is due to human causes versus natural sources of variability? Such questions are of enormous interest to those engaged in understanding, and perhaps restoring, the natural role of disturbances in ecosystem management (Perera et al. 2004). Comparing landscapes through time and space by accurate and consistent remote sensing is a first step, introducing questions of pattern significance and accuracy assessment. A remote-sensing approach may be the best way to confirm that detected differences in pattern are ecologically meaningful. Unfortunately, relatively new concepts, and even newer measures of fragmentation, may not yet be well-understood conceptually and quantified in existing software (e.g., Frohn 1998, Butler et al. 2004, Lele et al. 2008).

Hessburg et al. (2004) present a case study of present-day and historical forest patterns from the eastern Washington Cascades that will serve to illustrate the general approach (Table 4.3). Spatial patterns in forest land cover and structure (successional stages, stand development phases) were related to processes of forest disturbances,

Step	Action
1	Stratify by ecological subregions
2	Map historical vegetation from 1930s–1940s aerial photography
3	Reconstruct vegetation attributes in areas of timber harvesting
4	Analyze historical spatial patterns with landscape metrics
5	Define reference conditions
6	Set reference variation conditions
7	Estimate reference variation (including appropriate fire models)
8	Develop decision support model
9	Map current spatial patterns with current aerial photography
10	Compare metrics of current maps with historical reference

Source: Adapted from Hessburg et al. (2004).

TABLE 4.3 Steps in an Approach to Estimate the Departure of Present Forest Landscape Conditions from Historical Reference Conditions in USA (circa 1900)

including fires, forest diseases and insect outbreaks, unusual weather, and herbivory. Such processes are constrained by geological and climatic conditions. Mapping of spatial patterns was based on historical black and white photography, and current color infrared aerial photography, and a standardized interpretation protocol which identified key landscape attributes validated with field samples. The photointerpretation exercise yielded 18 different classification maps, including land cover, canopy layers, total crown cover, patches of forest age classes (e.g., old-growth), fuel loading, and crown fire conditions. Characterization of spatial patterns in the resulting classification maps was accomplished with five landscape metrics: total class area, patch density, mean patch size, mean nearest-neighbor distance, and edge density. At the landscape level, the mosaic was analyzed with nine metrics, among them patch richness, diversity, dominance, and contagion.

The range and variation in historical forest vegetation patterns and vulnerability to disturbance was evaluated through a combination of modeling and landscape reconstruction using Moeur and Stage's (1995) *most similar neighbor inference* (sometimes referred to as *backcasting* in the remote-sensing community—*see* Franklin et al. 2005). This enabled estimation of a *reference condition*, against which present day patterns could be compared. This comparison revealed vegetation changes that lay beyond the range of historical variation—*departures*—that could be explored in more detail to determine their potential ecological implications:

> The common finding among each of the 18 map evaluations is that the landscape had become highly fragmented, an observation that is consistently indicated by reduced contagion, elevated patch densities and mean nearest-neighbor distances, and reduced mean patch sizes. Nowhere is this better indicated than in the evaluations of fuel loading, crown fire potential, and fire behavior patterns...Current values of eight of the nine landscape metrics lie beyond these limits; historical values of five metrics also show evidence of departures"
>
> *(Hessburg et al. 2004: pp. 169–173)*

Currently, a lack of ecological context for the historical range of conditions to establish reference conditions and reference condition variance is a general, though local, problem to be overcome; often, reference conditions are estimated conservatively (Hessburg et al. 1999). Small sample sizes can also be an issue, however, powerful software tools to assist in the use of modeled or replicated landscapes are increasingly available (Fortin et al. 2003, Cardille et al. 2005).

CHAPTER 5

Remote Sensing in Wildlife Management

Just as our musical appreciation is increased greatly when more than one or two octaves are exploited, so also is our appreciation of the physical universe, through multiband spectral reconnaissance, which already can exploit more than forty "octaves."

—*R. N. Colwell, 1961*

Remote-Sensing Reviews, Case Studies, and Illustrative Examples

Recent Reviews

A large number of reviews have been published to document the use of remote sensing in wildlife management over the past few decades. One of the first and most comprehensive was conducted by Aldrich (1979) as part of a state-of-the-art review of "Remote Sensing of Wildland Resources" for the USDA forest service. Basic types of information needs were described, including the requirements for effective land-cover classification and mapping procedures defined for some broad forest, wetland, and rangeland management applications. Selected specific management information requirements were outlined, such as identification of an unusual stress condition, potential landslide, or water quality measure. One category of information needs was introduced as "observations and counts of occurrences" to encompass all manner of presence/absence information required in wildland resource management: buildings, cars, people, animals, roads, and so on. The review considered a number of sensor characteristics and imagery produced by photographic systems, multispectral scanners, thermal scanners, and microwave systems.

Although satellite applications were addressed, the technology was still very new. The emphasis was clearly on the photographic systems and the tremendous contribution aerial photography has made since widespread expansion and adoption of the aerial photo-interpretation approach to resource management in the decades preceding the review in the late 1970s. Several promising applications in wildlife management were identified:

1. Use of supplemental (70 mm) photography for stream trout habitat monitoring and wildlife census
2. Mapping bark beetle tree stress and animal detection using aerial thermal imagery
3. Tropical forest aerial surveys with active microwave imagery
4. Mapping rangeland productivity with Landsat multispectral scanner (MSS) imagery

Common problems associated with the digital use of Landsat data were thought to include the difficulty in finding people with the new skill set required to interpret the imagery, the relative high cost of the image data and associated image processing infrastructure, the complexity of handling georadiometric fidelity, registration of multi-temporal data for change analysis, and the influence of the topographic effect in digital imagery. A final challenge was perceived to exist with the relatively coarse spatial resolution (80 m) of Landsat data, compared to aerial photography, which would prevent certain applications of interest, even with synoptic acquisition over greater spatial extent than was possible with aerial platforms.

The contributions of remote sensing in wildlife management in those early decades [from the late 1950s to the publication of the Aldrich (1979) review] suggested a field dominated by aerial photointerpretation applications, with only a relatively modest impact by new sensors and digital aerial and satellite remote-sensing methods. The growing research agenda during the initial years of civilian terrestrial satellite remote sensing was beginning to take shape, but many observers maintained a cautious outlook for the technological and quantitative remote-sensing approach in wildlife applications (Braun 2005). In the most recent decade, though, literally dozens of individual aerial and satellite remote-sensing projects in wildlife management have been conducted (Table 5.1), with many implementing the suite of digital methods first identified as part of the new, innovative, quantitative remote-sensing approach.

Comprehensive examination of the current state-of-the-art suggests that enormous progress has occurred in developing critical remote-sensing contributions in the broad areas of habitat mapping,

Species of Interest	Location	Methodological Approach	Main Output and Result	Key References
Brushtail possum (*Trichosurus vulpecula*)	New Zealand	Satellite image classification, DEM, GIS model	Wildlife vector maps for TB caused by Mycobacterium bovis	McKenzie et al. 2002
Grizzly bear (*Ursus arctos*)	Alberta, Canada	Satellite image classification and empirical spatial models	Habitat suitability map, resource selection functions	Franklin et al. 2002a, McDermid 2005, Linke et al. 2005
Tibetan wild ass (*Equus kiang*)	Nepal, Tibet	Satellite image classification	Habitat suitability maps	Sharma et al. 2004
Kangaroo (*Macropus*)	Queensland, Australia	Aerial videography, thermal imaging	Animal census	Grierson and Gammon 2000
Rodents (*Muridae*)	USA and South America (Paraguay)	Aerial photography, satellite image classification, GIS model	Hantavirus distribution model	Goodin et al. 2006, Boone et al. 2005
Northern spotted owl (*Strix occidentalis caurina*)	Northwestern USA	Satellite image classification	Nesting habitat suitability, landscape structure mapping	Ripple et al. 1991, 1999
Emperor penguin (*Aptenodytes forsteri*)	Antarctica	Aerial and satellite image interpretation	Animal census	Sanchez and Kooyman 2004
Forest invertebrates (*Ant and beetle communities*)	Australia	Aerial multispectral vegetation index	Habitat suitability map, regression model	Lassau et al. 2005, Lassau and Hochuli 2008

TABLE 5.1 Sources of Wildlife Species and Habitat Studies Using Remote-Sensing Data

Species of Interest	Location	Methodological Approach	Main Output and Result	Key References
Wolves (*Canis lupus*)	Yellowstone National Park, USA	Satellite image analysis, GPS, GIS model	Spatial model, trophic cascade (willow)	Ripple and Beschta 2006, Kauffman et al. 2007
Saltwater crocodile (*Crocodylus porosus*)	Australia	Aerial photography, satellite image classification	Nesting habitat suitability map	Harvey and Hill 2003
Caribou (*Rangifer tarandus*)	British Columbia, Canada	Satellite image classification, GIS model	Habitat suitability map, landscape fragmentation	Hansen et al. 2001
Clouded leopard (*Neofelis nebulosa*), fishing cat (*Prionailurus viverrinus*)	Thailand, Malaysia, Borneo	GIS, image interprétation, camera traps, GPS	Animal census, habitat suitability map, species prediction model	Grassman et al. 2005
Forest songbirds; e.g., pine warbler (*Dendroica pinus*)	Northern Michigan, USA	Satellite image analysis, LiDAR and radar remote-sensing model	Habitat suitability map, vertical structure	Imhoff et al. 1997, Bergen et al. 2007
Reeve's muntjac (*M. r. micrurus*)	Taiwan	Camera traps, radiotelemetry	Population density, resource selection (diet)	McCullough et al. 2000, Lai et al. 2002
Capercaillie (*Tetrao urogallus*)	France	Satellite image classification	Habitat suitability map, species prediction model	Jacquin et al. 2005
Javan Hawk-Eagle (*Spizaetus bartelsi*)	Indonesia	Satellite image classification, regression model	Species occurrence and abundance model	Syartinilia 2008

Leopard (*Panthera pardus*)	West and central Asia	Satellite image vegetation index, DEM, regression model	Species distribution and habitat suitability	Gavashelishvili and Lukarevskiy 2008
Guanaco (*Lama guanicoe*)	southern Patagonia	Satellite image interpretation, DEM model	Species distribution map	Travaini et al. 2007
Black flies (*Simulium*)	Nigeria	Satellite image interpretation	Species distribution, human health model (Onchocerca volvulus)	Germade et al. 1998
Black-capped vireo (*Vireo atricapillus*)	Texas, USA	LiDAR image analysis	Vertical habitat structure, species distribution model	Leyva et al. 2002
Giant pandas (*Ailuropoda melanoleuca*)	China	Satellite image classification, GIS model	Habitat suitability map, landscape fragmentation	Mackinnon and De Wulf 1994, Hu 2001, Xu et al. 2006, Wang et al. 2008
Black-tailed prairie dog (*Cynomys ludovicianus*)	Kansas, USA	Satellite image classification, vegetation index	Habitat suitability map	Kostelnick et al. 2007
Spiny lobster (*Jasus edwardsii*)	New Zealand	Field-based videography, GPS, GIS model	Habitat suitability map, trophic cascade (kelp)	Parsons et al. 2004
Cholera (*Vibrio cholerae*)	Bangladesh	Ocean maps (e.g., sea surface temperature) and plankton blooms	Outbreak risk model	Beck et al. 2007

TABLE 5.1 Sources of Wildlife Species and Habitat Studies Using Remote-Sensing Data (*Continued*)

wildlife management, avian-habitat relations, epidemiology, and animal distribution studies (Table 5.2):

- Habitat mapping studies at local, regional, and national scales, with successful applications in:
  - The estimation of multiscale vegetation structure using aerial photography and satellite sensor technology
  - The characterization of vegetation and waterscape habitat with optical, microwave, thermal, and LiDAR remote-sensing spectral response patterns
  - Understanding the relationship between remotely sensed patterns and species observations or modeled distributions
  - Studies focused on parasitic diseases, such as malaria, or infectious diseases (through a search for remotely sensed *environmental correlates* for the conditions in which the diseases appeared)
- Remote sensing of terrestrial faunal diversity and animal distribution [including studies grouped by faunal taxa (mammals, birds, reptiles, amphibians, and invertebrates), and classes of surrogates available by remote sensing], such as:
  - Habitat suitability and prediction of animal density, species distribution (including rare or special species), and species richness estimation
  - Photosynthetic productivity and land-cover relationships
  - Multitemporal patterns and chemical constituents of vegetation (including forage quality)
  - Structural properties of habitat and landscape energy flows (genetics, hybridization)

The major trends in research and development over the past decade have included innovations closely associated with advances in sensor technology (e.g., continued improvements in spatial and temporal resolution), and the data developments (e.g., GPS, spatial models) that help ensure that remote sensing can be increasingly conducted systematically within the overall development of comprehensive wildlife *spatial information management systems* (Majumdar et al. 2005, Gottschalk and Huettmann 2006, McDermid et al. 2009). Within such systems, the image data and software tools to achieve wildlife management objectives are increasingly available and are continuing to improve. Overall, the selection of appropriate image data, matching temporal and spatial scales, and management of the complexity of modeling approaches, together with a myriad of human resource issues, are now recognized as among the most important critical success factors in remote-sensing wildlife applications (De Leuw et al. 2002, Leyequien et al. 2007):

Reference and Number of Studies Reviewed:	Broad Area of Review	Highlighted Applications Include:	Major Conclusions Relate to:
McDermid et al. (2005)—approximately 40 studies primarily in past decade	Remote Sensing in Land Cover and Habitat Mapping	Estimation of multiscale vegetation structure Mapping landscape heterogeneity	Large area applications Team collaborations Impacts of new sensor technology
McDermid et al. (2009)—approximately 30 studies primarily in past decade	Remote Sensing in Wildlife Ecology and Management	Animal density Species distribution Rare species Habitat suitability analysis Land cover and ecosystem process dynamics	Advances in sensor technology Interdisciplinary understanding Team dynamics Spatial information management systems
Gottschalk et al. (2007)—over 100 studies in past three decades	Remotely sensed Avian-Habitat Relations	Habitat conditions Modeling distribution or abundance Monitoring	Appropriate image data Matching temporal and spatial scales Complex modeling
Herbreteau et al. (2007)—89 studies in past three decades	Remote Sensing in Parasitic and Infectious Diseases Epidemiology	Parasitic diseases and vectors Infectious diseases and vectors Model development (e.g., risk factors)	Validation and scale Data availability and cost Modeling approaches Collaborative dynamics
Leyequien et al. (2007)—over 100 studies in past three decades	Remote Sensing and Faunal Distribution	Habitat suitability Productivity Multitemporal patterns Structural properties of habitat Forage quality	Increased synergy in remote sensing GIS data Ecological models Validation (traditional observations)

TABLE 5.2 Summary of Five Reviews of Remote-Sensing Applications in Broad Areas of Remote Sensing in Wildlife Management

- First, ongoing technological advancements in sensor capabilities, and image processing innovations, have made a significant and continuing impact on the development of remote sensing in wildlife management. Selection of image characteristics, and remote-sensing image availability, continue to impose certain limitations, but are no longer the major constraint in experimental design and analysis:
  - Image processing capability and analysis options have experienced a tremendous expansion. Digital image classification procedures, subpixel modeling, and mapping landscape heterogeneity, and challenges when implementing remote sensing over large areas, including image acquisition and temporal heterogeneity, multiple radiometric and geometric issues (e.g., image normalization, topographic effects), appear to have been largely resolved (or at least, to have achieved reasonable solutions for applications to proceed) (e.g., Aronoff 2005).
  - Ineffficient and incomplete accuracy assessment procedures are now rare (Foody 2002); importantly, standarized map evaluation procedures are commonly designed appropriately and early in project initiation (Cihlar et al. 1997, Wulder et al. 2006).
  - An emerging trend appears to be the need for increased synergy among remote-sensing products and GIS data within ecological models validated by traditional observation techniques (Madhavan et al. 2005, Zhu and Tateishi 2006).
- Second, however, *understanding* and *accessing* these data continues to be at a premium in many applications, which often do not appear to have developed as quickly as expected or hoped:
  - The skills required to implement geospatial solutions and remote-sensing approaches are not yet commonplace in the ecological community.
  - Apart from the technical skills, there is also a growing recognition of the importance of interdisciplinary understanding and team dynamics. Interdisciplinary and transdisciplinary science partnerships continue to develop, but have not yet reached a full state of maturity—many applications appear to proceed under collaborative arrangements suggested by a division of labor dualism, such as that promoted in a "tools and users" working environment.
  - Operational remote sensing, long considered a practical goal within reach, has consistently proven to be an elusive achievement.

Some gaps in the application record appear obvious; for example, a great deal of work experience has accumulated with remote sensing data from medium- and low-spatial-resolution satellite sensors, such as Landsat and the instruments associated with the NOAA-AVHRR series, but a smaller number of studies with high spatial resolution imagery have been reported than might have been expected based on the increasing availability of such data (Wulder et al. 2004). Very few waterborne disease studies appear to have been completed (Herbreteau et al. 2007), and multisensor fusion systems and object-based approaches are less well-developed and are employed only infrequently in large-area mapping situations to which they are ideally suited (McDermid et al. 2005, Barlow and Franklin 2007).

Recent experience confirms the importance of disseminating to a wider audience the key insights that have been accumulating as additional successful wildlife applications have been completed and the remaining technical challenges are identified and, ultimately, overcome (e.g., Zimmerman et al. 2007). Apparently, awareness of recent successful approaches and lessons learned elsewhere may not be as high as awareness of some earlier failures and current disappointments. Up-to-date information on successful remote-sensing approaches to these complex wildlife applications will be of particular interest and value to those embarking on a first-time remote-sensing approach to wildlife management, or those developing a new, more ambitious application on the foundation of a modest initial success story. Compiling a selective, though broadly ranging collection of specific examples to further describe the full potential of remote sensing in wildlife management may be useful; but even more importantly, what is needed is an appropriate *synthesis* to highlight and emphasize key themes and lessons learned. This synthesis may serve to guide and possibly lead the increasingly diverse wildlife management and conservation community to the more advantageous position of understanding the remote-sensing approach and the immediate and longer-term applications.

Remote Sensing of Habitat Indicators

Remote-sensing maps and statistical products are not habitat maps—instead, remote-sensing products can be used to develop habitat maps or models, perhaps designed for a species or guild or taxa of interest, in a process or series of steps that is initiated as the remote-sensing imagery are acquired and processed. One of the most powerful applications of remote-sensing imagery is in development of *habitat indicators* (Environmental Remote Sensing Group 2003) as surrogates to represent conditions at different scales (Weiers et al. 2004), suitability or quality (discussed in following sections), or perhaps designed to reveal species preferences (e.g., Van Bommell et al. 2006).

A comprehensive set of 11 remotely sensed, landscape-level habitat indicators was introduced by Tiner (2004) to quantify *natural habitat integrity* and assess the general ecological condition of watersheds. Natural habitats are typically defined with mature vegetation cover as the dominant attribute, such as forests and wetlands; such landscape features can be clearly recognized in imagery as sites that have not been developed for agriculture, transportation, or urban infrastructure. Identification of such nondevelopment and natural areas using remote sensing is relatively straightforward with visual interpretation or classification procedures, or possibly based on identifying vegetation types and specific plant communities. Natural habitat integrity can then be defined as the state of unbroken habitat, or the situation of a natural habitat cover (e.g., wetland) that is spatially intact over a relatively long temporal threshold. Areas with a high degree of natural habitat integrity are areas where natural habitat has existed for many years without physical disturbance by humans. Finally, a system to measure the extent to which a watershed was represented by these natural habitats of the area would indicate more or less habitat integrity.

Aerial photointerpretation of color infrared and black and white photos was used to derive the classification of natural habitat areas, which were then used to create the 11 habitat indicators based on three broad measures (Tiner 2004):

- Habitat extent (e.g., wetland extent index)
- Habitat condition (e.g., wetland buffer integrity index)
- Habitat disturbance (e.g., amount of drained wetland)

A composite index—*the natural habitat integrity index*—provided an overall score for watersheds (and sub-basins not examined here) based on a weighted sum of the difference between the habitat disturbance and habitat extent indices.

Examples of each developed indicator were provided from the Nanticoke River watershed, in Delaware, and a series of maps was produced that illustrate the values obtained on each of the indices (available at the National Wetlands Inventory home page, www.fws.gov/nwi web site accessed August 26, 2008). Some of the values computed for the indices are reproduced in Table 5.3. The indices were constructed so that deviations from 1.0 are readily interpreted as percentages; for example, an index value of 0.41 for natural habitat indicates that 41 percent of the vegetation cover was nondisturbed vegetation. The high value for the habitat fragmentation index, 0.71, indicates that 71 percent of the natural habitat was fragmented by roads. The composite index of 0.29 indicates that 29 percent of the landscape has been impacted, and this was interpreted to mean that this landscape is:

> ...a significantly stressed watershed due to human actions...significant degradation of water quality and aquatic habitat can be expected for the Nanticoke.
>
> *(Tiner 2004: pp. 235–236)*

(Compared to a maximum value of 1.0)	
Index	**Score**
Natural habitat (cover)	0.41
River stream corridor integrity	0.59
Vegetated wetland buffer	0.36
Wetland extent	0.41
Wetland disturbance	0.71
Habitat fragmentation (by roads)	0.38
Composite natural habitat integrity	0.29

Source: With permission from Tiner, R. W. 2004. Ecological *Indicators* 4:227–243.

TABLE 5.3 Scores for Natural Habitat Indices Derived from Remote Sensing for the Nanticoke River Watershed

This environmental stress was also expected to impact local species extinction rates and has specific and potentially devastating local impacts, particularly on bird abundance and distribution.

The natural habitat integrity approach is a landscape-level quantitative assessment that can be readily updated, and therefore, serve as the basis for a habitat monitoring system to detect variations from initial, expected, or assumed reference conditions (e.g., Table 5.4).

Indicator	Net Change Since Presettlement (%)
Wetland area	–38
Surface water detention	–39
Streamflow maintenance	–42
Sediment retention	–57
Shoreline stabilization	–23
Nutrient transformation	–41
Fish/shellfish habitat	–25
Waterfowl/waterbird habitat	–29
Other wildlife habitat	–41

Source: With permission from Tiner, R. W. 2004. http://www.aswm.org/calendar/integratingrest/tiner.pdf [March 3, 2009].

TABLE 5.4 Example Changes in Key Wetland Indicators Compared to Assumed Presettlement Reference Conditions Based on a Remote Sensing Mapping and Backcasting Approach

The strength of the system is in the development of scale-dependent habitat condition profiles for area units (such as watersheds, ecosystems, or sub-basins), and identification of areas of potential interest for restoration or conservation. The system enables quick and cost-effective assembling of coarse-filter variables, which can then be used in detailed correlations to field observations to reveal relationships between water quality, fish abundance, and land-cover changes. The maps, the imagery, and the illustration of simple relationships, are effective educational tools to increase awareness of human impacts on the natural environment.

Some improvements to this type of system can be immediately envisaged; for example, use of higher quality remote-sensing data, inclusion of other indicators of ecosystem integrity (e.g., indices for groundwater inputs, air pollution), and more complete integration of field- and image-data collected in different ways at different scales. Operational use of the system in watershed planning and management and continued development of the system was considered feasible (Tiner 2004).

Direct Animal Sensing and Counting

Aerial and satellite imagery have been used to sense animals and provide a basis for animal counting and census work (Gillespie 2001, Flamm et al. 2008). The wide range and innovative use of remote sensing in such surveys can only be hinted at here—for example, one novel study counted thousands of browsing animals in hundreds of pastures around the world, visible in high spatial resolution images freely available through Google Earth, and suggested animal body axis orientation may be partially determined by magnetic alignment (Minar 2008). Commonly, specific animal counts and estimates of populations have been developed using visual or computer processing of high spatial resolution aerial imagery and, recently, high spatial resolution satellite data. A wide variety of species in different settings have been enumerated this way, and there is no shortage of examples from which to draw a picture of the general image analysis and interpretation methods. Appropriate use of remote sensing with such counting of animals or population estimates requires a detailed understanding of the accuracy requirements in animal census, the options available for acquiring the necessary data, and, in implementing long-term monitoring of wildlife habitat, the conditions in which the census will occur.

Little known and harsh environments (e.g., Arctic, Antarctic) are particularly challenging to conduct animal surveys, but even in relatively well-known and easily accessible areas, animal counting by direct remote sensing must be based on a well-designed survey strategy. An important consideration is to match the spatial resolution with the animal characteristics, such as size and color. Typically, large animals with significant color contrast to surface materials in relatively open environments can be detected with maximum effectiveness. For example, initial surveys of Antarctic Emperor penguin (*Aptenodytes forsteri*) "huddles" were based on density data collected manually and then applied to large areas using imagery obtained from cameras carried aloft by balloons (Robinson 1990). Recently, Sanchez and Kooyman (2004) applied these specific estimates of the number of animals per square metre to imagery obtained by QuickBird sensors. Low light levels in July and August prevented optimal satellite image acquisition during peak huddle formation, and instead, QuickBird imagery acquired in October and September was used. Individual animals could be identified in high contrast situations (animals on ice), and smaller huddles of multiple animals were outlined manually.

The most effective methods of counting animals and estimating populations rely on observations in the field combined with appropriate remote-sensing approaches to assist in the development of management plans. Grierson and Gammon (2000) described a typical aerial remote-sensing application in counting and monitoring individual animals as part of Australia's kangaroo (*Macropus*) management program. The traditional approach was based on visual counting by observers in fixed-wing, ultralight, or helicopter aerial surveys, typically supplemented with field-based observations in smaller sample areas. These data required a correction factor (for time of day and year, for example) and were cost effective, though known to underestimate numbers, leading to uncertainty in management activities such as harvesting or culling.

For the comparison study, digital videographic imagery were acquired from airborne platforms at the same time as the implementation of traditional survey methods that depended on field and aerial observers and counting protocols. The digital video data were interpreted visually, and were found to be comparable in accuracy and precision to the traditional counts. The resulting advantages of the remote-sensing approach included:

1. Facilitating the integration of kangaroo counts in specific habitat models
2. More accurate species identification and density estimation
3. A greater ability to support modeling exercises (e.g., for allocation of harvesting tags)

Thermal imagery, in particular, was found to be an excellent data source early in the morning when the animals were most active and the surroundings at cooler temperatures, providing the highest contrast at the most advantageous flying altitudes and times. Aerial videography and thermal scanning were considered to have introduced a degree of accountability unavailable in the traditional, existing monitoring systems.

The *Clouded Leopard Conservation Project* (www.cloudedleopard.org web site accessed August 29, 2008) is designed to understand the conservation status and ecology of the clouded leopard (*Neofelis nebulosa*), an endangered felid in southeast Asia. The clouded leopard skull is unlike any other cat today, and possibly a modern example of a cat in the early stages of divergence from now extinct saber-tooth cats (Christiansen 2006). Also of interest in the region are the fishing cat (*Prionailurus viverrinus*), Asiatic golden cat (*Catopuma temminckii*), and marbled cat (*Parodfelis marmorata*)). All four cat species are threatened by habitat loss, persecution and illegal hunting, but little is known about their natural history beyond observations of animals in captivity. This conservation project has used remote sensing in three different ways, including digital cameras in field documentation of track sets, the deployment of camera traps, and satellite image habitat modeling in a GIS environment with radiocollar data:

1. Digital images of pugmark track sets permit detailed analysis of individual characteristics; four individual clouded leopards were identified by their pugmarks in repeated transects in one survey area in Sabah, Borneo (Wilting et al. 2006). A rough estimate of nine individuals per 100 km^2 was developed with the CAPTURE software package (Rexstad and Burnham 1991) using estimated probabilities of capture-recapture, the effective surveyed area, and sampling confidence intervals. Based on existing protected habitat areas and reserves, a Borneo population estimate of clouded leopards was derived (1500–3200) and found to be much lower than previously published. The population estimate may provide some guidance in setting protected reserve size, although in reality, such estimates "can only be a working hypothesis for future research" (Wilting et al. 2006, *see also* Gordon et al. 2007).
2. A habitat analysis using satellite imagery and GIS was used to identify areas of critical habitat, currently threatened habitat, the amount and protection status of the remaining habitat, and areas with the greatest potential for felid conservation (Howard et al. 2002). Annual ranges, activity patterns, diet- and prey-use of four clouded leopards were analyzed with radiocollar information (Fig. 5.1) and then interpreted with SPOT satellite image classification of vegetation types (Grassman et al. 2005); certain habitat preferences were revealed (selection of open- and closed-forests, streams, and hunting activity in grasslands and orchards) that may be important in understanding response to forest harvesting activity and other environmental changes.

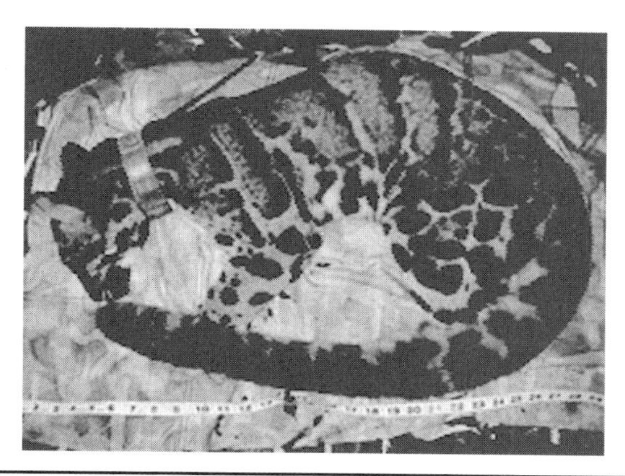

FIGURE 5.1 Live capture and collar fitting of a clouded leopard in Thailand. (*With permission from* Grassman et al. 2005. *Journal of Mammalogy* 86:29–38.)

3. A field survey with camera traps (also deployed to help identify areas for live trapping) enabled individual animal identification and supported interpretation of habitat selection in some circumstances—for example, clouded leopards appeared to travel on the ground more than expected, to hunt arboreal prey, and to avoid confrontations with other predators [e.g., dholes (*Cuon alpinus*)]. Failure to photograph the marbled cat despite intensive sampling efforts "...suggests that future surveys and field studies are vital to clarify the status of this cat in the wild" (Grassman et al. 2005).

A multiple-step capacity building component is an integral part of the Clouded Leopard Conservation project. Local field officers were trained in traditional and the innovative new approaches, which included work with digital cameras, satellite remote-sensing imagery, and GIS, to monitor habitat and determine animal density. The training opportunities also served to increase general public awareness and were featured as a component of outreach programming. A related study of skull morphology, and the application of molecular genetic methods, enabled the re-evaluation of clouded leopard subspecies partitions based on patterns of population genetic variation (Buckley-Beason et al. 2006). Analysis of 109 clouded leopards of known geographic origin (captive and museum animals) suggested a critical new species-level discrimination and the separation of the Borneo population (*Neofelis diardi*), which itself may be comprised of two subspecies, from the mainland clouded leopard population (Wilting et al. 2007). An accurate and formalized taxonomy is critical to the conservation of the clouded leopard—for example, to support planned reintroductions in areas of local extirpation or subspecies extinctions, and in captive breeding programs.

Field placement of cameras and camera traps is an increasingly valuable approach to acquire crucial data in difficult situations related to animal distribution, behavior, and population characteristics (Rowcliffe et al. 2008). This technology has been used in many parts of the world, but has proven to be of considerable value in dense forest environments, such as occur in tropical forests. In southern Taiwan, for instance, McCullough et al. (2000) deployed camera traps combined with animal capture and marking, radiotelemetry, and fecal nitrogen analysis to test hypotheses concerning the home range and dietary behavior of wild Reeve's muntjac (*Muntiacus reevesi micrurus*), a small, secretive, and little-known forest ungulate. For four months, 16 cameras were positioned in 3 ha of the core study area occupied by two radiocollared muntjacs, and 14 photos were acquired of muntjacs over the survey period. At the same time, a further 19 cameras were placed in similar environments without radiocollared animals; 16 photos of muntjacs were acquired there. Estimates of population size were then calculated using the program NOREMARK (White 1996). These data suggested that the two areas had similar abundance of muntjacs; camera pictures of the radiocollared muntjacs were subsequently used to estimate density. Detailed interpretation of these photos in context of the other data acquired (e.g., on diet) led to a deeper understanding of foraging behavior, female social organization, and territorial behavior by males (McCullough et al. 2000, Lai et al. 2002).

Remote Sensing of Habitat Resources

General Habitat Suitability Mapping

A vast literature has developed to document and explain the use of remote sensing in habitat suitability and habitat quality analyses. Actual species observations—where animals occur—are an essential element of habitat suitability approaches and on their own likely will comprise one of the best indicators of the significance of habitat (e.g., Dique et al. 2004). But for a number of obvious reasons—cost and scale, for example—these data are not always available, and identifying distributions of preferred habitat or suitable habitat using remote sensing is an approach with considerable potential. Increasingly, software tools and resources to enable habitat suitability modeling with remote sensing are widely available and relatively straightforward to apply in a wide range of environmental settings (e.g., Congalton et al. 1993, Hansen et al. 2001, Juntii and Rumble 2006).

A general remote-sensing habitat suitability mapping approach typically involves field observations (sightings to confirm map predictions, or on which to base habitat interpretations), an understanding of habitat conditions or attributes of importance, and a remote-sensing mapping protocol. This general approach was applied to habitat suitability mapping for a forest grouse, the Capercaillie (*Tetrao urogallus*), in France (Jacquin et al. 2005). This forest- and grassland-dwelling bird is an umbrella species of particular biodiversity and ecological importance because of its unique genetic character and distribution. The Capercaillie is listed as threatened under the European Birds Directive, and significant declines have been documented over the past decade. The known habitat requirements were reviewed, seasonal differences in habitat use, and interseasonal preferences were summarized, and then vegetation patterns associated with high quality Capercaillie habitat were identified on satellite remote-sensing imagery. These patterns included large patches of conifer forest, presence of deciduous tree fruits (heath), and grassland patches within an elevation range of approximately 1200 to 2300 m asl.

Fuzzy classification of SPOT-5 VHR sensor data (10 m spatial resolution) produced two measures that distinguished suitable Capercaillie habitat over small (10 ha) areas:

- Indicators of relative amounts of coniferous and deciduous forest, grassland, and heath; for example, based on observations and earlier studies, nesting habitat area was defined as composed of more than 40 percent forest, approximately 22 percent heath, 25 percent grassland, and minor amounts (1 percent) of exposed soil; winter habitat had comparatively larger amounts of forest and soil, and reduced amounts of heath and grassland.
- A forest-cover indicator (incorporating a forest-density level); for example, the proportion of dense forest to nonwooded areas, and the distances or adjacency to areas of high quality defined by the first set of indicators.

These indicators were calculated over a 10 ha area to represent reference conditions for suitable habitat. The actual conditions mapped in the SPOT imagery were then compared to field estimates using a histogram matching technique, suggesting that 83 percent of the area mapped represented suitable Capercaillie habitat, which agreed well (over 95 percent) with the areas where the bird was known to exist based on earlier sightings and field observations.

This general habitat suitability approach can be improved with more careful consideration of human impacts, and there are numerous ways in which to quantify and include more explicit human disturbance information. Readily mapped are linear features, such as major roads, that can be identified with automated software techniques in imagery with appropriate spatial resolution characteristics (Chen et al. 2005). Depending on the type of human activity in an area, other linear developments (seismic lines, roads, pipelines, powerlines) might be of interest; these conditions have been mapped manually with IRS imagery (5 m panchromatic) to determine their influence on endangered caribou habitat use over large areas in west-central Alberta (Oberg et al. 2000), for example, and were determined to be important in the grizzly bear project described later in this chapter (Linke et al. 2005). Even more detailed human activity linear features, such as networks of human paths and trails may be of interest; for example, trails in a park environment in northern Spain were mapped by Díaz Varela et al. (2006) using specially acquired very high spatial resolution aerial photography. The intention was to quantify the negative habitat effects of different levels of human use (e.g., soil compaction).

This level of mapping human activity-related features has proven to be at the very limit of current aerial and satellite remote-sensing capability, and is described more fully in later sections. The general question of the extent to which remote sensing can detect or be useful in analysis of human use or disturbance features is an outstanding research topic of considerable interest in wildlife management (He et al. 2009).

Other types of human impacts may not be as readily discerned even with the increased technological capabilities resulting from recent sensor developments. For example, one study of the Tibetan wild ass (*Equus kiang*), a poorly known species whose conservation status is uncertain and habitat requirements are not well-known (Schaller 1998), in the Surkhang, Upper Mustang region of Nepal, was based on concerns of potential conflicts with pastoralists and livestock (Sharma et al. 2004). Supervised classification of ASTER imagery, DEM data, and distance metrics (e.g., distance from communities and temporal pastures), produced a *kiang* habitat suitability model, which delineated areas of primary habitat (areas interpreted to offer food, shelter, and water), and secondary habitat (conditions suitable for some specific needs, such as reproduction). However, suitable habitat in one valley was apparently avoided by *kiang*, and this was determined to be a result of human activity not considered in the habitat suitability analysis—religious pilgrimage. More detailed investigation determined that the valley was considered to be of local religious significance, and was frequented by pilgrims during a portion of the year that conflicted with *kiang* habitat use (Koirala and Shrestha 1997). These finding suggested that management and conservation of *kiang* should be coupled with an increased understanding, and possible regulation, of local human activities.

Two other improvements to general habitat suitability models based on remote sensing involve the use of historical pattern analysis (Thompson and Angelstam 1999, Obade 2008), and the use of more complex modeling techniques that consider species distribution data on both presence and absence (Beauvais et al. 2006). Statistical models of species distribution will require an understanding of spatial autocorrelation to reduce the possibility of overestimating the importance of some environmental correlates. Species abundances are positively autocorrelated, meaning that nearby points in space tend to have more similar values than would be expected by random chance. Some of this autocorrelation is a result of the patchiness of the landscape or habitat, but typically, uneven distribution of sampling effort will introduce spatial autocorrelation.

Absence data were deployed by Syartinilia (2008) in a study of habitat suitability and Javan Hawk-Eagle (*Spizaetus bartelsi*) distribution in Indonesia after an intensive field sampling effort. An autologistic regression function significantly outperformed other forms of the general habitat suitability model. The logistic regression models had improved overall accuracy, and higher predictive power in estimating Javan Hawk-Eagle distribution based on habitat attributes, and were therefore superior in helping identify new areas for detailed field surveys, possible connectivity among existing suitable habitat, and in all aspects of conservation planning. Absence data are difficult to acquire and are typically estimated from areas in which the species of interest was not observed (depending on sampling intensity and protocols), suspected to be absent (Beauvais et al. 2006), or modeled as *pseudo-absence* (often from background areas in which species data are missing) (Pearce and Boyce 2006).

Gavashelishvili and Lukarevskiy (2008) hypothesized that leopards (*Panthera pardus*) in west and central Asia were more likely to occur in areas with milder climate, higher vegetation productivity, and greater distance from urban areas. Again, absence data were difficult to obtain and instead were estimated in background areas of suitable but unoccupied habitat. With regional climate data (e.g., snow cover), Spot Vegetation NDVI (e.g., surrogate for herbivore food and cover), and Shuttle Radar Topography Mission DEM data (e.g., altitude ranges, terrain-adjusted proximity to urban areas), a logistic regression model was developed to predict leopard distribution and habitat suitability in Armenia, Azerbaijan, Georgia, Iran, Russian Federation, and Turkey. Habitat connectivity and the presence of isolated leopard populations in marginal or near-marginal habitat were evaluated as model outputs. These are critical elements in a long-term leopard survival and conservation plan based on protecting source populations, and at the same time, providing dispersal opportunities by linking these populations.

Two advantages in the logistic regression modeling approach can be highlighted. Typically, there will be:

1. A reduction in the number of independent variables per regression analysis
2. Improved model accuracy by decreasing the likelihood of multicollinearity and random variations

Species Distribution Models

Employing historical patterns of species occurrence is one approach to improve general habitat suitability models, and such an approach may also lead to the development of *predictive species distribution models* (Franklin 1995, Kelly et al. 2001, Guisan and Thuiller 2005). These models attempt to quantify resources, limiting factors, and disturbances, and are built systematically from basic species biology and environmental patterns developed from a remote-sensing image or existing base map. The model development process includes early conceptualization, data preparation and model fitting, spatial predictions, evaluation, and assessment. One species distribution model output may be similar or even equivalent to the simpler general habitat suitability model discussed in previous sections. In essence, this first step can be characterized as a prediction of general habitat suitability using environmental correlates. The general habitat suitability output will often be in the form of a habitat suitability map, but the process can be expanded to include a species distribution prediction. This second step will provide a much greater opportunity to build relevant ecological theory into the modeling process than is possible with general habitat suitable mapping approaches (Guisan and Thuiller 2005). For instance, model recursion will permit systematic selection of causal environmental predictors, and model design will enable ample opportunity to consider the impact of initial data-driven decisions, such as extent and grain size (spatial resolution) (Guisan et al. 2008).

Remote-sensing input to species prediction models may be driven by the need to characterize current understanding of the relevant ecological processes or gradients operating to limit or track species or habitat changes. Examples include species response to fragmentation, competition, seed dispersal, niche position, and population dynamics (Catling and Coops 2004, Guisan and Thuiller 2005, Letcher et al. 2007). Species prediction models are adaptable; for example, Travaini et al. (2007) outlined an integrated framework to generate species distribution models and maps for guanaco (*Lama guanicoe*) in southern Patagonia. Innovative geospatial technologies were incorporated into each step of the process, including:

1. Wildlife census work, based principally on GPS technology
2. Predictive distribution modeling, based principally on high quality remote-sensing data and DEMs produced from diverse sources, such as the Spaceborne Imaging Radar (SIR) sensor on board the Space Shuttle
3. Generation of dynamic and informative maps in a GIS and statistical modeling environment

The initial output was a general habitat suitability map, but additional model output included maps showing "relative probability of contact with a guanaco group per unit area" on the steppe habitat delineated by remote sensing, and "predicted guanaco density class."

A key feature of the final guanaco species distribution model was the reliance on free remote-sensing imagery, such as those available over the Internet from Google Earth, for the environmental land-cover predictor. Such an approach is possible when habitat conditions are simple and can be reliably interpreted visually on available imagery. For scientific remote sensing (such as would be required in digital classification analysis), such data are inadequate for a variety of reasons (for example, uncertain seasonality, lack of radiometric control, or inadequate recourse to the original spectral response patterns). For more complex and scientific analyses, however, many other free remote-sensing data sets have been made available through the Landsat archive and other sources (e.g., Woodcock et al. 2008).

Standardized data sets, such as those available through the *Global Land Cover Facility* (http://glcf.umiacs.umd.edu/aboutUs/), are a valuable global resource. The Global Land Cover Facility data set is based on the MODIS Vegetation Continuous Field (VCF) products, themselves available free over the Internet (Hansen et al. 2003, 2006). These VCF data estimate the percentage of trees, bare soil, and herbaceous vegetation for each 500 m spatial resolution pixel. These data were used recently in a study of landscape-scale sexual segregation of elephants in Kruger National Park (Smits et al. 2007). Traditional helicopter census of elephant herds analyzed with VCF-derived habitat information indicated differences in distribution patterns of mixed herds and bull groups, and consequently, differences in habitat impact of herbivory and water-resource use. Scale differences in observations and inherent uncertainty in the remote-sensing methods were thought sufficient reason to consider the VCF-based models as "... more valuable in an exploratory way, indicating which variables drive the landscape scale and sexually distinct dry season distribution patterns of elephants, rather than being accurate models for predicting fine-scale localized patterns" (Smits et al. 2007: p. 232).

In a different example, a system called *Rapid Epidemiological Mapping of Onchocerciasis* (REMO) was developed based on available topographic maps in Nigeria (Germade et al. 1998). A team of epidemiologists, geographers, and entomologists identified the human population living in areas of high risk to human of *onchocerciasis* (river blindness), which is caused by the prelarval (microfilaria) and adult stages of the filarial nematode *Onchocerca volvulus*. The disease is transmitted by the bite of certain species of female *Simulium* flies (black flies) that are active by day and are found near rapidly flowing rivers and streams. Distance buffers around streams, and proximity measures to villages, were extracted from the topographic maps and used with GPS data to build a simple predictive model of *Simulium* species distribution. The species distribution maps based on these models were able to guide medical teams conducting detailed physical examinations and treatments in the area.

More complex applications, such as those contemplated in *landscape epidemiology* and pathogen prediction models, will require purpose-produced remote-sensing data at specific scales to achieve successful model building and use (Panah and Greene 2005). For example, Sharma and Srivastava (1997) constructed both exploratory and explanatory spatial models to understand and forecast remotely sensed habitat relations of mosquito bearing *malaria* for prevention and control. Such scale-dependent predictive distribution models of disease vectors require significantly higher spatial precision and accuracy of model inputs than can be obtained from existing topographic maps or coarse spatial resolution imagery (described in a later section, Animal and Human Health Studies).

Species prediction models for insects, mammals, and birds have been developed at multiple scales with remote sensing. A few additional examples will serve to more fully illustrate the general species distribution modeling approach; these examples, like those highlighted above, have selected elements in common, typically comprised of species observations, image data, and remote-sensing/mapping systems:

- *Small mammals.* In Australia, Catling and Coops (1999) used trapping grids for small mammals across a range of abundances and found similar relationships between field-observations and predicted species abundance from a remotely sensed surrogate of habitat complexity derived from 2 m spatial resolution aerial videography imagery. These results suggested the development of species distribution models, coupled with habitat quality simulations, which incorporated time-since-disturbance mapped by remote sensing (Coops and Catling 2000).
- *Birds.* Similarly, in the United States, observations of occurrences of black-capped vireo (*Vireo atricapillus*) and LiDAR data were used to develop a predictive species distribution model based on predefined habitat quality classes (Leyva et al. 2002). The LiDAR data were used to describe the vertical structure of suitable habitat classes of particular shape and land cover; however, field confirmation showed that linear habitat features, known to be important, were only poorly classified (less than 10 percent accuracy, not much better than a random distribution). A new approach based on visual interpretation, or digital fusion with high spatial resolution multispectral or comparable imagery, was recommended.
- *Insect communities.* Insect species community distribution patterns can be predicted using remote-sensing data in a spatial modeling approach (Coops and Catling 1997, Lassau et al. 2005). Beetle communities, for example, are often highly related to land cover and land use (Eyre et al. 2005a,b). In one study, a patchy mosaic of eucalypt-dominated sandstone woodlands on nutrient-poor soils in Australia was examined for diverse beetle communities and the relationships to remotely sensed measures of habitat complexity (Lassau and

Hochuli 2008). In the field, habitat complexity was visually assessed by characterizing tree, shrub, and herb cover, soil moisture, leaf litter, and amount of downed woody debris. Beetle pitfall traps were deployed and beetles subsequently identified to family or genus, then categorized as morphospecies. In high spatial resolution aerial spectrographic imagery, habitat complexity was associated with a vegetation index (NDVI). Analysis of the patterns in beetle species richness and abundance confirmed that high spatial resolution aerial remotely sensed NDVI was an accurate predictor of beetle species community patterns and beetle species composition in these relatively open canopy forests. A similar situation was observed by Lassau et al. (2005) in prediction of ant species communities in forest areas with differing NDVI values (Table 5.5).

These habitat suitability and species distribution studies suggest the existence of a larger, more fundamental and very general environmental conservation problem, already revealed in a variety of environmental settings with species ranging from China's giant pandas (*Ailuropoda melanoleuca*) (Mackinnon and De Wulf 1994) to American martens (Hargis et al. 1999) to Mexican palm trees (Arroyo-Rodriguez

Ant Species/ Morphospecies	Abundance in Southern Sites Predicted by Remote-Sensing NDVI Value		Abundance in Northern Sites Predicted by Field-Based Habitat Complexity Values	
	Low	**High**	**Low**	**High**
Aphaeonogaster longiceps	1 (8%)	167 (92%)	104 (36%)	94 (36%)
Iridomyrmex sp.1	779 (83%)	31 (42%)	321 (86%)	174 (86%)
Melophorus sp.1	104 (92%)	0 (0%)	95 (71%)	8 (21%)
Meranoplus sp.1	132 (83%)	131 (17%)	166 (79%)	18 (29%)
Monomorium sydneyens	409 (83%)	19 (58%)	169 (86%)	3 (7%)

Source: With permission from Lassau, S. A., G. Cassis, P. K. J. Flemons, L. Wilkie, and D. F. Hochuli. 2005. Ecography 28:495–504.

*The proportion of sites in which each of the species was trapped (from each treatment) is presented as percentages in brackets.

Table 5.5 Abundance of Five ant Species/Morphospecies between Sites with Low and High Remotely Sensed NDVI Values (Southern Sites) and Field-Based High- and Low-Habitat Complexity in Northern Sites*

et al. 2007). The general problem is that species viability may be threatened not only by habitat loss, but by *habitat fragmentation* and the existence of small population sizes in scattered and isolated habitat areas. Analysis of species distribution models that incorporate the habitat structure and patterns of landscape change will facilitate accurate and reliable modeling with remote sensing.

Multiscale Habitat Structure

Habitat Fragmentation

Measures of fragmentation and connectivity are powerful diagnostic tools that require the quantification of multiscale landscape structure, typically through the application of remote sensing. Fragmentation of landscapes has been shown to impact wildlife species in a variety of ways (Fahrig 2003). Of the wide variety of possible examples in wildlife management, two habitat fragmentation studies are selected for presentation here:

1. The design and management of reserves for China's giant panda
2. Management impacts on the spatial structure of grasslands following conservation planning in the midwestern United States

Remote Sensing in Giant Panda Reserve Analysis

Pandas are one of the world's most iconic species at risk. The past two decades has seen the implementation of a remote sensing-based approach to understand impacts and changes in China's remaining panda habitat, and to identify new reserves and linkage opportunities. The approach has evolved to rely increasingly on remote sensing to understand panda habitat impacts because of three essential findings:

1. The recognition of the tremendous power of a broader and multiscale perspective available only from satellite image analysis [first articulated in panda conservation work by Mackinnon and De Wulf (1994)]
2. The role of field and remote-sensing data working together in models that include animal movement and population predictions (e.g., abundance)
3. The important new ability to monitor and understand the impacts of change in habitat resources and human-landscape use over time provided by synoptic and repetitive remote-sensing approaches

The conservation and management situation in China in the 1980s was strongly influenced by the urgent need to understand the apparent decline of the giant panda, which stimulated a great deal of research and management activity to understand three related habitat aspects:

1. Habitat fragmentation and connectivity
2. Habitat suitability based on food sources
3. A myriad of human disturbance factors (Xu et al. 2006, Wang et al. 2008)

The following brief summary describes the successful use of remote sensing in specific work in these aspects to identify additional areas for panda reserves, which are one important component supporting giant panda conservation and management.

The Qionglai mountain range contains six nature reserves and is currently home to about 30 percent of the entire wild panda population (Hu 2001). Spatial distribution of habitat in these reserves and between reserves was examined using satellite imagery and field observations by Xu et al. (2006). Habitat was defined by forest-cover attributes, bamboo availability and distribution, and elevation and slope factors using a niche-model approach further refined in a later study of pandas in the Pingwu County area, Sichuan Province (Wang et al. 2008). In Qionglai, human activities known to cause habitat degradation—such as roads, human settlements, and agriculture—were incorporated using a distance function, but data on other human factors (such as tourism) that could affect the quality and distribution of habitat were not available, and therefore not included in the model (Xu et al. 2006).

Cloud-free Landsat TM imagery were classified using supervised maximum likelihood decision rules and training data with an 80 percent accuracy over general forest/shrub/herb classes. Analysis of panda habitat suitability included the distance calculations, the DEM data, and a Chinese government bamboo understory distribution map of unknown accuracy (but thought to be an overestimate). Small areas of habitat on the habitat suitability map were removed with a 390 ha area-filter based on previous panda research that suggested the average home range size of individuals was between 390 and 620 ha. On this map, specific habitat conditions in certain configurations were identified:

1. *Key areas* containing large areas of suitable habitat with little human disturbance
2. *Linkage areas*, containing habitat suitable for panda movement between existing areas of reserves, or suitable habitat patches

Linkage habitat areas were identified based on environmental criteria that corresponded with expert knowledge about panda movements, and avoidance and foraging behavior.

The assessment confirmed a broad and ongoing fragmentation pattern of panda habitat in the Qionglai mountains. The total amount of suitable panda habitat was calculated to exceed 500000 ha, but approximately 80 percent of this area was represented by marginal habitat significantly fragmented by roads and human activities. Approximately 36 percent of the total habitat area was determined to be under some form of environmental protection, but at least five key areas of suitable habitat were identified without current protection. Such areas are valuable because they have potential to establish additional reserves. Four other areas were identified as linkage habitats between isolated or potentially isolated key habitat areas. Only one of these linkage habitat areas occurred inside an existing nature reserve. The uncertainty of the final habitat distribution map (with a relatively poor representation of food resources) and the lack of full consideration of the human activities (neglect of disturbances caused by tourism, for example, and no seasonal variation in human activity) in the final habitat suitability model, suggested that the key and linkage habitat areas were only broadly defined. However, the broad pattern showed the unmistakable and deliterious effects of fragmentation and habitat loss and helped support obvious and urgent management conclusions—the creation of new panda habitat reserves and movement or linkage areas (Xu et al. 2006).

Solutions ultimately advanced to reduce the rate of panda decline and restore viable populations included strengthening environmental protection, but also rethinking the design and configuration of natural reserves of panda habitat. The actual environmental processes now known to be responsible for a precipitous panda population decline—habitat loss, edge effects, and fragmentation associated with human encroachment—and the extinction risk associated with isolated populations, were identified and monitored using satellite remote sensing as part of clear management objectives to conserve the species in the wild (Xu et al. 2006). The identification of these patterns then encouraged positive management actions and decision making to create new reserves, extend existing reserves, and restore and reconnect critical habitat; panda reserves have increased dramatically (from fewer than 15 to more than 50 in two decades—and panda numbers may even be increasing (http://www.panda.org/about_wwf/what_we_do/species/about_species/species_factsheets/giant_panda web site accessed January 15, 2009).

However, with such small absolute numbers [an alarmingly low wild population thought to consist of about 1500 individuals (Xu et al. 2006)], and fragile habitat conditions, subject to influence by still larger processes (e.g., national economic development), there is little time available, for example, to apply a long-term monitoring approach to determine actual panda resource use, or to model alternate habitat attributes to understand panda population viability.

Conservation Reserve Program for Great Plains Grasslands

The Conservation Reserve Program (CRP) was developed in 1985 to reduce soil erosion and improve water quality throughout the midwestern United States. This program provides an excellent opportunity to review the use of satellite imagery to examine landscape structure changes and the effect on wildlife caused by the widescale reconversion of agricultural lands to grasslands and woodlands, a central policy directive in the CRP (Egbert et al. 2002).

Earlier studies (e.g., Heske et al. 1999) had documented the significant decline, and increase in fragmentation, over previous decades in natural habitats for a number of wildlife species of interest and concern in the midwestern United States. Changes were typically related to the large-scale clearing and conversion of grasslands, wetlands, and woodlands to agricultural croplands. The impact of grassland and woodland habitat fragmentation, and increasing edge density, for example, was to decrease avian diversity (McIntyre 1995) and the observed numbers of interior-dwelling bird species (Bender et al. 1998). Edge-dwelling bird species increased in abundance, and some wildlife species were observed to be more susceptible to predation following such landscape change. Fragmented landscapes dominated by agriculture "may concentrate predators and act as ecological traps for nesting birds because they attract high densities of breeding birds that are subjected to high rates of nest predation" (Heske et al. 1999: p. 345). Edge, degree of isolation, patch size, and shape are all important components of landscape spatial structure in these environments (Helzer and Jelinski 1999).

Egbert et al. (2002) used a satellite-derived map of CRP lands in Kansas to compare pre- and post-CRP implementation landscape structure. Nine multiseasonal Landsat images acquired in 1987 and 1992 were classified into land-cover classes using an unsupervised clustering approach. An accuracy evaluation using an independent sample confirmed that the maps were more than 90 percent correct. Postclassification change detection was based on map differencing to produce a final change map for spatial analysis using FRAGSTATS software (McGarigal and Marks 1995). Nine landscape metrics were calculated: total landscape area, percent of area in grassland, number of patches, mean patch size, patch density, edge density, mean shape index, nearest neighbor distance, and interspersion/juxtaposition (Table 5.6). The changes documented suggested the landscape structure had changed, following the introduction of the CRP, to a landscape with increased amounts of grassland in fewer isolated and larger patches, less edge, more complex shapes, and increased (or "more equitable") interspersion and connectivity. These trends confirmed a reversal of the grassland habitat fragmentation process that had long been in effect prior to the CRP influence on land management practices (Park and Egbert 2008).

Landscape Metric	1987	1992
Grassland area (ha)	72690	97670
Percent area in grassland	22.1	29.6
Number of patches	1035	930
Mean patch size (ha)	70.7	115.7
Patch density (ha)	0.31	0.28
Edge density (m/ha)	11.7	11.3
Mean shape index	1.62	1.52
Nearest neighbor distance (m)	326.2	335.1
Interspersion/juxtaposition index	76.6	81.9

Source: With Permission from Egbert, S. L., S. Park, K. P. Price, R.-Y. Lee, J. Wu, and M. D. Nellis. 2002. *Computers and Electronics in Agriculture* 37:41–156.
*(Conversion of agricultural lands to grasslands, wetlands, woodlands, as mapped in Landsat imagery)

Table 5.6 Summary of Grassland Landscape Metrics in Kansas Illustrating the Impact of the Conservation Reserve Program*

This study is a powerful illustration of the importance and capability of mapping landscape structure for large-area wildlife habitat following management or policy changes that create differences in landscapes over time and space, but requires careful consideration of a number of influencing factors. Three are highlighted here:

1. The land-cover classes were mapped as single, homogenous classes (e.g., grassland, woodland) (Egbert et al. 2002). Such macrohabitat classes may contain spatial variability (e.g., density differences) that, while beyond the sensor discrimination capability, is important in wildlife habitat analysis (microhabitat). For example, in Kansas, the CRP-converted grasslands were created with seed mixtures that may or may not be similar to adjacent and historical natural grasslands; nevertheless, natural grasslands and CRP-converted grasslands were all mapped as a single grassland cover type. Some management practices, which are known to impact wildlife (e.g., treatment for weeds), continued on these CRP-converted lands, but were not applicable to natural grassland areas.
2. Classification accuracy is a critical feature of multitemporal studies identifying macrohabitat land-cover classes and landscape structure from remote sensing (Shao et al. 2001, Shao and Wu 2008). Misclassification of agricultural areas as native prairie grasslands in the Landsat-derived land cover map is a

relatively common error and can be a significant source of uncertainty in subsequent analysis; for example, in one comparative study on CRP and other lands in Kansas, black-tailed prairie dog (*Cynomys ludovicianus*) colonies were predicted to occur in areas later determined to be unsuitable habitat (Kostelnick et al. 2007). Clearly, additional follow-up work to validate land-cover accuracy and the actual impact of landscape structure changes on wildlife are required (Ribaudo et al. 2001).

3. Finally, the landscape metric approach is strongly scale- and technique-dependent. There are significant issues when comparing landscape metrics over time (Davidson 1998). This problem can become acute when using maps prepared with differing geospatial technologies; when comparing a satellite-derived land-cover classification map to a land-cover classification based on aerial photointerpretation, for example (Franklin et al. 2002a), or when using sensors of different spatial resolution (Benson and MacKenzie 1995). Few guidelines exist for analysts faced with difficult choices that accompany landscape metric uncertainty and sensitivity to input map quality or consistency. One approach is to create simulations and compute confidence intervals for landscape changes to better understand the significance of expected or observed differences. Critical to understanding metric sensitivity to change is the landscape configuration (autocorrelation) and composition (proportion) (Gergel 2007). Significantly, the concept of stationarity, as a requirement for "homogeneity" across the process shaping the landscape, appears in only vague forms in the ecological literature (Remmel and Csillag 2003).

Spatial Heterogeneity

Variability of land-cover classes, and variability within land-cover classes, may introduce spatial heterogeneity of importance to wildlife. For example, predator-prey systems are thought to persist over the long term due to heterogeneity in predation rates caused by prey refugia in space or time (Fryxell et al. 1988, Kareiva and Wennegren 1995, Ellner et al. 2001). In one study, Kauffman et al. (2007: p.696) found a "striking degree of spatial variability in predation at the landscape level" when studying sites where wolves (*Canis lupus*) killed elk (*Cervus elaphus*) in Yellowstone National Park. Most of the variability was caused by the physical features of the landscape, represented by snow conditions, slopes, vegetation openness, and road- and stream-proximity features, where prey and predator interacted. Changes in behavior following wolf reintroduction in Yellowstone may have stimulated changes in elk behavior, contributing to a "trophic cascade" influencing willow (*Calix* spp.) community growth, and increasing spatial heterogeneity (Ripple and Beschta 2006).

Accurate maps of landscape structure and spatial heterogeneity are increasingly required to determine the habitat preferences of species, particularly at the landscape-scale or -level (Wiens 1995).

Vertical and horizontal landscape structure are a result of underlying dynamic processes that vary to create spatial heterogeneity; mapping these variations and quantifying the impacts is a significant issue in remote sensing and ecology. In one early study of Northern spotted owl (*Strix occidentalis caurina*), Ripple et al. (1991, 1997) clarified linkages between spatial heterogeneity, landscape structure, and owl population dynamics (e.g., reproductive rates). Recommendations were developed concerning the amount and quality of old growth and mature forests surrounding nesting sites. Recently, Brosofske (2006) presented a case study of multiresolution analysis of vegetation transects embedded within Landsat TM-based land-cover classifications maps to explore patterns in plant diversity in Wisconsin. Four types of landscapes were assessed by sampling ground plants and analyzing patterns using wavelet analysis. One landscape was influenced by fire disturbance, and other landscapes were influenced by timber harvesting; consequently, spatial heterogeneity caused by these disturbances in these landscape conditions was determined to be significantly different from each other and from premanagement conditions. The structure of plant diversity, in turn, was strongly influenced by patch size, patch structure, edge effects, and landscape context (e.g., distance to road) across a range of scales.

These effects of habitat spatial heterogeneity documented in plants and animals in coastal and terrestrial ecosystems have been mapped more consistently with continuity in satellite remote-sensing mapping technology. The Landsat sensors provide an impressive base mapping foundation upon which to build an analysis of habitat changes over the past three decades. In one study of British Columbia mountain caribou (*Rangifer tarandus*), Hansen et al. (2001) used Landsat to document habitat suitability changes as a result of a reduction in spatial heterogeneity indicated by remote-sensing landscape metrics calculated in 1975 (using Landsat MSS imagery), prior to extensive forest harvesting activity, and 1997 (using Landsat TM imagery), following two decades of local forest management activities. The different forest stands that represented spatially heterogenous habitats for caribou were discriminated on the basis of a remotely sensed structural complexity index comprised of measures of forest age, biomass, and crown volume characteristics. Patch density and edge density had increased, and interior core area had decreased, which together with the reduction in old growth and mature forest stands, reduced overall caribou habitat suitability by more than 8 percent. Spatial analyses prior to the Landsat era may be accomplished with access to earlier satellite imagery (e.g., the Corona archive), aerial photography, and historical maps (Franklin et al. 2005).

Spatial heterogeneity may be an important feature of coastal habitats. In the Leigh Marine Reserve, and at many coastal sites throughout northeastern New Zealand, survey work over two decades had suggested that "urchin barrens" habitat replaced areas of kelp forest following increased fishing of urchin predators (Parsons et al. 2004). A trophic cascade in the recovery of kelp forest in relatively shallow water was related to predator recovery [e.g., snapper (*Pagrus auratus*) and spiny lobster (*Jasus edwardsii*)]. Changes in the subtidal reef habitats were mapped by divers on video transects approximately 1.5 above the substratum and positioned by radiotelemetry. Additional mapping data were obtained with imagery obtained through the hull of a glass-bottomed boat. Visual interpretation of the video imagery was based on the recognition of nine benthic habitat types associated with habitat-forming algal species, sponges, coralline turf, and sediments. A GIS proximity interpolation routine generated the final, spatially continguous map of the entire reef area from the available (relatively sparse) image data. This map was compared to a diver-transect benthic habitat map of the same area constructed 22 years earlier. These map comparisons were difficult because of the different technologies used to produce the maps; the earlier map was much smoother and with larger, fewer patches because of the manual application of a minimum mapping unit and visual interpolation of the original mapping data over the study area.

The total number of patches for each map was 113 for 1978 and 274 for 2000, but the percentage of kelp forest area was about the same (approximately 35 percent of the area). The area of urchin barrens habitat had declined from around 26 percent of the reef in the early map to being nonexistent in the map produced following reduced predator fishing with the creation of the reserve. Conversely, turfing algal habitat increased from 6 percent to almost 38 percent. In deeper water, the extent of kelp forest and sponge flats was found to decrease by 25 and 33 percent, respectively, while turfing algal habitat had increased by 50 percent. This increase in turfing algal habitat had not been previously documented due to the spatial scale of traditional sampling methods and the inability of visual analysis or image interpretation to document spatial heterogeneity (or within class variability).

The most obvious changes over the two decade mapping period were the almost total disappearance of urchin barrens across all depths, and the recovery of the kelp forest and shallow water algal assemblages caused by a trophic cascade related to predator recovery (i.e., consistent with the cessation of fishing within the reserve and higher predation of sea urchins by lobster and snapper). The changes within habitat classes were consistent with trends in community structure identified at permanent sites located in the study area. However, the differing mapping technologies prevent more comprehensive comparison of habitat structure changes—or changes in spatial heterogeneity—over the time interval of interest.

Vertical structure is another dimension of spatial heterogeneity. Vertical structure may be calculated as a measure of vegetation vertical complexity, and has multiple direct and indirect effects on species-habitat relationships. For example, vegetation vertical distribution influences the distribution of animals that consume plants, and may alter foraging behavior. Microhabitats associated with microclimates are typically more diverse in more vertically complex forests, which may also contain more diverse food sources. Birds as a group have been most frequently associated with vertical vegetation structure detected by remote sensing (Imhoff et al. 1997). However, a number of faunal examples can be cited, including the co-occurrence of squirrels in Africa that differ in their use of vertical vegetation, and the co-existence of species of *Anolis* lizards in Puerto Rico associated with different perch types stratified by forest height (e.g., Brokaw and Lent 1999).

A key requirement is the availability or use of remote-sensing imagery that are capable of capturing the necessary structural dimensions of interest. SAR imagery and LiDAR data are suitable for this task:

- *Multidimensional vegetation structure mapping with SIR-C radar and Landsat imagery*. In northern Michigan, a SAR image bird habitat analysis was based on significant niche associations between the northern songbirds of interest and the configuration of habitat patches, patch area, edge effects, and vertical complexity surrogates, such as forest-stand age or biomass (Bergen et al. 2007):
  - Landsat-derived land-cover maps were combined with biomass estimates from the SIR-C data obtained by backscatter model inversion.
  - Biomass prediction accuracy was determined to exceed 1.4 kg/m^2 dry weight in a biomass range of 0 to 30 kg/m^2.
  - Bird surveys and work with the Genetic Algorithm for Rule-Set Production (GARP) (Stockwell and Peters 1999) revealed three species—pine warblers (*Dendroica pinus*), chipping sparrows (*Spizella passerina*), and red-eyed vireos (*Vireo olivaceus*)—that differed significantly in vertical vegetation structure habitat requirements.
  - Habitat suitability maps were produced for these three species for analysis with spatial variables, and the results suggested that species predictions were more precise with vertical structure (such as understory conditions and tree biomass) included in GARP models of bird occurrence.
  - Predictions for birds that required fine spatial heterogeneity (chipping sparrow) were not as accurate as GARP models for birds with more distinct vertical structure habitat

characteristics that could be mapped at the scale of the satellite remote-sensing imagery.
    - Creative ways of incorporating new vertical structure habitat variables thought to be of importance, such as within-canopy vertical structure, will depend on developing sensor synergy between high spatial detail optical imagery and interferometric radar (e.g., BioSAR™) or aerial LiDAR data (Bergen et al. 2007).

- *LiDAR and aerial photography for vertical structure.* In South Africa, Levick and Rogers (2008) provided an example of optical and LiDAR data synergy with aerial photography (black and white, and multispectral imagery) and aerial LiDAR data in monitoring horizontal and vertical spatial heterogeneity in savannah ecosystems in Kruger Park. This environment is controlled at the landscape scale by climate and geology, and spatial variations (or microhabitat) over small areas are expressed by two main vegetation life forms (woody trees and grasses) controlled primarily by rainfall, topography, soil type, fire, and herbivory. A multistep approach was implemented:
    - Object-based image analysis, incorporating contextual and ancillary data in image classification, was used with the high spatial detail aerial photography to identify woody plants.
    - Confusion between trees and dark soils was resolved with a multiresolution image segmentation technique creating successively finer divisions of image objects for classification. Over 97 percent accuracy in tree recognition was obtained.
    - The vertical component was derived from 25 kMz frequency first/last pulse ALTM 1225 Optech LiDAR data processed to create a vegetation canopy model (spatial depiction of tree height).
    - Each tree identified in the first step was attributed a specific vertical structure and spatial context. Validation revealed that the provider-LiDAR interpolation method had introduced a spatial interpolation error in nontree vegetation areas, but with the tree recognition map, these data were corrected prior to interpretation of spatial heterogeneity.
    - A significant R^2 of 0.851 between tree height and field observation of height was obtained. This approach provides detailed individual tree recognition, woody cover estimates, and tree height distribution necessary to depict both horizontal and vertical spatial heterogeneity of savannah ecosystems.

Comparable synergistic results with fusion of data from active and passive aerial optical systems have been reported in other spatially heterogeneous environments, including riparian zones (Dowling and Accad 2003), mixed-deciduous woodlands (Koukoulas and Blackburn 2005, Hill and Thomson 2005), conifer forest stands (Hyde et al. 2006), and Australian eucalyptus (Turner 2007). In all these applications using high spatial detail imagery, a common image processing step—segmentation to isolate individual plants—is essential to depict micro-habitat spatial heterogeneity and land-cover spatial variability at the required level of detail (or scale) in the resulting map products. A general introduction to methods designed for the integration of LiDAR and satellite data from Landsat and QuickBird imagery would include traditional pixel-based classification approaches as well as contextual processing and image segmentation (Wulder et al. 2008).

Anthropogenic Change and Disturbances

Linear Features

Landscapes are increasingly recognized as cultural landscapes, shaped over time, in an interactive process linking human needs with natural resources in a specific topographic and spatial setting (Bürgi et al. 2007). Anthropogenic change can take many forms and has been classified according to a multitude of criteria depending on the need to understand the importance of changes in the landscape. Possible groupings of forest change could be based on the cause of change (natural and anthropogenic), rate of change (gradual vs abrupt), levels of impact on vegetation health (normal change variability vs catastrophic change), and permanence (reversible and irreversible landscape changes) (Gong and Xu 2003). Within this wide assortment of change types, many different processes and outcomes are apparent.

Direct Impact of Linear Features

Roads have diverse and negative effects on wildlife and biodiversity, and are a complex management issue in many environments (Findlay and Bourdages 2000, Forman et al. 2003). Roads are a major contributor to the proportion of *impervious surface cover* in an area (Weng 2007) which, when increased, can indicate habitat loss, significant process changes (e.g., hydrology), and decreased habitat quality. Impervious surfaces influence landscape configuration and fragmentation. Changes in these characteristics, in turn, are strongly predictive of a large number of ecological conditions, such as stream health and water quality in certain habitat settings (Snyder et al. 2005). Specific road impacts on wildlife in tropical rainforests are associated with habitat loss and alterations, edge and disturbance effects, invasions, road mortality, and barriers to movement (Goosem 2007). A 2-year census of one 3.6 km two-lane paved road section adjacent to the Big Creek National Wildlife Area in the Long Point region of Lake Erie recorded mortality of more than 32,000 individuals (mainly juvenile frogs, but also adult amphibians, reptiles, mammals, and birds) (Ashley and Robinson 1996).

One of the most challenging types of change to detect by remote sensing, and of great interest in many habitat applications, is the detection of these linear features. Of interest are not only roads, but trails, seismic lines, pipelines, and other types of urban and transportation infrastructure. Two broad approaches are possible:

- *Image processing to extract roads.* Remote-sensing methods to directly map roads and other linear features are plentiful and increasingly accurate when applied to specific remotely sensed data, including high spatial resolution satellite imagery (Mena and Malpica 2005) and aerial LiDAR (Clode et al. 2007). Many analysts prefer to interpret images visually and digitize the results in a GIS environment. Typical digital image processing steps include image fusion, texture analysis, and segmentation (sometimes called skeletal extraction based on mathematical morphology). This type of software is often a standard offering in many commercial image processing systems (Mather and Koch 2004, Aronoff 2005). More specialized software packages are increasingly available (*see*, for example, the December 2004 issue of *Photogrammetric Engineering and Remote Sensing*, which contains more than a dozen original papers describing a comprehensive suite of road detection software approaches, evaluation techniques, and applications). One software system called LINDA (linear feature network detection and analysis), originally developed by Wang (1993) for relatively coarse or medium spatial resolution satellite imagery, now works with a wavelet extraction tool to increase effectiveness with higher spatial resolution imagery (Chen et al. 2005).
- *Modeling road locations.* The progress in effectiveness of linear feature extraction software has been impressive, but in very dense road networks (urban areas), or in areas with roads or other linear features that are difficult to extract because of low spatial resolution imagery or low contrast to surrounding features, *contextual* or *spatial modeling* approaches may be useful. In forested areas with active forest management activities, for example, it may be reasonably straightforward to search for road locations based on land cover (forest) and topography (valleys and stream occurrences). This approach could be tailored to consider road building technology and regulatory environments that might constrain road construction and other infrastructure developments.

A version of this modeling approach was implemented by Huettmann et al. (2005) in a study to determine the likely future impact of roads and other linear features on landscape structure in western Alberta, an area then experiencing rapid economic development associated with resource extraction industries. The main objective was to create a number of *future landscape habitat scenarios* showing features strongly associated with resource extraction; roads are typically one of the first and lasting changes in such areas. Assumptions on future road locations were based on forest harvesting economics, and a realistic footprint for other linear features expected in the area, such as pipelines, was derived from adjacent areas that had already experienced similar development. The analysis suggested that road and pipeline linear features typically increased edge density and decreased large patch dominance and mean patch size characteristics. Such changes in landscape configuration and composition can have important impacts on wildlife (Riitters et al. 1995).

Roads and pipelines are not the only linear features introduced by human resource development activities that can influence landscape configuration and composition. The impact of *seismic cutlines* on habitat and landscape structure in this type of environment could be significant and of some importance for certain species of interest, such as grizzly bears (Linke 2003). In the process of conventional oil and gas exploration in western Alberta, a dense network of seismic cutlines—typically ranging from 2 to 5 m in width—is created that dissect the landscape. These cutlines may provide food or movement corridors for individual animals, and may introduce a level of fragmentation to landscapes at a finer scale than had previously been mapped with satellite remote-sensing data. These smaller linear features are not portrayed on forest inventory maps or medium spatial resolution imagery such as those obtained by Landsat for land-cover classification purposes (Color Plate 5.1; Franklin et al. 2002a, He et al. 2009). Typically, the impact of cutlines could be hypothesized to increase landscape heterogeneity, increase habitat fragmentation by dissecting large patches and convert interior habitat to edge habitat, and influence patch context (e.g., isolation) (Linke et al. 2008).

Roads are often avoided by grizzly bears, but few observations have been made to support the idea that bears avoid seismic lines or other linear features (McLellan and Shackleton 1988, 1989). Instead, the introduction of seismic lines into this landscape may alter the landscape structure, and through changes in landscape structure, influence grizzly bears. Linke et al. (2005) set out to explore these relationships in the Yellowhead Ecosystem of western Alberta (described more fully later in this chapter, Case Study: Landscape Change and Grizzly Bear Health in Alberta) in order to understand the meaning of landscape structure, and changes caused by linear features such as seismic cutlines, to grizzly bear populations:

1. Grizzly bear GPS data were collected on movement and habitat use following capture and collaring of 19 and 20 individuals, in 1999 and 2000, respectively; the number of sample points ranged between 131 and 300 locations per individual throughout the early summer season.
2. Home ranges were computed for each bear and the distance at which strong autocorrelation existed was determined; these statistics were used to calculate a landscape unit or scale of movement, and all GPS data within these areas (approximately 49 km^2 hexagons) were used to yield densities of bear landscape use.
3. Landscape structure was computed using FRAGSTATS on a satellite-derived land-cover map which displayed 11 cover types and was determined to be approximately 83 percent correct; seismic lines were manually mapped using a high spatial detail IRS image (5 m spatial resolution, mapping accuracy 88 percent), and together with streams and an index of vertical ruggedness from the DEM, were included in the spatial analysis.
4. Fourteen independent variables (landscape metrics such as mean patch size, shape index, diversity measures, and proportion of land-cover classes, including proportion of seismic lines) were related to the grizzly bear landscape use density data.

These results confirmed that while seismic lines did not have a direct relationship to population-level landscape use, an *indirect* influence was observed because landscape spatial heterogeneity, which increased in areas with more seismic lines, was important in explaining bear landscape use (Linke et al. 2005). *Increased* bear landscape use was indicated by increasing mean patch size, and seismic lines demonstrably reduced mean patch size in an effect similar to the road effect reported by others (e.g., McGarigal et al. 2001). Bears appeared to use areas more when the spacing between landscape patches was consistent; seismic lines caused the spacing between landscape patches to become more variable (Linke 2003, Linke et al. 2008). Male and female bear landscape use also appeared to differ with spatial heterogeneity, and the majority of bear landscape use variance remained unexplained at this scale of analysis.

Partial Land-Cover Change

Spatial heterogeneity may be expressed in partial land-cover changes created by processes that are not sufficient to induce a change in land cover. Instead, the impact is in the creation of a change in other attributes that are associated with land cover and are known to vary within a specific land cover. For example, in Scandinavia, most forest stands are subject to between one and three thinnings, in which approximately 20 to 50 percent of the basal area is removed, before the final clearcutting (Olsson 1994). The material left on the ground, and the gaps created, present a spectral response pattern that can be detected using appropriate remote-sensing data. In some studies, analysts have noted the importance of partial harvesting in an area, perhaps because of field knowledge, or regulatory and policy changes affecting forestry practices, but have not been able with available sensors to detect those changes and understand their influence on fragmentation processes and habitat spatial heterogeneity (e.g., Zheng et al. 1997).

Betts et al. (2003) have provided an example of partial land-cover change analysis using remote sensing in the development of habitat indicators influenced by silvicultural activities (e.g., thinning) and partial (or selective) forest harvesting practices in New Brunswick, Canada. Changes to species composition of forest stands and age distribution result from these forest management activities—changes which, in turn, influence habitat characteristics, including number of cavity trees, shrub layer composition and development, and landscape spatial heterogeneity. Available Landsat imagery were acquired over a 5-year time period (1992–1997), georadiometrically corrected, transformed to brightness/greenness/wetness indices, and subjected to image differencing techniques (Franklin et al. 2000). This differencing procedure discriminated silvicultural changes and partial harvest (typically between 20 and 30 percent commercial stem removal) with 71 percent accuracy and no significant errors of omission. The final change layer was combined with clearcuts on both Crown land and local harvesting operations mapped in the field with GPS. Growth of new habitat was inferred through analysis of regeneration and young stands mapped in the forest inventory.

Three categories of landscape metrics were selected to analyze the landscape changes in five habitat types associated with indicator species (Table 5.7, Betts et al. 2003):

1. *Habitat area.*The amount of landscape that meets the requirements of a selected indicator species.
2. *Patch size metrics.* Maximum, minimum, mean, and frequency distribution of contiguous habitat patches within a landscape.
3. *Landscape configuration metrics.* Include proximity of habitat patches within a landscape, and frequency distributions of nearest forest patches for five habitat types.

The habitat criteria were developed based on existing expertise and knowledge of home range size, dispersal distances, and population viability analysis, under an indicator species approach in which it was assumed that if the habitat requirements of a set of mature forests

Habitat Type	Indicator Species	Spatial Requirements
Old tolerant hardwood (e.g., American beech (*Fagus grandifolia*), sugar maple (*Acer saccharum*), yellow birch (*Betula alleghaniensis*	Barred owl (*Strix varia*)	Patch > 20 ha, interpatch distance > 3 km
	White-breasted nuthatch (*Sitta carolinensis*)	Patch > 40 ha, interpatch distance < 1 km
Old hardwood (tolerant hardwoods combined with some intolerant species (e.g., trembling aspen (*Populus tremuloides)* and conifers)	Northern goshawk (*Accipiter gentiles*)	Patch > 20 ha, interpatch distance > 3 km
	Hairy woodpecker (*Picoides villosus*)	Patch > 30 ha, interpatch distance < 1 km
Old spruce-fir mixed forest (e.g., red spruce (*Picea rubens*) and balsam fir (*Abies balsamea*) with other conifers and some tolerant and intolerant hardwoods	American marten (*Martes americanus*)	Patch > 375 ha, width > 1 km
Old pine forest (pure stands of eastern white pine (*Pinus strobus*) or jack pine (*Pinus banksiana*)	Pine warbler (*Dendroica pinus*)	Patch > 15 ha, interpatch distance < 1 km
Old mixedwood forest (typically tolerant or intolerant hardwoods mixed with red spruce and balsam fir)	Northern flying squirrel (*Glaucomys sabrinus*)	Patch > 60 ha, interpatch distance < 1 km

Source: With permission from Betts, M. G., S. E. Franklin, and R. Taylor. 2003. *Canadian Journal of Forest Research* 33:1821–1831.

TABLE 5.7 Spatial Requirements and Indicator Species of Five Habitat Types Used in Wildlife Fragmentation Analysis

specialists was met, populations of other, more generalist species, would also be maintained.

Changed forest conditions represented 4.5 percent of the total land area and 5.6 percent of the forested land area of the study area selected over the 5-year period. Two habitat types were reduced by approximately 1 percent per year over the time period captured in the Landsat imagery, but mixedwood habitat declined by 9.7 percent. Mean patch size declined for most habitat types, and the number of "large" patches decreased markedly. Area and patch size distribution for all five habitat types decreased, largely as a result of the introduction of new roads and some harvesting activities. The distance between patches decreased, and some differences in the changes observed were recorded in different jurisdictions known to be influenced by different management decisions (e.g., Crown land vs private woodlots). In most habitat types, the rate of fragmentation exceeded the rate of habitat loss.

Overall, 42 percent of the area studied represented habitat in patches sufficient to meet the needs of the selected local indicators species, a decline from 44 percent at the beginning of the study time period. Management implications following this analysis include suggested changes to forest harvesting practices to ensure appropriate characteristics of species composition and spatial heterogeneity are maintained. The study confirmed that partial changes within a land-cover class are significant, in addition to the class-by-class changes in land cover; *both* are important in habitat analysis. Sustainable management for timber volume will need active planning to ensure mean patch size and other spatial requirements of indicator species are met.

A cautionary note may be appropriate. Substantial gaps in knowledge exist for effective management decisions related to landscape structure and change; for example, the interpretation of the ecological meaning of landscape metrics and landscape change remains a significant challenge in virtually any natural or managed environment (Franklin and Dickson 2001, Gergel 2007). Detailed landscape configuration and composition sensitivities of many species—and, particularly, the indicator or special species of interest—remain largely unknown. Patch size criteria for a given species may either be underestimated or considered inappropriately as diagnostic when there are many uncertainties. For example, in New Brunswick, selected indicator species may be engaged in *habitat supplementation* across a gradient of habitat conditions (Dunning et al. 1992, Betts et al. 2003). Relating the landscape changes observed to individual field observations of species abundance and behavior must be a priority. Population effects are even more uncertain and presently are typically only vaguely understood.

Animal and Human Health Studies

Advances in pathogen, vector, and reservoir and host ecology have allowed assessment of a greater range of environmental factors that promote disease transmission, vector production, and the emergence and maintenance of disease foci as well as risk for human-vector contact (Beck et al. 2000, Brown et al. 2008, Tran et al. 2008). Remote-sensing approaches to *landscape epidemiology* have continued to improve (Panah and Greene 2005), with successful examples recently reported in domestic animal and wildlife disease studies, such as tuberculosis (TB) in New Zealand (McKenzie et al. 2002), Lyme disease in northeastern United States (Beck et al. 2000), bluetongue in the Mediterranean Basin (Guis et al. 2007), chronic wasting disease in wild deer populations (Farnsworth et al. 2005), foot and mouth disease (Curtis et al. 2005), and hantaviruses (Boone et al. 2005, Goodin et al. 2006). Table 5.8 lists a number of potential links between remotely sensed environmental conditions and disease compiled by Beck et al. (2000). A few of these applications are described in greater detail in the remainder of this section.

Hantaviruses are zoonotic, aetiological agents maintained by rodents of the family *Muridae*. The occurrence of hantavirus in rodent hosts has been correlated to a number of climatic and environmental factors, including landscape structure derived from Landsat imagery and aerial photography (Langlois et al. 2001). In Paraguay, Goodin et al. (2006) used the Global Land Cover 2000 data (spatial resolution of 1 km^2) to map six land cover types (Eva et al. 2002). Following a rodent capture protocol, a total of 362 rodents from ten species known to host hantaviruses were tested for the presence of hantavirus antibodies, resulting in 27 seropositive individuals. The relationship between serostatus and remotely sensed land cover type was examined in nonparametric tests of proportions and qualitative comparison of observed and expected values. A significant difference in habitat association between seropositive and seronegative rodents was observed; seropositive rodents were found with disproportionately high frequency in areas where human disturbance in the form of intensive and mosaic agricultural landscapes had occurred.

Such analyses are indicative, but do not reveal cause-and-effect explanations as to why agricultural land cover is more likely to harbor seropositive rodents. However, the relationship discovered in Paraguay was thought realistic and important enough to help inform a health monitoring system designed to relate land cover change to potential viral outbreaks in rodents and humans (Goodin et al. 2006). A similar study in western United States concluded that the level of connectivity among host populations distributed across the landscape could be monitored using remote sensing; there, however, increasing fragmentation of habitat reduced the likelihood of hantavirus infections in deer mice because of their relative isolation from neighboring populations (Boone et al. 2005).

Mapping Lyme disease transmission risk in New York State required an understanding of the relationship between forest patch size and deer distribution. White-tailed deer serve as a major host and mode of transportation of the adult tick (*Borrelia burgdorferi*) that transmits Lyme disease. Canine seroprevalence rate (the assumption being that dogs were more likely to acquire tick bites on or near their owner's property) was compared to land cover obtained by classification of Landsat-derived indices of brightness/greenness/wetness,

Factor	Disease	Mapping Opportunity
Vegetation/ crop type	Chagas disease Hantavirus Leishmaniasis Lyme disease Malaria Plague Schistosomiasis Trypanosomiasis Yellow fever	Palm forest, dry & degraded woodland habitat for triatomines Preferred food sources for host/ reservoirs Thick forests as vector/reservoir habitat in Americas Preferred food sources and habitat for host/reservoirs Breeding/resting/feeding habitats; Crop pesticides vector resistance Prairie dog and other reservoir habitat Agricultural association with snails, use of human fertilizer Glossina habitat (forests, around villages, depending on species) Reservoir (monkey) habitat
Vegetation green-up	Hantavirus Lyme disease Malaria Plague Rift Valley fever Trypanosomiasis	Timing of food sources for rodent reservoirs Habitat formation and movement of reservoirs, hosts, vectors Timing of habitat creation Locating prairie dog towns Rainfall Glossina survival
Ecotones	Leishmaniasis Lyme disease	Habitats in and around cities that support reservoir (e.g., foxes) Ecotonal habitat for deer, other hosts/reservoirs; human/vector contact risk
Deforestation	Chagas disease Malaria Malaria Yellow fever Yellow fever	New settlements in endemic-disease areas Habitat creation (for vectors requiring sunlit pools) Habitat destruction (for vectors requiring shaded pools) Migration of infected human workers into forests where vectors exist Migration of disease reservoirs (monkeys) in search of new habitat
Forest patches	Lyme disease Yellow fever	Habitat requirements of deer and other hosts, reservoirs Reservoir (monkey) habitat, migration routes

Table 5.8 Potential Links between Remotely Sensed Environmental Conditions and Disease

Factor	Disease	Mapping Opportunity
Flooding	Malaria Rift Valley fever Schistosomiasis St. Louis encephalitis	Mosquito habitat Flooding of dambos, breeding habitat for mosquito vector Habitat creation for snails Habitat creation for mosquitoes
Permanent water	Filariasis Malaria Onchocerciasis Schistosomiasis	Breeding habitat for Mansonia mosquitoes Breeding habitat for mosquitoes Simulium larval habitat Snail habitat
Wetlands	Cholera Encephalitis Malaria Schistosomiasis	Vibrio cholerae associated with inland water Mosquito habitat Mosquito habitat Snail habitat
Soil moisture	Helminthiases Lyme disease Malaria Schistosomiasis	Worm habitat Tick habitat Vector breeding habitat Snail habitat
Canals	Malaria Onchocerciasis Schistosomiasis	Dry season mosquito-breeding habitat; ponding; leaking water Simulium larval habitat Snail habitat
Human settlements	Diseases	Source of infected humans; populations at risk for transmission in general
Urban features	Chagas disease Dengue fever Filariasis Leishmaniasis	Dwellings that provide habitat for triatomines Urban mosquito habitats Urban mosquito habitats Housing quality
Ocean color	Cholera (Red tides)	Phytoplankton blooms; nutrients, sediments
Sea surface temperature	Cholera	Plankton blooms (cold water upwelling in marine environment)
Sea surface height	Cholera	Inland movement of Vibrio-contaminated tidal water

Source: With permission from Beck, L. R., B. M. Lobitz, and B. L. Wood. 2000. Emerging Infectious Diseases 6:217–227. Available from http://www.cdc.gov/ncidod/eid/vol6no3/beck.htm [September 8, 2008].

Table 5.8 (*Continued*)

which were also correlated to tick abundance (Beck et al. 2000). Results suggested that human-host contact risk was positively correlated to deciduous forest cover and forest-edge characteristics. Beck et al. (2000) also highlighted a case study involving remote sensing of temporal and spatial patterns in the Bay of Bengal associated with increased risk of cholera (*Vibrio cholerae*) outbreak in Bangladesh. TOPEX/Poseidon and NOAA AVHRR imagery at relatively low spatial resolution were used to map sea surface temperature, sea surface height (for tidal prediction), sediment loads, and plankton blooms, an important marine reservoir of cholera.

High spatial detail satellite imagery and landscape metrics were used to determine the conditions of emergence of bluetongue in Corsica, a French Mediterranean island where the disease occurred for the first time recently (Guis et al. 2007). Bluetongue is a viral disease of ruminants transmitted by some species of *Culicoides* (biting midges). The approach was to identify environmental parameters related to bluetongue occurrence in the neighborhood of 80 sheep farms. A logistic regression model computed within three subsequent buffer distances of 0.5, 1, and 2 km revealed the role of landscape metrics, particularly those characterizing land cover and land-use units, such as grasslands and woodlands, as well as farm type, latitude, and solar radiation to explain the presence of bluetongue. Internal and external validation both indicated that the best results were obtained with the 1 km buffer size model (area under Receiver Operating Characteristic curve = 0.9 for internal validation and 0.81 for external validation). The results showed "...that high spatial resolution remote sensing and landscape ecology approaches contribute to improving the understanding of bluetongue epidemiology."

The brushtail possum (*Trichosurus vulpecula*) is a wildlife vector for TB caused by *Mycobacterium bovis* in New Zealand (McKenzie et al. 2002). The spatial distribution of possums is a critical health issue; possum habitat and denning sites mapped with remote sensing represents an important breakthrough in the possible control and risk reduction for farming productivity and human health. In one study, SPOT HRV imagery were classified into nine land cover classes with an overall accuracy of 93 percent. The full extent of habitat heterogeneity was analyzed with DEM data on slope, and farm boundary data, obtained from a GIS database. Models of TB risk were produced at the "individual possum scale" based on a health database from animal trapping; TB hot-spot risk information included estimates of possum habitat, denning likelihood, and farmer's knowledge of where they believed cattle were becoming infected. Spatial variables based on FRAGSTATS were included (habitat patch metrics, edge metrics, diversity and proximity measures) and showed that habitat spatial heterogeneity was strongly associated with a high TB risk. However, additional training was required to help farmers (and control operators) understand and use the risk-maps to implement effective decision making and control programs in the field (McKenzie et al. 2002).

Case Study: Landscape Change and Grizzly Bear Health in Alberta

Land Cover, Structure, and Resource Selection

This case study illustrates a wide range of ways in which remote sensing, GIS, and spatial modeling activities are used to support the development of knowledge of grizzly bear resource use, health, and management in Alberta. First, the technical details of the work are briefly summarized; second, the hypothesized links between grizzly bear health and landscape change are described. This work is ongoing and only preliminary results illustrative of the methods employed are provided here. And third, the final section of this case study presentation highlights some of the management implications suggested in this 10 year research project. Interested readers are encouraged to visit the Foothills Research Institute Grizzly Bear Research Project web site for more comprehensive material and publications (http://foothillsresearchinstitute.ca/pages/Programs/Grizzly_Bear.aspx web site accessed on January 17, 2009).

Resource extraction activities (forestry, mining, and oil and gas development), agriculture, and an increasing demand for recreational activities have resulted in alteration of landscapes along the eastern Rocky Mountain slopes of Alberta, traditional habitat of remaining grizzly bear populations (Stenhouse and Graham 2008). Grizzly bears are widely recognized as an indicator of ecosystem health. Bears are resource generalists with large home range, and catholic resource requirements, suggesting major challenges in understanding the best approaches for wildlife management particularly in areas experiencing rapid landscape change. A key to developing management plans for this species has been the production of a seamless remote-sensing habitat map (and associated data layers) for use with observational data and models (Color Plate 5.2; McDermid 2005, Stenhouse and Graham 2008).

Annual spring capture and collaring has been conducted since 1999 to obtain data necessary to build resource selection functions (RSFs). These functions are powerful spatial analysis tools (Boyce and McDonald 1999) to quantify the association between bear movement data and remotely sensed habitat, and permit comparative analyses relative to ongoing disturbance (e.g., mountain pine beetle infestation and associated forest harvesting activities). Approximately 15 to 20 GPS radio-collars were deployed on grizzly bears each year following bear captures by trap, snare or helicopter-darting. Bears were immobilized using a drug combination of Telazol and xylazine administered by rifle/pistol; atipamazole was used to reverse the xylazine after handling procedures were completed (Cattet et al. 2003). Other sampling (e.g., to acquire necessary data on tooth, blood, and tissues) was conducted to understand population dynamics, genetics, and health. Bears were weighed and measured and monitored during processing. Radio-collars and ear tag transmitters (with remote release mechanisms) were fitted and programmed to provide location data every hour during the active period and every two hours during the denning period.

The GPS data were collected and used to design some survey activities. For example, field observation data collected at some GPS points included bear (e.g., scat) and den information, and vegetation conditions (understory, forest composition, presence of bear foods). Some activity inference was possible (e.g., feeding or bedding) and more detail on specific activities was developed with newly designed bio-logging GPS/digital camera collar data (Hunter et al. 2005). Other relevant data were collected during berry abundance surveys, road use intensity (from digital trail cameras), and scat and hair samples (for DNA analysis).

All of these data have a spatial component when analyzed with land cover, crown closure, species composition, and vegetation phenology produced from Landsat and MODIS imagery for all of known grizzly bear range in Alberta (McDermid 2005). The remote sensing mapping procedures are described in detail elsewhere (Stenhouse and Graham 2008), and included object-based segmentation, maximum likelihood decision rules, continuous variable modeling using NDVI and other vegetation indices, and a comprehensive accuracy assessment protocol (Franklin and Wulder 2002, McDermid et al. 2005). Annual changes in landscape conditions (Table 5.9) were obtained by modeling (Fig. 5.2)

Year	Road Density (km/km^2)	Cutblock Proportion (%)	Wellsite Density (#/km^2)	Pipeline Density (km/km^2)	Mine Area (km^2)
1998	0.35	2.04	0.09	0.126	1.10
1999	0.36	2.33	0.10	0.130	1.10
2000	0.37	2.64	0.11	0.136	3.50
2001	0.39	2.97	0.12	0.138	6.58
2002	0.42	3.03	0.11	0.127	10.64
2003	0.44	3.36	0.12	0.129	12.20
2004	0.44	3.68	0.13	0.131	14.26
2005	0.45	4.04	0.14	0.132	15.73

Source: Adapted from Stenhouse and Graham 2008.

TABLE 5.9 Example of Changes in Road Density, Cutblock Proportion, Wellsite Density, Pipeline Density, and Mine Area (1998–2005) in One Specific Area of Grizzly Bear Habitat in Western Alberta Mapped Using Satellite Remote Sensing and GIS Methods*

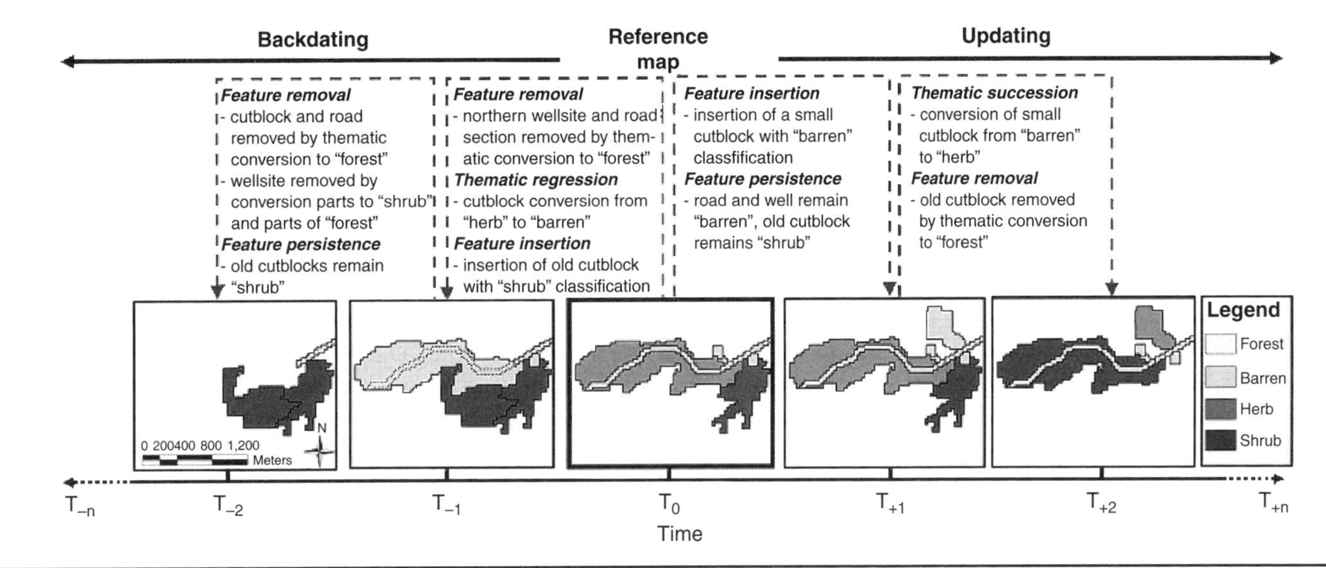

FIGURE 5.2 Map change detection and update procedure in grizzly bear habitat mapping application in Alberta. (*With permission from* Linke et al. 2009. *Photogrammetric Engineering and Remote Sensing*, in press.)

and from best-available satellite remote-sensing imagery, including MODIS, which provided reasonably accurate detection of changes greater than approximately 15 ha in size (Pape and Franklin 2008).

A model of grizzly bear density that included selected landscape metrics (e.g., patch density, contiguity index) was developed; the model was significant and explained approximately 35 percent of the variance in bear density across a wide variety of landscape conditions. Interpretation of this model suggested that grizzly bear use of an area increased with increasing patch density, increasing amounts of large, contiguous patches, and increasing variation in the distances between similar patches (Linke 2003). Using a limited sample of points to test the model performance, prediction of grizzly bear presence was 87 percent correct, and the overall prediction accuracy, including both presence and absence, was 71 percent, when compared to observed values (the GPS locations). A link was confirmed between decreased grizzly bear landscape use and increasing agricultural activity and habitat fragmentation (Nielsen and Boyce 2002).

The distribution of 12 critical grizzly bear food resources was estimated using the developed remotely sensed data layers (e.g., land cover, leaf area index), terrain variables (e.g., topographic position), and field observations acquired in plots across the landscape (Nielsen et al. 2009). Seasonal weights were used to simulate changes in grizzly bear diet in different parts of the landscape and over time. The performance of the food-based habitat model was evaluated based on radiotelemetry data and also compared to the resource selection models and to bear occupancy estimates developed from the DNA studies. Compared to random locations, for example, grizzly bear radiotelemetry locations were almost 20 times more likely to occur in sites predicted to contain known grizzly bear foods.

Health and Change

Grizzly bear response to human alteration of the environment was defined by a suite of physiological and behavioral measures with long-term stress hypothesized to be a linking mechanism. The duration of a stressor is important; for example, short-term stress (lasting seconds to hours) encountered during the normal activities and experiences of daily life rarely pose a threat to healthy animals, but long-term stress (lasting days to months) as can occur with human alteration of the environment, can exceed an animal's ability to cope (Moberg and Mench 2000). Long-term stress may lead to a loss of capacity to sustain normal biological function (i.e., growth, reproduction, immunity, activity), and impaired health (termed "distress") including reduced growth, impotency, infection, and sometimes premature death. Population-level effects occur depending on the proportion of the population that is distressed, although the time lapse between human alteration of the environment and reduction in wildlife population performance can take many years (Findlay and Bourdages 2000).

A number of methods have been developed for populations and individual animals to understand the impact of long-term stress on wildlife management strategies. Health function scores, for example, are comprised of suites of variables associated with different aspects of health; included may be measures of stress (e.g., cortisol and certain heat shock proteins), growth (e.g., body condition index), immunity (e.g., white blood cell count), and movement (e.g., mean daily movement rates). By constructing weighted linear functions of individual health measures, such scores for the population of bears inhabiting the study area could potentially reveal significant variation year-to-year, with age, food quality, and other habitat conditions, and with bear location. Typically, grizzly bears with a home range in areas of high human activity display higher levels of stress and relatively low immunity measures.

The analysis of the relationship between landscape and health data, in the form of scores of other measures, is challenging for many reasons; for example, there are influences on health measures caused by the live capture protocol (Cattet et al. 2008). Additionally, confounding variables (or covariates) such as age and sex, and small sample sizes, complicate analyses. The sheer number of variables involved (more than 30 health measures, more than 20 environmental variables) across biological, temporal, and spatial scales required that complex statistical models be applied. One powerful approach is to match health data and landscape data by home range through principal components analysis followed by ordination and an analysis of covariance (ANCOVA mixed models); in essence, the technique is searching the health and landscape data for patterns to clarify the relationships between bear demography, health, and a changing environment.

Principal components analysis is an appropriate technique to reduce data redundancy and create indices appropriate when comparing large numbers of health and landscape variables. Following tests of significance and developing procedures for handling missing values, an analysis of covariance reveals potential gradients and relationships in the extracted components; for example, the form of the relationship between body condition index (BCI) (dependent) and environmental variables (independent predictors) in one environmental strata was determined to be:

$$\begin{aligned}\text{BCI} = {} & \beta_0 + \beta_1{}^*\text{sex}^*\text{age} + \beta_2{}^*\text{month} + \beta_3{}^*\text{year}^2 + \beta_4{}^*\text{canopy closure} \\ & + \beta_5{}^*\text{roads} + \beta_6{}^*\text{roads}^*\text{yr} + \beta_7{}^*\end{aligned}$$

Interpretation of this index model in the context of the overall set of relationships (Fig. 5.3; actual values shown are for illustrative purposes only) revealed some general patterns of interest in management applications. For example, the index suggested that road density (*roads*) influenced bear size in a counterintuitive way because higher road density was positively correlated with larger body size. This

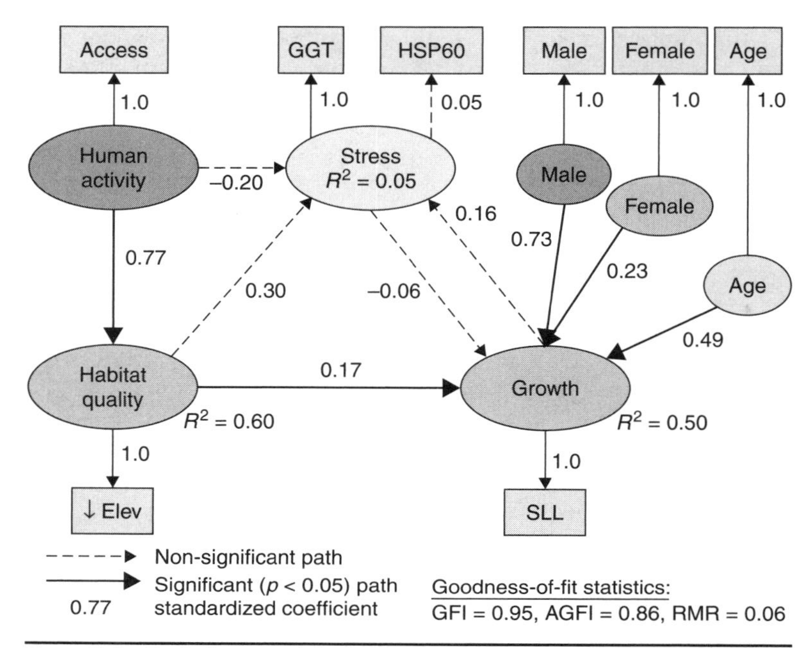

FIGURE 5.3 Preliminary model relationships between grizzly bear demography, health, and landscape change. Values of relationships shown are for illustrative purposes only. (*Courtesy of* Dr. Marc Cattet, University of Saskatchewan.)

may be a result of availability of better habitat resources, such as game and other meat sources, whose densities are typically higher in areas of disturbance and younger forests which occur in areas with higher road density. In contrast, bear body size was lower in higher elevation areas, which is potentially due to lower habitat quality at higher elevations. This general result suggested that while mortality pressure may be higher in areas with higher road density and consequently higher anthropogenic effects, the increased availability of food resources may actually allow bears living in these areas to grow faster and larger.

A similar index model development predicting a stress protein, heat shock protein (*HSP60*), was determined to have the form:

$$\text{Log}(HSP60) = \beta_0 + \beta_1{}^*\text{change} + \beta_2{}^*\text{change}^*\text{year} + \beta_3{}^*\text{roads} + \beta_4{}^*\text{roads}^*\text{year} + \beta_5{}^*\text{proximity}$$

Interpretation of this protein index as a stress indicator suggested that bears with more landscape change (*change*) and roads (*roads*) within their home range experienced higher levels of stress. Generally, linear access features (*roads*), canopy closure, edge contrast, and elevation were significantly associated with grizzly bear health when age-sex class were controlled. A stress gradient may be suggested with higher road density, lower canopy closure, and higher edge contrast. Variation in mean patch size was significantly related to reproductive variables (e.g., levels of testosterone). Certain of these gradients could potentially be explained by decreased survival rates in areas with higher road density shifting the distribution of ages to younger bears. For example, subadult males were more likely to have higher road densities within their home range areas; apparently, decreased survival rates in these areas potentially eliminated many of the adult males, compared to areas with a lower road density and higher protection status.

Management Implications

Management of bears and human activity in grizzly bear habitat requires an understanding of the changes in bear health that may be associated with landscape change. Analysis of reproductive rates can reveal a link between environment, health, and population demography. If only postchange effects are monitored, there may be few management options available. Instead, an analysis of stress may provide an early warning system that can suggest management approaches to achieve wildlife management goals before significant commitment occurs in the field.

This project investigated the relationships between landscape structure, human-caused landscape change, grizzly bear health and population status through combined use of remote-sensing technology, GPS, radio-telemetry, wildlife health evaluation, and molecular techniques in proteomics. The work has confirmed strong linkages exist between environment, long-term stress, and grizzly bear health, a finding which has significant utility and importance in the management of grizzly bear habitats in Alberta and are of global significance in wildlife and biodiversity management and research applications.

A significant management outcome from this long-term research effort was the creation of new *Grizzly Bear Conservation Areas* along the eastern slopes of Alberta in areas experiencing rapid landscape change (Nielsen et al. 2009; (http://srd.alberta.ca/fishwildlife/wildlifeinalberta/grizzlybearmanagement/pdf/GrizzlyRecovPlan_FINAL_March_31_2008.pdf web site accessed January 17, 2009).

CHAPTER 6

Remote Sensing in Biodiversity Monitoring

People rarely see the irony inherent in the idea of preserving wilderness. "Wilderness" cannot be defined objectively; it is as much a state of the mind as a description of nature.

—*Yi-Fu Tuan, 1974.*

A Wide Range of Examples and Case Studies

Biodiversity Reviews

A significant number of individual studies and reviews have been conducted over the past two decades to assess the contributions of remote sensing to biodiversity management and monitoring (Table 6.1). Such work has emphasized the diverse use of remote sensing in specific biodiversity projects, and has tended to focus on the important habitat mapping and monitoring aspects of this complex endeavor (Stoms and Estes 1993, Nagendra 2001). Recently, powerful spatial modeling approaches have been introduced, focused, for example, on mapping productivity and the linkages between productivity, spatial heterogeneity, and biodiversity (Turner et al. 2003, Muchoney 2008). Presently, mapping and modeling biodiversity, habitat, spatial heterogeneity, productivity, and special species-environment relationships, are appropriately considered separate, though linked, components within the overall framework of a *biodiversity monitoring system* (Franklin and Dickson 2001, Crosse 2005, Duro et al. 2007, Henry et al. 2008).

Typically, components within a monitoring system or individual monitoring projects commit to use remote sensing to contribute to biodiversity assessment because participants have concluded that scattered or systematic plots in which plant or animal data are collected are simply not able to provide a complete understanding of

Biodiversity Indicator of Interest	Location	Methodological Approach	Main Output and Result	Key References
Threatened fish species in streams, fish species diversity	Western Ghats, India	Satellite image classification, correlation model	Maps of stream habitat, environmental correlates	Sreekantha et al. 2007
Dynamic habitat index	Canada	Satellite image vegetation index	Estimates of ecosystem productivity	Duro et al. 2007
Endangered species: Sonoran pronghorn (*Anilocarpra americana sonoriensis*)	Western United States	High spatial detail imagery, radiotelemetry	Habitat suitability model, geostatistical model of habitat quality	Wallace et al. 2002, Wallace and Marsh 2005
Rare species: Mountain bongo (*Tragelaphus euryceros isaaci*)	Kenya	Multiscale satellite imagery, regression model, data fusion	Species prediction model, resource selection	Estes et al. 2008
Invasive species: Leafy spurge (*Euphorbia esula*)	Western United States	Satellite image classification, field-spectra, hyperspectral imagery	Detection maps	Everitt et al. 1996, Hunt et al. 2004, Lawrence et al. 2006
Plant species diversity	contiguous United States	Satellite vegetation index and ecosystem productivity models	Environmental correlates of plant and bird species richness	Waring et al. 2006, Coops et al. 2004
Plant species diversity	Inner Mongolia	Satellite vegetation index	Maps of plant species richness	John et al. 2008
Plant species diversity	Mount Heron, Israel	Multiple spatial resolution satellite vegetation indices, DEM	Prediction model of annual and perennial plant richness	Levin et al. 2007

Table 6.1 Biodiversity Studies Using Remote-Sensing Data

Biodiversity Indicator of Interest	Location	Methodological Approach	Main Output and Result	Key References
Forest canopy biodiversity	Hawaii	Hyperspectral imagery, regression model	Leaf chlorophyll, N, water; maps of species richness	Carlson et al. 2007
Bird species richness	Finland	Satellite image classification, landscape metrics	Agriculture/forest bird species richness patterns	Luoto et al. 2004
Bird species richness	New Mexico, USA	Digital orthophotography, texture analysis	Grassland bird species richness patterns	St. Louis et al. 2006
Bird species richness	Maryland, USA	LiDAR data analysis, vertical complexity	Forest and wetland bird species richness patterns	Goetz et al. 2008
Bird species richness	South Dakota, USA	LiDAR data analysis, foliage height diversity	Forest bird species richness patterns	Clawges et al. 2008
Reef fish species richness	Diego Garcia (British Indian Ocean Territory)	Satellite image analysis	Maps of reef habitat complexity	Purkis et al. 2008
Fragmentation	Barsey Rhododendron Sanctuary, Sikkim, India	Satellite image analysis	Maps of habitat change and disturbance index	Behara et al. 2005, Kushwala et al. 2005
Fragmentation	Andaman and Nicobar Islands, India	Satellite image analysis	Habitat disturbance, correlation models	Roy et al. 2005
Ecosystem productivity	Global	Satellite image analysis	Correlation models	Turner et al. 2003

TABLE 6.1 Biodiversity Studies Using Remote-Sensing Data (*Continued*)

biodiversity or change in biodiversity over time. For example, the total number of species present in even a relatively biodiversity-poor region is very large, and the interactions between those species and with the environment immense (Lee et al. 2005). There are few options available in monitoring such complex phenomena over any appreciable area or time period. Additionally, a small but increasing number of these species may be special in some important way (Thompson and Angelstam 1999); for example, a keystone or umbrella species, or perhaps a species-of-concern, at-risk, vulnerable, threatened, or endangered. Understanding even a small number of such species-environment interactions and habitat changes over multiple scales of interest using field data alone is a daunting, virtually impossible, task (Molleman et al. 2006). What a remote-sensing approach does offer is support for a monitoring system in which *the overall variation in biodiversity, habitat, and special species-environment relationships within the whole landscape may be considered at appropriate spatial and temporal scales.*

Biodiversity monitoring is best considered as an integrated part of, rather than separate from, environmental research and management applications (Canter 1993, Miller 1994, Dutton 2001). A strong rationale for the investments required to initiate and maintain biodiversity monitoring programs is support to *adaptive ecosystem management* and policy development. A strong link between monitoring and management can serve to counteract a general tendency for long-term monitoring to become detached over time from both research and policy (Lee et al. 2005). Such a detachment or separation may have significant consequences, including the threat of uncertainty or possibly a real reduction in the high-level support often required to maintain monitoring system investments. A second and related challenge to the integrity of a biodiversity monitoring program sometimes occurs, despite initial designs, when the results of monitoring are thought to have a modest or even no discernable impact on environmental policy and decision making (e.g., Kummer 1992, Nagendra and Gadgil 1999b, Phinn et al. 2003, LeDrew and Richardson 2006). The lack of effective feedback from monitoring programs to management may weaken the resolve to continue intensive and extensive monitoring efforts.

Perhaps the first and most influential review of the use of remote sensing in biodiversity monitoring focused principally on vegetation species richness as an indicator of biodiversity, and outlined selected broad areas of research activity thought necessary to ensure effective remote-sensing contributions to emerging monitoring programs (Stoms and Estes 1993):

- Mapping and inventory of patterns of species richness, resource quality and quantity, and spatial heterogeneity [and species (or community) assemblages—*see also* Nagendra and Gadgil 1999a,b]
- Modeling species-habitat relationships and determining biophysical controls of species richness patterns and other

indicators of biodiversity (including direct sensing or mapping of individual plants and animals—*see also* Nagendra 2001, Leyequien et al. 2007)

- Mapping and modeling the effects of anthropogenic processes and environmental change (and understanding and quantifying energy/productivity relations—*see also* Turner et al. 2003, Phillips et al. 2008)

As earlier discussion in this book shows remote-sensing monitoring approaches in terrestrial ecosystems have long sought an *operational status* in many of these critical areas, but a dispassionate view would suggest that, though enormous progress has been achieved, this work is far from complete. At present, much remote-sensing biodiversity work might be considered *quasi operational* (Andréfouët et al. 2008). However, moving to fully operational remote sensing is considered of the utmost importance in establishing a baseline understanding of terrestrial and marine biodiversity, and landscape and waterscape monitoring and management (Crosse 2005, Strand et al. 2007, Muchoney 2008).

What is needed is a concerted effort to employ the remote-sensing approach and ensure continued added-value in biodiversity and wildlife applications of remote sensing. Such efforts must strengthen the important progress in biodiversity monitoring using remote sensing, evident in many types of specific habitat assessment (Kushwaha 2002) and species distribution prediction modeling approaches (Guisan and Thuiller 2005, Bradley and Fleishman 2008), for example, discussed in the preceding chapters of this book. An exhaustive review of the state-of-the-art confirms that linking remote sensing and other geospatial data to terrestrial or marine habitat information, or general biophysical conditions and processes, and relating these to biodiversity, is a powerful *conceptual* approach to identify more accurately the ways in which the environmental status of the waterscape- or landscape-influences and determines biodiversity (Innes and Koch 1998, Nagendra 2001, Turner et al. 2003). This critical and urgent work has only just begun.

The complexity of developing a comprehensive biodiversity monitoring system or network involves three related issues or trends, which are outlined here with reference to marine biodiversity monitoring:

1. *Habitat mapping.* Coral reef environments and habitat are among the world's most diverse and threatened marine ecosystems (Knudby et al. 2007). Remote sensing of coral reef macro- and micro-habitat variables, such as seagrass cover and live coral on reefs, is widely considered an essential part of an effective approach to understanding marine biodiversity (Gower 2006).
2. *Data and modeling approaches.* Many individual remote-sensing studies of marine habitat, such as those identifying seagrass

spectral characteristics, have emphasized that the near-future interdisciplinary challenge lies in global systematization of the data acquisition and processing protocols, archiving, and cataloguing and modeling approaches (Andréfouët et al. 2008).

3. *Technological innovation.* Monitoring systems must consider continued or even accelerating advancements in technology. Recent progress is partially a result of accelerating improvements in remote-sensing capability (e.g., spatial and temporal resolution improvements), impressive advances in modeling functionality, and the development of marine biodiversity indicators, which require significant focus and investment (e.g., *OzEstuaries* 2003).

The role of *biodiversity indicators* is a crucial one to consider in development of waterscape and landscape biodiversity monitoring systems (Strand et al. 2007). Much experience has been gained through empirical remote-sensing biodiversity studies in marine (Newman et al. 2006), grasslands, rangelands, and forest environments (Warren and Collins 2007). Virtually all of these studies were based principally on the idea of developing land cover or habitat maps and spectral indices at the landscape scale from satellite or aerial optical remote sensing, and then relating these maps in simple regression models to field observations of species richness, variability, or other similar species-level indicator. Although an emphasis on species richness indicators has clearly developed for practical reasons, significantly, *managing* for biodiversity is understood to require an understanding of processes and relationships, at multiple spatio-temporal scales (Noss 1990, Hansen et al. 1993, Franklin and Dickson 2001, Robinson 2008). Management approaches typically require a well-understood process of identifying clear objectives relative to biodiversity and remote sensing, and will typically involve setting certain target levels of species richness or abundance to be monitored.

Species richness monitoring, and the biodiversity indicator approach in general, are sometimes judged inadequate if the purposes for monitoring are not well-defined. *Indicator design criteria* can be decisive in determining the success of monitoring systems—but the question of *purpose* is a critical one. Biodiversity monitoring to detect vague or poorly understood conditions or trends will be difficult to implement, let alone maintain, over the long time periods and spatial scales necessary to establish credibility (Coops et al. 2008c). Experience suggests that a balance must be achieved among the agreed purpose(s), sampling scheme, data collection and analytical steps; otherwise, there is a high probability that little or no data analysis and interpretation will be performed despite the investment of substantial resources. Additional issues to consider, in development of the biodiversity indicator monitoring approach, will include:

- *Indicator selection.* Multiple measures are typically necessary in addition to species richness—a selection of biodiversity measures, from the large number of broadly based suite of indicators that can be quantified and monitored over time by remote sensing, will need to be carefully considered (Strand et al. 2007). Selection of too few indicators (e.g., only species richness with no attention to the status of invasive, threatened, or endangered species), or the misplaced adherence to a single biodiversity "index" to value biodiversity, measure achievement, or monitor status after management treatments, will jeopardize the system. Such reliance on single or simplistic indicators appears to have failed (or, at least, is not widely recommended) mainly because species, ecosystems and processes are neither commensurable (i.e., capable of being measured by a common standard) nor fungible (i.e., capable of mutual substitution) (Lee et al. 2005).
- *Indicator validation.* Newly constructed or proposed indicators should link with existing data, and reliance on untested technology without adequate backup or testing should be avoided. Indicators must be selected with scientific rigor. A substantial challenge to the integrity of monitoring systems has also been noted in the definition of *reference conditions* for biodiversity monitoring programs; this is a similar problem earlier discussed in studies to monitor or predict habitat integrity and change (e.g., Tiner 2004). Recently, one or two promising empirical approaches have been developed based on statistical ranges and comparison with observed occurrence and abundance or *biodiversity intactness indices* (e.g., Hessburg et al. 2004, Nielsen et al. 2007). Future progress in these initiatives is required to help alleviate this potential weakness in current design of biodiversity monitoring programs.
- *Scale.* The importance of scale and the need for biodiversity assessment at multiple spatio-temporal mapping scales and hierarchical levels (or ecological scales) has been frequently highlighted (Stoms 1994, Peterson and Parker 1998, Turner et al. 2003, Chen et al. 2006). In essence, this insight has developed because biodiversity indicators at one mapping scale may have limited applicability or predictive power at higher or lower levels of the ecological hierarchy. Remote sensing—an inherently multiscale observation approach—has much to offer to those interested in ensuring that monitoring across ecological scales is accomplished in ways that will facilitate more comprehensive understanding of biodiversity and change in biodiversity over time. However, great care in remote-sensing monitoring program design must be exercised. The

demands of a multiscale remote-sensing information system, with multiple platforms and sensors, could easily overwhelm even the most ambitious biodiversity monitoring data and information protocols (Franklin and Dickson 2001).

- *Collaboration.* Increasingly, commentary in the literature suggests that the quality of interaction between the diverse contributors (sometimes formalized as communities) to environmental science questions must improve (Clark et al. 2000, Tompkins 2005, Pohl 2005). In fact, there is evidence of a growing recognition that boundaries between scientific communities are essentially artificial and can be counterproductive (Lattuca 2001, Wilson 2006, Payton and Zoback 2007). Early gentle reminders of the limits of disciplinary, and some multidisciplinary and even interdisciplinary approaches, have grown to become a rising chorus of those urging more expansive approaches to complex environmental science research and management (e.g., Pintea et al. 2002, Xiao et al. 2007, Murakami et al. 2008).

Thoughtful and influential reflections by researchers and senior scientists, building on decades of disciplinary progress and increasing interdisciplinarity, highlight a growing passion for *consilience* (Wilson 1999) and *transdisciplinarity* (Naveh 2007). Environmental philosophy has begun to integrate these passions with earlier development of philosophical and epistemological concepts such as an "environmental ethic" and "reverence for life" (Sarkar 2005, Evans 2005). The vast literatures on disciplinary history and development are an essential resource to understand the dynamic and evolving "map of the intellectual landscape" (Gregory 1994). For those traversing this landscape and seeking new insights in search of solutions, there are some interesting waypoints previously identified or philosophically erected to provide guidance along the way:

- *Biology.* "Leading researchers are coming increasingly to agree that the future of biology depends on interdisciplinary studies within and beyond biology" (Wilson 2006).
- *Human geography.* "Modern human geography has been defined through a series of strategic encounters with anthropology in the eighteenth century, sociology at the turn of the nineteenth and twentieth centuries, and economics in the middle of the twentieth" (Gregory 1994).
- *Ecology.* "An understanding of ecology does not necessarily originate in courses bearing ecological labels; it is quite as likely to be labeled geography, botany, agronomy, history, or economics" (Leopold 1949).
- *Physical geography.* "It would be good if we could again approach the earth with unhampered curiosity and attempt

to satisfy that curiosity by whatever means the problems we encounter suggest" (Leighly 1955).

- *Environmental public policy*. "Regardless of the subject—biodiversity, genetics, endangered species, population growth, oceans, fisheries, forests, parks and reserves, the atmosphere, outer space, or land- and water-use—every discipline and standpoint has something to contribute in the examination of problems and solutions. Learning how to integrate and synthesize specialized knowledge and contributions is central" (Clark et al. 2000).

The evidence for a new and growing *transdisciplinary* understanding in natural sciences, and specifically in remote sensing and biodiversity work, can be discerned increasingly in the curricula of environmental studies/science training programs (e.g., Kalluri et al. 2003) and in graduate education across the university (e.g., Clark et al. 2000). Such learning environments increasingly emphasize the role of leadership, team dynamics, and integration. These trends are coupled with continued innovation in interdisciplinary and transdisciplinary undergraduate programming in the areas of conservation, environmental studies and environmental science, and remote sensing (e.g., Estes and Foresman 1996, and *International Center for Remote Sensing Education* http://www.icrsed.org/proj.html web site accessed January 21, 2009).

Many observers have highlighted the need to increase connections and pay attention to multidisciplinary, interdisciplinary, and transdisciplinary team dynamics in biodiversity work (Innis and Koch 1998, Ramanujan 2004, Warren and Collins 2007). The underlying theme is the growing understanding that *working across disciplines* is essential to improve the quality of biodiversity and conservation science and management. Joint efforts must be carefully nurtured and consciously managed. Over time it is likely that the obstacles presented by the continued existence of disciplinary-based structures will become more obvious, and the collaborative situation may improve as understanding increases about what is needed to monitor biodiversity successfully. Such a pattern is visible in many areas of research and practice requiring the development of a common vision with collaborative approaches to some of society's most significant and long-standing problems (e.g., understanding the link between environment, public health, and public policy) (Brewer 1988, Clark et al. 2000, Smith and Carey 2007, Hirsch-Hadden et al. 2008). The importance of establishing true collaborations of understanding and implementing a shared vision in the interdisciplinary and transdisciplinary approach to biodiversity monitoring and management, discussed earlier in this book, should not be underestimated.

Remote Sensing of Freshwater Fish Biodiversity

A successful approach to monitoring multiscale biodiversity in the Western Ghats, India, linked landscape-scale remote-sensing information to species richness information (Sreekantha et al. 2007). This environment is one of the world's *biodiversity hotspots* (Myers et al. 2000) with over 4000 species of flowering plants, 330 butterflies, 289 fishes, 135 amphibians, 156 reptiles, 508 birds, and 120 mammals (Daniels 2003). The Western Ghats form an important watershed with a number of catchments known to have experienced significant hydrological, land use, and land-cover change in recent decades, and with several known introduced and invasive fish species expected to impact the local fish fauna. The spatial extent of the Western Ghats and the pervasive changes suggested a remote-sensing approach linked with detailed field sampling protocols would be an effective strategy to biodiversity monitoring.

At the broad regional scale, multispectral IRS 1C imagery were used. These image data have a spatial resolution of 23.5 m. A maximum likelihood decision rule classified six broad land-cover categories (e.g., evergreen forest, deciduous forest, plantations, agricultural land, urban infrastructure). Certain of these forest areas had long been considered sacred forests (or *kans*), traditionally sustaining more endemic tree species, but recent forest management practices had resulted in some changes, in addition to fuel wood collection and cattle grazing, even in these forests. Once the broad land-cover map was created, detailed mapping of the areas surrounding fish sampling sites in streams was conducted using high spatial detail satellite imagery, available aerial photography, and field survey techniques (Sreekantha et al. 2007). Fish sampling sites were located within areas identified with differing land-cover characteristics, and then sampling with nets and lines occurring at locations in streams and reservoirs based on a stratified random sample of sites with representation of reservoir and stream microhabitats (riffles, pools, cascades, backwater, etc.). Vegetation sampling in the vicinity of these water sample locations was based on *in situ* transects in the catchment areas.

A comparison of endemism and threatened freshwater species revealed a high percentage of threatened species (16 of 64), and lower numbers of exotic species in streams compared to reservoirs at these sites. Areas with greater natural vegetation cover with higher levels of plant endemism were positively correlated with fish species richness, fish species endemism, and the presence of fewer endangered fish species. Streams whose catchments were dominated by deciduous forest, agricultural lands, or otherwise degraded vegetation cover with a low degree of plant endemism, had few endemic fish species with wider distribution ranges; such areas had declining fish species richness, and a higher number of endangered or threatened species.

The overall interpretation of the detailed survey data and relationships to species information suggested that preserving and restoring natural evergreen forests in this moist forest ecoregion was vital to maintain and improve stream perenniality and fish species richness in the Western Ghats catchment:

> ...this study highlights that endangered and endemic fish species are precariously clinging onto stream habitats where patches of primeval forests, though degraded substantially, still persist
>
> *(Sreekantha et al. 2007: p. 1602)*

This fish species richness analysis of streams and reservoirs is one component of a national initiative to monitor regional biodiversity in the whole of the Western Ghats and the west coast of southern India, comprised of a network of regular monitoring stations (2–3 year intervals), field transect surveys by volunteers (typically college teachers and students), and multiscale remote sensing and GIS database development (Nagendra and Utkarsh 2003; http://www.gisdevelopment.net/application/environment/conservation/becp0001a.htm web site accessed January 18, 2009). This work recognizes the importance of transdisciplinarity by linking economic and ecological information in a common vision to understand overall environmental impacts.

The monitoring goals of this network were developed with earlier initiatives to use remote sensing, GIS, and landscape ecology in the analysis of land cover and land-use change, estimation of deforestation rates and rates of forest fragmentation, examination of the spatial correlates of forest loss and socioeconomic drivers of land cover and land-use change, climate change and changes in the distribution of biodiversity, biomass estimation, gap analysis of the effectiveness of the protected area network, and conservation planning in the Western Ghats (Menon and Bawa 1997).

Biodiversity Indicators

Trends in Habitat and Ecosystem Extent

Land-cover conditions, and ecosystem extent and integrity, are considered necessary inputs to provide indicators of biodiversity based on assumed or modeled area relationships and trends over time (Coops et al. 2007). Land-cover biodiversity indicators can be simple—for example, *area of a particular land cover or habitat over time*—or complex, comprising of multiple variables in an index or function (Strand et al. 2007). Since most such indicators represent a defined and measurable land attribute or condition, there are of course an enormous number of possible indicators; recall the discussions in earlier chapters of this book (*see also* Strand et al. 2007), outlining the wide assortment of biome-specific indicators that have previously been studied or recommended. However, a coarse-filter approach to selection of indicators will typically begin by testing the assumption that multiple variables of interest in biodiversity monitoring are highly related to a few simple measures, such as land cover, that can be readily determined using remote-sensing data sets (Franklin and Dickson 2001).

Four such simple *coarse-filter biodiversity indicators* were recommended by Duro et al. (2007) in the development of BioSpace, a proposed Canada-wide national biodiversity monitoring program (Color Plate 6.1). Measures of productivity, disturbance, topography, and land cover were thought to integrate the climatic and structural response of vegetation and faunal species, and, when monitored over time across large areas (e.g., over decades and all of Canada), provide a unified scheme or foundation for biodiversity monitoring that would capture overall national trends and highlight issues within changing regional landscapes. Implementation of the BioSpace approach in Canada occurs at the strategic level and is used to identify, nationally, broad changes in the underlying landscape factors that are believed to strongly influence biodiversity. Insights into areas that require more detailed investigations are suggested in the identified broad trends, which when validated across large areas (e.g., Waring et al. 2006) serve to substantiate the overall approach (Ahern 2007, Coops et al. 2008a).

The BioSpace approach is briefly reviewed here, and in later sections of this chapter additional monitoring program initiatives are discussed in more detail. The overall interpretation is based on the highlights from a large number and wide variety of studies and biodiversity monitoring programs in different environmental settings.

BioSpace Productivity

A *dynamic habitat index* summarized landscape productivity to provide an environmental correlate to biodiversity at the national scale (Coops et al. 2008b). The habitat index was based on separate measures of fPAR (fraction of photosynthetically active radiation) derived from 16-day or monthly MODIS composites at 1 km spatial resolution. One successful strategy to use such composite data considers the observed annual and monthly variations in MODIS fPAR. These MODIS-based productivity variations were used by Mackey et al. (2004) in Australia to estimate biomass partitioning and availability of food and other habitat resources for animals; BioSpace modified this approach for the Canadian landscape by comparing the monthly variations to longer-term fluctuations observed in the preceding years (since 1999, when MODIS data were first acquired routinely).

BioSpace Disturbance

The disturbance patterns of interest in biodiversity monitoring over the whole Canadian landscape were assessed using MODIS vegetation indices following the image processing procedures recommended by Mildrexler et al. (2007). Again, the BioSpace approach considered comparisons of the long-term (since 1999) variations to monthly and annual variations contained in the indices; major disturbance events, such as forest fires, insect damage, flooding, and land conversion, could be identified in these variations with a series of thresholds specific to Canada's ecozones developed following tests of index sensitivity (Coops et al. 2008c).

BioSpace Topography

Available Shuttle Radar Topographic Mission (SRTM) data were used to develop 90 m spatial resolution topographic data over the Canadian landmass. Such data were useful in correcting the MODIS image data for topographic and view angle effects, but were also considered broadly predictive of biodiversity variables when compiled in various topographic indices (such as terrain roughness). The BioSpace approach assumed that large-scale topographic variability can synthesize climatic variables across gradients that influence biodiversity.

BioSpace Land Cover

The vast majority of Canada had been mapped using Landsat ETM+ imagery to circa 2005 land-cover conditions in a nation-wide land-cover mapping program, which preceded the BioSpace program, called Earth Observation for Sustainable Development (EOSD) (Wulder et al. 2003). The initial EOSD land-cover map was produced by unsupervised classification (hyperspectral clustering) and labeling with a total of 23 land-cover classes, specifically tuned to ecozone legends developed locally. The BioSpace land-cover map product was a resampling of this EOSD map to 1 km spatial resolution for broad cover types and landscape structure metrics (e.g., patch shape and size). These land-cover patterns were thought to directly influence major ecological dynamics and processes related to biodiversity.

Threatened Species Habitat

Changes to special species habitat—for example, endangered or threatened species habitat—is widely accepted to be an indicator of biodiversity, because such habitat is a strong indicator of species abundance and health. Focusing on special species habitat, such as threatened or endangered species habitat, is a recognition of the importance of certain species for various values, including biodiversity; all species are not equally important and interchangeable, as might sometimes be assumed in roll-up measures or indicators of species richness, or other broadly based species indicators (Burbridge 2004, Lee et al. 2005).

Remote sensing is an important approach to monitoring habitat, as has been shown in earlier sections of this book, in which many mapping and monitoring habitat issues have already been discussed from the remote-sensing perspective. There are some additional challenges that may more frequently arise when dealing with the habitat conditions and preferences of a special species as part of a biodiversity monitoring project.

Species Habitat Preferences

Detailed threatened or endangered species habitat-monitoring approach is optimal where the threatened or endangered species habitat preferences are known and are amenable to remote sensing, but there are many situations in which such conditions are not in place. In one instance in Kenya, Estes et al. (2008) used multiscale remote-sensing image (SPOT, ASTER and MODIS) with sparse field data to develop habitat maps for the rare mountain bongo antelope (*Tragelaphus euryceros isaaci*). Almost nothing was known about bongo habitat use, but important microhabitat features, such as canopy and understory structure, were identified through logistic regression models of vegetation structure data collected in plots (0.04 ha) associated with bongo presence ($n = 36$) and absence ($n = 90$). Spectral mixture analysis and image texture analysis of the remote-sensing data identified the broad macrohabitat conditions selected by bongo. This input enabled maps of these conditions over large areas to be constructed to form the basis of a microhabitat bongo distribution prediction model that was finely tuned to resource selection characteristics obtained by field observation.

Substitution of Species Habitat Preferences

In other situations, an analogue might be found in studies of a similar or related, though more plentiful, species; one such innovative study in Arizona of the endangered Sonoron pronghorn (*Anilocarpra americana sonoriensis*) shows some of the possibilities in the development of remote sensing for critical threatened species habitat monitoring applications in which alternatives are sought for species preferences.

The habitat requirements of the endangered Sonoran pronghorn are not well known, but are thought to include distinct landscape preferences related to vegetation type and structure associated with summer and winter differences in forage quality, cover/exposure, and predator avoidance behavior (Wallace and Marsh 2005). Some of these preferences may be similar to those displayed by more numerous, and better-known, related pronghorn species located in adjacent environments throughout the southwest United States.

In New Mexico, the general pronghorn landscape preferences appeared to suggest that spatial heterogeneity and landscape productivity influenced the population viability of pronghorn. This insight was incorporated into the preparation of a habitat suitability model for the endangered Sonoran pronghorn. Separating spatial heterogeneity at the scale of pronghorn habitat use required analysis of pronghorn sightings, high spatial detail imagery, and texture image processing techniques based on geostatistical semivariograms (Wallace and Marsh 2005):

- Pronghorn sightings were obtained of radio-collared animals by field personnel from fixed-wing aircraft.
- Digital orthophotos and IKONOS imagery (1 m spatial resolution panchromatic and 4 m spatial resolution multispectral) were acquired, georegistered, and processed for vegetation indices.
- Local semivariance was computed for these imagery (texture analysis) at each pronghorn sighting location, with a window size of 25 × 25 pixels and over multiple lag distances greater than the expected autocorrelation in this landscape.
- Geostatistical measures of range, nugget, sill, and slope were correlated with pronghorn sightings and a sample of random locations, to create a map of habitat suitability (expressed as landscape preferences).

The mapping process revealed that areas of high habitat suitability for the endangered Sonoran pronghorn were found in low relief terrain with characteristic vegetation structure and spatial heterogeneity that could be mapped with aerial remote-sensing imagery.

The habitat suitability map for endangered Sonoran pronghorn suggested that pronghorn occupy more structurally homogeneous landscapes compared to random landscapes, and landscapes without pronghorn sightings. Using the more general pronghorn preferences as indicative of the endangered species of pronghorn preferences resulted in better maps and models of habitat suitability for the endangered species than relying on methods that did not consider any information on environment-species relationships (such as simple coarse-filter land-cover mapping).

The geostatistical measures derived from high spatial detail remote-sensing imagery comprised a sensitive suite of indicators that were able to predict potential locations of an endangered species in a large and diverse landscape. Validation of the usefulness of model based on the general species preferences relied on relatively few samples and anecdotal evidence rather than a rigorous test. The approach was recommended as a foundation to monitor changes in landscape structure (e.g., reduced shrub cover) and fragmentation in future (Wallace et al. 2002).

These threatened or special species habitat monitoring and species distribution prediction projects highlight the importance of habitat in biodiversity monitoring, and represent a growing awareness and understanding of the importance of special species habitat mapping initiatives in biodiversity monitoring systems.

Trends in Invasive Alien Species

Successful remote sensing of invasive alien species is based on detection of biochemical, phenological, or structural characteristics or modifications of landscapes associated with a change in ecosystem functioning (Inoue et al. 2000, Stow et al. 2000, Underwood and Ustin 2007). Mapping of invasive animal species includes some high profile remote-sensing success stories, including mapping of forest-insect damage (Hall et al. 2007), and feral animal impacts (Simberloff et al. 2005). However, mapping other animal invasion phenomena typically are less commonly attempted with remote-sensing approaches than mapping plant invasions. A few examples of these plant invasive alien species mapping projects are reviewed briefly here to supplement those highlighted in Chapter 5 of this book.

A summary of some of the many invasive plant species that can be identified by *distinctive flowering* or other coloring, and selected remote-sensing approaches used in mapping these invasive plant species, is provided in Table 6.2 (*see* Underwood and Ustin 2007).

Species and Setting	Unique Feature	Imagery or Approach
Broom snakeweed (*Gutuierrezia sarothrae*)	Phenology	AVHRR
Zebra mussels (*Dreissena polymorpha*)	Turbidity	AVHRR
Water hyacinth (*Eichornia crassipies*)	Greenness and emergent properties	SPOT imagery, aerial hyperspectral
Jubata grass (*Cortaderia jubata*)	Cellulose and lignin	Aerial hyperspectral
Fire tree (*Myrica faya*)	Nitrogen and water content	Aerial hyperspectral
Salt cedar (*Tamarisk* spp.)	Habitat suitability criteria	MODIS NDVI
Locust infestation (*Chortoicetes terminifera*)	Soil type and condition, breeding sites	Landsat

Source: With permission from Underwood, E., and S. Ustin. 2007. In Strand, H., et al., eds, *Sourcebook on Remote Sensing and Biodiversity Indicators*. Technical Series No. 32. Secretariate of the Convention on Biological Diversity, Montreal. 163–179.

Table 6.2 Examples of Invasive Species Mapped with Remote Sensing

A general paucity of good invasive species mapping examples exists, but three recent remote-sensing success stories that identified distinct flowering phenomena include:

- *Mapping silver wattle (Acacia dealbata) in Chile.* In this environment, wattle is an invasive plant closely associated with disturbances (such as roads). A successful remote-sensing approach identified silver wattle using color aerial photography based on the conspicuous flowering during the winter season. A distinctive yellow pattern near roads was relatively easy to interpret manually or digitally when the photography was acquired when the plants produce peak yellow flower patterns (Pauchard and Maheu-Giroux 2007).
- *Mapping perennial pepperweed (Lepidium latifolium) in western United States.* A native of Eurasia found in wetlands and recently disturbed sites; pepperweed was accidentally introduced into the United States in the 1930s as containment in seed. Flowering occurs in late spring and summer; flat, dense clusters of white flowers develop at the apex of the flowering stem. The white flowers were accurately discriminated on California roadsides and rangelands using hyperspectral digital image analysis (Andrew and Ustin 2006).
- *Mapping leafy spurge (Euphorbia esula) in midwestern United States.* Leafy spurge is an adventive, perennial weed that infests over one million hectares of land in North America. Early work had established that leafy spurge negatively influenced biodiversity in grassland ecosystems (Scheiman et al. 2003), and remote-sensing solutions to detection and mapping were needed:
    - The potential of aerial photography and videography in mapping the distinctive yellow-green color of leafy spurge flower bracts was quickly identified (Francis et al. 1979). The flower bracts were shown to be spectrally unique when compared to co-occurring green vegetation in laboratory and field conditions (Everitt et al. 1995).
    - Detailed *in situ* spectroradiometer analysis by Hunt et al. (2004) compared leafy spurge leaf and flower bract spectral response patterns. The bracts had lower reflectance at blue wavelengths (400 to 500 nm), greater reflectance at green, yellow, and orange wavelengths (525 to 650 nm), and approximately equal reflectances at 680 nm (red) and near-infrared wavelengths (725 to 850 nm).
    - The biochemical characteristics of leafy spurge were examined; pigments from leaves and flower bracts were extracted in dimethyl sulfoxide, and the pigment concentrations were determined spectrophotometrically (Hunt et al. 2004).

> Carotenoid pigments were identified using high-performance liquid chromatography. Flower bracts had less chlorophyll *a*, less chlorophyll *b*, and less total carotenoids than leaves. These chemical differences suggested that absorptance by the flower bracts would be less, and reflectance greater at blue and red wavelengths. The carotenoid to chlorophyll ratio of the flower bracts was approximately 1:1, explaining the hue of the flower bracts, but not the value of reflectance. The primary carotenoids were lutein, β-carotene and β-cryptoxanthin in a 3.7:1.5:1 ratio for flower bracts and in a 4.8:1.3:1 ratio for leaves, respectively. There was 10.2 $\mu g\ g^{-1}$ fresh weight of colorless phytofluene present in the flower bracts and none in the leaves. The fluorescence spectrum indicated high blue, red, and far-red emission for leaves compared with flower bracts. Fluorescent emissions from leaves may have contributed to the higher apparent leaf reflectance in the blue and red wavelength regions.

The spectral characteristics of leafy spurge leaves and flowers associated with chemical constituents confirmed the cause of the general patterns visible in color-aerial photography and aerial videographic data. The next step was to develop a linkage to aerial and satellite remote-sensing data to serve as a long-term and large area leafy spurge mapping system:

- First, the spectral characteristics were used to construct a spectral library (endmembers) for use with aerial hyperspectral digital remote-sensing imagery (Hunt 2004). AVIRIS imagery were acquired in northeastern Wyoming; the results of the spectral mixture analysis with these endmembers detected leafy spurge canopy cover across broad areas imaged by the hyperspectral sensor. The resulting model predictions of leafy spurge canopy cover by aerial remote sensing had an R^2 of 0.69, and leafy spurge presence/absence was estimated with an overall accuracy of 95 percent. This level of accuracy for leafy spurge detection was also obtained in an independent leafy spurge hyperspectral aerial image classification by Lawrence et al. (2006).
- Second, mapping leafy spurge with aerial photography and hyperspectral imagery formed the basis for a regional- or landscape-level mapping approach based on Landsat and SPOT satellite imagery (Williams and Hunt 2004). Using a vegetation index approach, Landsat and SPOT imagery acquired at approximately the same time as the hyperspectral aerial imagery were found to have a maximum classification accuracy of 66 percent using relatively straightforward supervised classification methods. Higher classification accuracies may be achieved with a more sophisticated classification approach (for example, using more complex decision rules), as earlier discussed in this book.

Overall, the key output of this multiscale, multispectral, hyperspectral satellite, aerial and field-based survey approach provided an invasive alien species assessment for use in biodiversity monitoring and in management applications.

Patterns of Biodiversity

Plant Species Richness

Empirical remote-sensing models of plant species richness have continued to improve and now are used frequently at multiple scales in a variety of environmental conditions. Such studies relating spectral patterns to plant species patterns are sometimes referred to as specific instances of the *spectral variation hypothesis* (SVH) (Rocchini 2007). In essence, the central idea is to use spectral response patterns in aerial or satellite images to predict species richness, controlled by ecological scales of investigation, and different spatial and spectral resolutions (Fig. 6.1, *see* Rocchini et al. 2004, Warren and Collins 2007).

For large areas, imagery from satellites such as MODIS have been used in prediction of plant species diversity across regions as diverse as the contiguous United States (Waring et al. 2006) and Inner

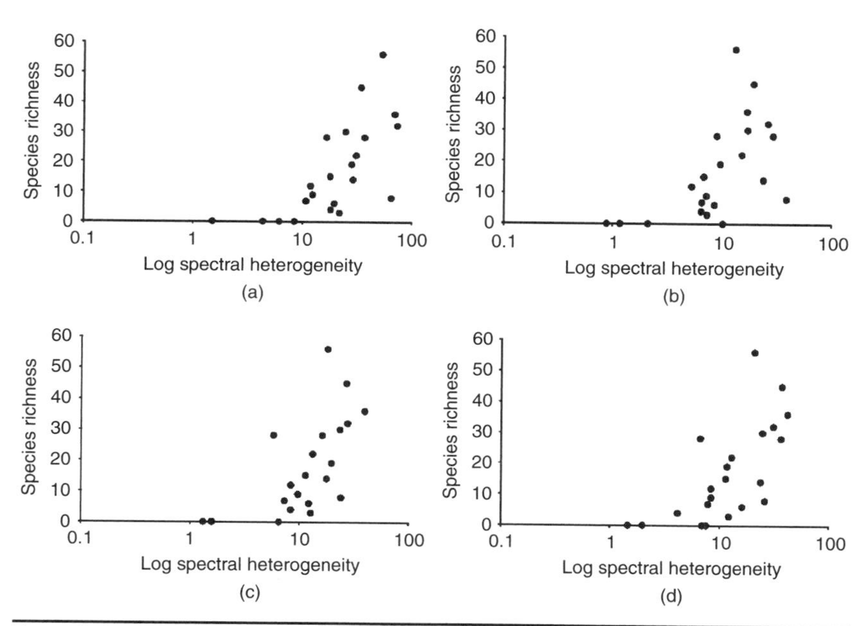

Figure 6.1 Scatterplot of species richness vs. spectral heterogeneity measured as mean distance from spectral centroid in different resolution satellite imagery: (a) QuickBird; (b) Aster; (c) Landsat ETM+; and (d) resampled Landsat ETM+. (*With permission from* Rocchini, D. 2007. *Remote Sensing of Environment* 111:423–434.)

Mongolia (John et al. 2008). Over smaller areas, medium resolution (e.g., Landsat data) and high spatial resolution imagery have also been tested and, typically, predictive relations are found which vary significantly, but predictably, in form and strength (Levin et al. 2007). Two promising approaches to develop the relationships between satellite-derived spectral response patterns and species richness are based on the many observations of a strong correlation between remotely sensed data and estimates of productivity (a reflection of energy relations) and spatial heterogeneity. A typical approach is recommended in linking species-area curves to models of habitat spatial heterogeneity to improve prediction of overall biodiversity patterns (Botkin et al. 2007). Understanding these relationships has greatly enhanced knowledge of biodiversity patterns and effective biodiversity management at all relevant scales (John et al. 2008).

A multiscale landscape analysis at Mount Hermon, Israel, a biodiversity hotspot, illustrates the general approach and outcome using field data and multiscale optical satellite imagery (Levin et al. 2007). Landsat TM, ASTER and QuickBirddata were acquired over this region over several seasons in a nested sample design. The imagery were georadiometrically corrected and used to generate texture variables, NDVI, and other vegetation indices:

- Thirty-four 0.1 ha quadrats were surveyed along an elevation gradient from 300 to 2200 m in particular topographic settings away from roads and other infrastructure. Total, perennial, and annual vascular plant species richness were calculated along with species indicators such as range size rarity (calculated as the sum of the inverse of the range sizes of all the species occurring in each quadrat).
- A strong relationship was found between image NDVI and plant species richness at Mount Hermon. Figure 6.2 shows the relationship between species richness and Landsat NDVI, which was strongly determined by the elevation gradient.

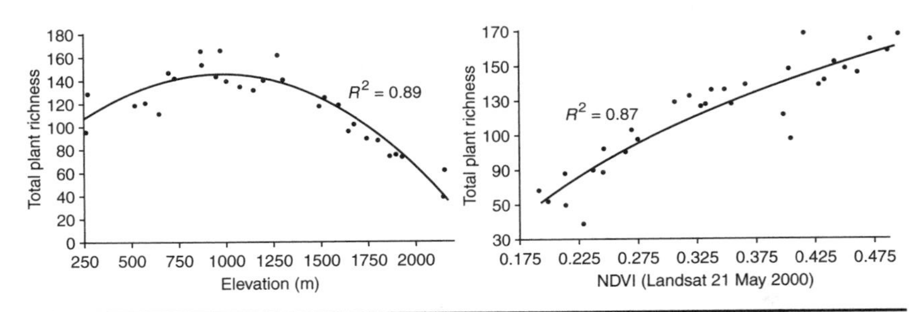

FIGURE 6.2 Plant species richness relationships to elevation and Landsat NDVI at Mount Hermon. (*With permission from* Levin et al. 2007. *Diversity and Distributions* 13:692–703.)

NDVI explained up to 87 percent of the variation in total plant species richness in this environment, and this relationship was even stronger (by an average of 4 percent) when only perennial plant species richness was examined. NDVI was negatively correlated with range size rarity (as was plant species richness). The spatial variation of spectral response within each quadrat as measured by QuickBird NDVI and texture measures were significantly correlated with species richness ($R^2 = 0.74$).

- Vegetation productivity (as represented by NDVI measures), rather than habitat variability (as represented by texture or spatial heterogeneity), appeared to be a more important factor in explaining species richness and rarity in this mountain environment. Higher levels of rarity and endemism were found at higher elevations, and these observed relationships could be understood with reference to current ecological understanding of energy relations, gradients, productivity, and climatic influences. The study showed differences in relations between NDVI and annual versus perennial plant species richness, but this may have been a consequence of sample selection (fewer sites with higher tree cover).

The Mount Herman study concluded with a confirmation of the value of remote sensing "...in evaluating and predicting richness and rarity patterns in mountain areas at regional and global scales, which is an important component in biodiversity estimation" (Levin et al. 2007: p. 701). Such a conclusion has been found in many other landscape-level studies of species richness, which have confirmed the approach. In addition, these efforts have highlighted the importance of explaining the observed correlation with an understanding of the underlying ecological processes independent of the particular characteristics of the ecosystem. For example:

- In Burkina Faso, Landsat-derived vegetation indices (to represent landscape productivity), and Shuttle radar topographic data (to represent landscape spatial heterogeneity), improved species richness predictions compared to predictions based on climate data alone, and were useful in understanding the relationships between plant diversity, soils, and water availability (Schmidt et al. 2008).
- In Saskatchewan boreal forests, mixedwood boreal forest species richness was predicted with Landsat spectral response patterns (Warren and Collins 2007). A simple multivariate linear model used independent variables derived from remote sensing and digital elevation model data that were indicative of light, moisture, and nutrient conditions in the study area

(canopy species type and stem density, distance from the nearest ridgeline, time since the last fire). Statistically significant estimates of species richness at both the plot and stand scales were generated and validated with field observations.

- On Horn Island, Mississippi, in the northern Gulf of Mexico, a strong predictive relationship between high spatial resolution hyperspectral remotely sensed indicators of soil exposure and plant species richness was identified in mesic habitats (Lucas and Carter 2008).

The development of these insights into the vegetation productivity/spatial heterogeneity/species richness relationships over diverse environmental conditions (Mount Herman, Burkina Faso, Saskatchewan, Mississippi) is strengthened by the results of a hyperspectral remote-sensing study of canopy biodiversity in Hawaiian lowland rainforests (Carlson et al. 2007):

- Hyperspectral data spanning the 400 to 2500 nm wavelength range acquired by the NASA Airborne Visible and Infrared Imaging Spectrometer (AVIRIS) were analyzed at 17 forest sites with species richness values ranging from 1 to 17 species per 0.1 to 0.3 ha.
- Spatial variation (range) in the shape of the AVIRIS spectra (derivative reflectance) in wavelength regions associated with upper-canopy pigments, water, and nitrogen content were well correlated with species richness across the 17 field sites.
- An analysis of leaf chlorophyll, water, and nitrogen content within and across species suggested that increasing spectral diversity was linked to increasing species richness by way of increasing biochemical diversity.
- A linear regression analysis showed that species richness was predicted by a combination of four biochemically distinct wavelength observations centered at 530, 720, 1201, and 1523 nm ($r^2 = 0.85$, $p < 0.01$).

Relationships between plant chemistry, plant productivity, and species richness indicators can be used to map biodiversity; in lowland forest reserves throughout the Hawaiian study region, this mapping was accomplished at approximately 0.1 ha resolution (Carlson et al. 2007). One of the most important results of this study was to confirm that leaf-level variation in pigment concentrations, nitrogen and water concentration scaled predictably with the number of species analyzed. This finding forms an important connection between the hyperspectral reflectance-derivative data, vegetation productivity, spatial heterogeneity, and species richness.

Animal Species Richness

Animal species richness prediction using remote sensing has improved significantly as spatial resolution and modeling approaches have advanced in recent years. For instance, certain studies have successfully related animal species richness, in a variety of taxa, to remotely sensed measures of productivity (e.g., NDVI) and land-cover condition and variability (e.g., Catling and Coops 1999). Accuracy in predicting animal species richness typically improves as land-cover detail (and microhabitat definition) increases (Leyequein et al. 2007). Bird species richness has been frequently considered from a remote-sensing database of land cover or productivity measures (Seto et al. 2004). Three relevant animal species richness studies have recently been completed to illustrate the overall remote-sensing approach at the landscape scale. Such work illustrates the trend to increasingly finer detail in understanding and mapping these relationships than can be obtained by field methods alone.

Regional Scale Bird Species Richness in an Agricultural Landscape

There are many bird species richness studies at the landscape scale in forest environments, but a relative paucity of such work has been identified in key environments subject to long-term human disturbance, such as agricultural-forest mosaics (Luoto et al. 2004). Major cultural processes which influence biodiversity in such mosaics may be present, including:

1. Conversion of forest lands to agriculture
2. Implementation of changing agricultural practices, and the intensification and industrialization of agriculture, which can often lead to increased landscape fragmentation
3. The decline of agriculture activities in marginal areas, which may reduce open habitats and create different niche habitat

Overall, in recent decades, observers have recorded a steep decline in abundance and distribution of many bird species, hypothesized (and empirically) to be strongly linked to habitat fragmentation and habitat loss in wintering areas, and fatal pesticide use (Stutchbury 2007). The determinants of bird species richness in such complex environments are often too complex to be modeled using broad macrohabitat land-cover types and pattern alone; instead, other predictive variables that capture fine microhabitat variations, such as measures of landscape structure or texture, and spatial heterogeneity, may be required.

Luoto et al. (2004) set out to examine fragmenting landscape patterns relationships to bird species richness in the agriculture-forest mosaic of southwestern Finland. Land-cover mapping using available Landsat imagery identified different types of agriculture crops and pasture, a predominately natural or relatively undisturbed grassland cover, a wetland cover class, and forest cover represented by different age classes (e.g., older conifers, regeneration or plantation). Habitat structure variables (edge length, Shannon's diversity index, mean patch size, and largest patch index) were calculated using FRAGSTATS, and DEM variables related to topographic position and soil moisture indices were included. Bird occurrence data were assembled from field observations for use in a 600 km^2 grid spatial model.

Individual habitat relationships were identified; for example,

1. Mean patch size was negatively correlated to bird species richness.
2. The areas with highest biodiversity were the areas with the highest topographic variability, high variation in soil moisture, highest cover type variability, and low amounts of intensive agricultural land cover.

General linear models of observed bird species occurrence explained more than 66 percent of the variance in species richness, and when extrapolated to other areas, identified a number of bird species hotspots (river valleys) and coldspots (flat, intensively cultivated plains). Scale variability was highlighted as a potential problem to be considered in all such studies; biogeographical variation on the scale of large landscapes versus ecological variation, which should include consideration of landscape structure and composition, but also inter-specific interactions such as competition and predation.

Overall, the issue of species richness patterns operating at different scales than those on which they are observed was overcome (Luoto et al. 2004: p. 1954):

> A very auspicious feature in our study is that both ends of the gradient of richness of bird species richness are modeled reasonably well. This indicates that this kind of predictive modeling approach may be useful in a priori exclusion of the most uninteresting areas, and provides a cost-efficient procedure to complement bird surveys in the area under investigation.

This approach was recommended for use together with species distribution models, which can consider more detailed ecological habitat conditions that may be critical in a particular species occurrence. This finer level or scale of remote sensing to predict bird species richness has been accomplished elsewhere using two separate approaches based on image texture analysis and structural interpretations using LiDAR data. At these finer mapping scales, and increased image spatial resolutions over small areas, prediction of bird species richness using remote sensing will typically require consideration of several factors, including spatial heterogeneity (Goetz et al. 2007).

Using Different Types of Aerial Imagery to Predict Bird Species Richness

An emphasis has developed on situations in which focusing on a small patch or component of the landscape provides insufficient insight or explanatory power of bird biodiversity—what is important are such relationships when considered within the spatial variability of the landscape as a whole. For example:

- Biodiversity monitoring has successfully been based on high spatial resolution *image texture* derived from aerial multispectral imagery. Earlier studies had linked image texture to land cover and habitat heterogeneity and species diversity indicators; for example, spatial heterogeneity has been analyzed using image texture measures in aerial imagery (Coops and Catling 1997) and in satellite imagery (Hepinstall and Sader 1997) to assess biodiversity.
- This texture analysis approach was updated recently by St.-Louis et al. (2006) in a New Mexico grassland/herb/forest complex. First- and second-order texture measures based on spatial co-occurrence matrices were derived from digital orthophotos at 1 m spatial resolution and related to data of 24 bird species in over 100 plots; an overall adjusted $R^2 = 0.76$ was obtained. The best single texture measures (which increased from grassland to forest, as expected) accounted for 57 percent of the variation in species richness; this finding provided strong evidence for the high degree of accuracy with which texture measures could represent spatial heterogeneity in this environment. However, window sizes were not conclusively tested—this texture parameter might have to be related to species characteristics, such as territory sizes and other habitat features, such as food presence, habitat quality, and distance of escape cover.
- An even more detailed bird species richness and spatial heterogeneity analysis was conducted by Goetz et al. (2008) using LiDAR data acquired in Maryland forests. LiDAR metrics, such as canopy height and the vertical distribution of canopy (including understory), were used to build a model of *canopy structural diversity* to predict observed bird species richness. The best models accounted for approximately 45 percent of the variability in bird species richness, a much higher percentage than could be predicted using land-cover variables or two-dimensional spectral information alone. Accuracy of the models improved when the bird species richness data were stratified into guilds dominated by forest, scrub, suburban, and wetland species. Additional work—to identify the ability of LiDAR to detect conditions

preferred by multistage habitat specialists, for example—was recommended, confirming the approach recently implemented by Clawges et al. (2008) in a study of foliage height diversity and avian diversity, density, and occurrence in South Dakota forest environments.

Prediction of Fish Diversity in Coastal Reef Environments

A similar approach and outcome has been documented in studies of fish diversity and spatial heterogeneity in coastal reef environments using high spatial detail remote-sensing imagery. Generally, the diversity, abundance, and distribution of reef fish are related to the spatial heterogeneity and physical complexity of benthic habitat across reefscapes. To confirm these relationships, IKONOS imagery were used to map substratum type and seabed topography for a reef system in Diego Garcia (British Indian Ocean Territory) (Purkis et al. 2008). Replicate fish counts were made at seven measurement stations across the study area using visual census. Monte Carlo simulation revealed that species richness and abundance of several guilds and size groupings of reef fish appraised *in situ* were correlated with the satellite-derived seabed parameters over relatively large seafloor areas (5030 m^2). This work suggests that high spatial resolution satellite remote sensing is capable of predicting reef habitat complexity at a scale relevant to fish. Larger size classes of fish were better predicted with the satellite habitat complexity or texture data. New management applications to predict fish stocks, and identify potential sites for marine protected areas where intensive field surveys are not practical, were recommended.

Biodiversity and Fragmentation

The impact of fragmentation has been frequently studied in several environments that are of special concern for biodiversity and habitat loss. The general applicability of the relationship between habitat fragmentation and biodiversity has been noted in political environments as diverse as Asia (Behara et al. 2005), Australia (Saunders et al. 1998), and North America (Wang et al. 1998), and of course in virtually every kind of natural environment (Fahrig 2003, Strand et al. 2007). In India, for example, fragmentation and biodiversity relationships have been identified using remote-sensing approaches in two diverse ecosystems of interest for their conservation value:

1. *The Barsey Rhododendron Sanctuary in Sikkim (Kushwala et al. 2005).* This region supports 26 of the 36 reported Rhododendron species in this part of India, and is also the site of numerous rare fauna, such as the kaleej pheasant (*Lophura leucomelanos*), leopard and leopard cat (*Felix bengalensis*), red panda (*Aliurus fulgens*), and palm civet (*Paguma larvata*). Medium spatial

resolution IRS imagery were georadiometrically corrected and interpreted for different land cover types, (over 90 percent classification accuracy), and analyzed using an ArcInfo extension (customized Bio_CAP software package), which computes landscape metrics such as porosity, juxtaposition, and interspersion. A total of 70 field plots, with the size of the plot determined by species-area-curve analysis, were established for species identification and classification (endemism, vulnerability, rarity, economic importance, etc.). The analysis confirmed that the Sanctuary is relatively undisturbed, with low levels of fragmentation; however, within this environment, cover types associated with higher levels of fragmentation showed lower levels of species richness and biodiversity. A disturbance index was constructed to simplify the map interpretation, which revealed that areas in which cattle grazing had occurred were degraded and displayed low species richness.

2. *Andaman and Nicobar Islands (Roy et al. 2005).* Vegetation and land cover were mapped using IRS satellite data for comparison to plant species richness and other biodiversity measures acquired in the field. Biological richness was estimated as a function of six biodiversity attributes (i.e., spatial, phytosociological, social, physical, economical, and ecological), which identified forest vegetation in different phytogeographically distinct groups on the two tropical island groups—Andaman and Nicobar Islands. Fragmentation was expressed as a calculated *disturbance index* (weighted sum of porosity, patchiness, interspersion, and juxtaposition); the spatial variability in this index clearly indicated that Nicobar Islands land cover types were less disturbed compared to Andaman Islands. The plant communities of Andaman showed high plant diversity in terms of number of species (523 species) and more heterogeneity compared to those of Nicobar (347 species); Andaman Islands also recorded more threatened and rare species in a variety of covertypes (Table 6.3).

Location	Endemic Species	Rare or Threatened Species
Andaman Islands	182	69
Nicobar Islands	103	24

Source: Modified with permission from Roy, P. S., H. Padalia, N. Chauhan, M. C. Porwal, S Gupta, S. Biswas, and R. Jagdale. 2005. *Ecological Modeling* 185:349–369.

Table 6.3 The Distribution of Unique (endemic, rare, and threatened species) over All the Vegetation Types Mapped with IRS Data in Andaman and Nicobar Islands

> The effect of fragmentation in the Andaman Islands was reflected by the decrease in species diversity ($r^2 = 0.97$) and species richness ($r^2 = 0.82$). Such a comparative study under relatively controlled conditions is rare and valuable; the interpretation of these results can be expected to help guide biodiversity and special species management planning and monitoring programs on these islands and in other environmental settings.

The following section introduces specific examples drawn from diverse jurisdictions to illustrate the key design principles used in several current, planned, or developing biodiversity monitoring programs. The role of remote sensing in relation to modeling and field observations is highlighted. The presentation introduces examples of regional (e.g., Alberta Biodiversity Monitoring Institute, New Zealand Biodiversity Monitoring Program, Tropical Savanna Biodiversity Monitoring Program), national (Australia, Canada), and international scale (Circumpolar Biodiversity Monitoring Program) biodiversity monitoring and assessment. Only a brief introduction to each program is provided here, but many references and web sites exist to support these programs, which continue to evolve rapidly in response to changing information needs and monitoring capabilities. Many more such programs are anticipated in the near future to support biodiversity monitoring systems, management, and research.

Biodiversity Monitoring Programs

Alberta Biodiversity Monitoring Institute

In development for the past decade, the *Alberta Biodiversity Monitoring Institute* has implemented operational biodiversity measurement based on *a cumulative-effects monitoring approach* and a broad suite of indicators (www.abmi.ca web site accessed October 26, 2008). The goal of this biodiversity monitoring program is the ability to achieve a 90 percent probability of detecting a 3 percent change per year in biodiversity (measured as combined and individual species, habitat, and human footprint indices) after four samples over time acquired at more than 1600 field sites on a 20 km grid covering the province. In the initial design, each of the 1600 sample sites is surveyed once every 5 years. Reference conditions are observed in areas currently managed for natural ecosystem processes (e.g., National Parks), historical landbase conditions, and an estimate of reference conditions obtained in an *ecosystem-intactness* modeling approach (Nielsen et al. 2007).

The remote-sensing component of the ABMI program was conceived to include several key information products or indicators (Franklin and Dickson 2001):

- Human footprint (percent area in agricultural land cover, farmsteads, well sites, open mines, industrial facilities, roads, railways, pipelines, power lines, cutlines, cutblocks)
- Landscape structure and habitat change maps
- Wetland and aquatic systems (rivers and lakes, intermittent streams) extent and condition
- Canopy models (based on LiDAR and high spatial detail imagery)

The program employs *information pyramids* (Fig. 6.3) as a framework for aggregating and simplifying biodiversity measurements.

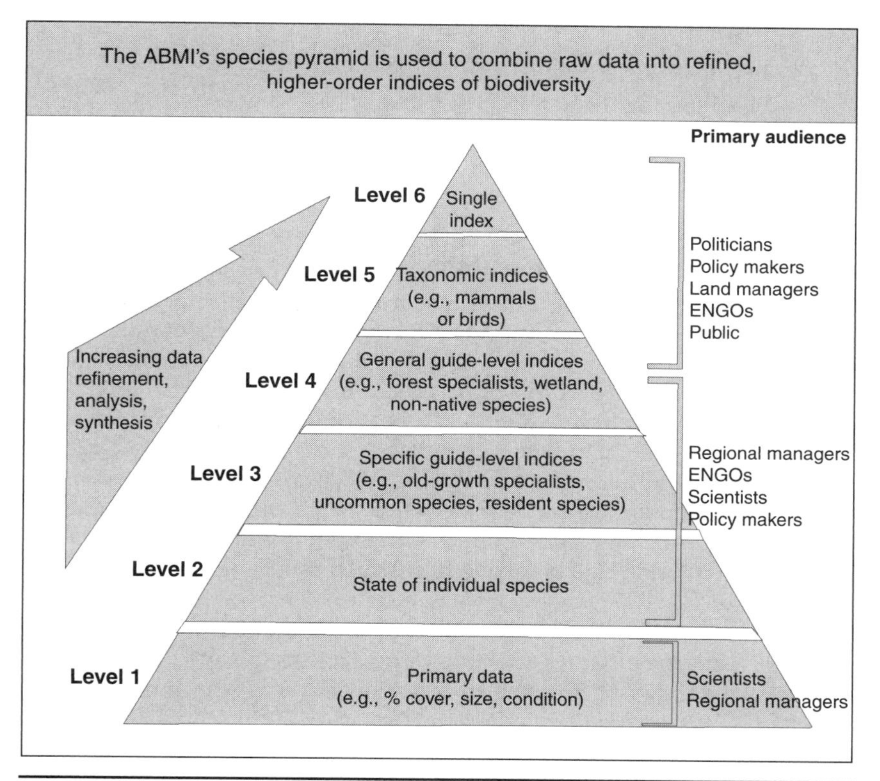

FIGURE 6.3 Alberta Biodiversity Monitoring Institute "Information Pyramid" (*With permission from* ABMI Index Development Group. 2006. *The Alberta Biodiversity Monitoring Program's Species Pyramid*. The Alberta Biodiversity Monitoring Program, Alberta, Canada. Report available at: www.abmi.ca [February 19, 2009].)

The information pyramids are an aggregation of the observational data based on the following principles:

- Data used in information pyramids are scientifically sound.
- Data analysis methods are transparent, scientifically rigorous, and peer-reviewed.
- Information products are flexible and responsive to the diversity of stakeholders needs.
- Products are easy to interpret, timely, and relevant.

Information products give insight into the underlying reasons for change in biodiversity.

Some key advantages to the information pyramid approach are thought to include:

1. Consistency across relatively large geographic areas with a wide diversity of ecosystems over time
2. Responsiveness to the needs of decision makers
3. Flexible design and adaptive protocols
4. An arms-length management that emphasizes relevant and independent knowledge

New Zealand Biodiversity Monitoring Program

Biodiversity decline was identified in February 2000 as "the most pervasive environmental issue" in New Zealand. With significant political support, subsequent developments included a comprehensive biodiversity strategy based on principles of resources management, conservation, governance, community participation (and a special relationship with Maori indigenous peoples), and biosecurity (http://www.biodiversity.govt.nz/picture/doing/nzbs/contents.html web site accessed September 14, 2008). A wide range of programs and initiatives were recommended aimed at ten environmental themes including monitoring land biodiversity, coastal, freshwater, and marine biodiversity.

A comprehensive review of the available literature on a large number of specific biodiversity monitoring projects in New Zealand and elsewhere was part of the preparation in designing the *New Zealand Biodiversity Monitoring Program* (Lee et al. 2005). Many of the reviewed projects were comprised of local or regional surveys focused on vegetation communities, threatened plants, birds, animal pests, weeds, and invertebrates in sensitive or protected environments. Typically, the objectives of these surveys was to determine species distributions and population trends in conjunction with specific management plans, the regulatory environment, or actual decisions (e.g., intervention, research activities). However, as an overall New Zealand biodiversity monitoring program, these activities suffered from a number of challenges (Lee et al. 2005):

- Poor linkages between groups working on different aspects of biodiversity management
- Lack of standardized reporting
- Limited data context
- Poor or no quality control
- Differential, or no, program effectiveness evaluation

An emphasis on *ecological integrity*, defined as the full potential of indigenous biotic and abiotic features, and natural processes, functioning in sustainable communities, habitats and landscapes, had developed in many of these surveys. But the end result of summarizing all of the results of these surveys was at best, a biodiversity assessment for New Zealand that could be characterized as being comprised of literally dozens of different indicators and uncertain comparisons to baseline conditions. Very possibly, the overall effect would to reduce the effectiveness of the individual monitoring activities themselves *as constituent components of a national biodiversity monitoring system*.

The development of remote sensing was appropriately considered an essential contributor to the New Zealand Biodiversity Monitoring Program. The program identified particular remote-sensing contributions in the development of specific biodiversity indicators, including land cover, productivity (e.g., NPP), status of threatened, weed and pest species, pollutants, ecosystem composition, structure and function, water quality (e.g., chemistry), ecosystem disruption (e.g., hydrological change, mass erosion), status of protection (e.g., habitat integrity), biological responses to climate change (e.g., mortality), human use (e.g., road density), and socioeconomics (e.g., eco-vandalism). But again, lack of coordination, data context, quality control, and standardization among a large number of individual surveys, compared to a centrally managed and nationally designed remote-sensing approach, would likely lead to reduced effectiveness in monitoring biodiversity over larger areas and the full diversity of environments. One program design response to help provide coherence to biodiversity monitoring over diverse environmental conditions was *thematic specialization*.

Thematic Biodiversity Monitoring

One of the most advanced program components, or themes, of the overall structure of the New Zealand Biodiversity Monitoring Program was a forest biodiversity monitoring component (Allen et al. 2003). Nested within the overall biodiversity monitoring program, the essential design elements of a *New Zealand Forest Biodiversity Monitoring Program* were created to complement specific national and international reporting commitments associated with climate change, carbon storage, biodiversity, and forest health agreements. Developing a forest monitoring system to support such a wide range of information needs at multiple spatial and temporal scales required a clear understanding of nine principles underlying the proposed monitoring method (Table 6.4, *see* Allen et al. 2003). The final system design was "...to include a combination of remote sensing and point-based sampling and modeling." The sampling design did not include forest or ecosystem prestratification. Instead, a design was introduced that employed a grid-based, nested, systematic design for plot locations aimed at guilds and using broad indicators. Such an approach was also implemented in the *Alberta Biodiversity Monitoring Institute* program in western Canada described in the previous section (Stadt et al. 2006).

Five broad biodiversity indicators were proposed in the *New Zealand Forest Biodiversity Monitoring Program* (Allen et al. 2003):

- Forest area
- Tree mortality and recruitment
- Exotic plants
- Browsing impacts of introduced animals
- Quantity of dead wood.

Within these broad indicators, a partial listing of proposed forest health monitoring indicators included:

- Tree species diversity, regeneration, mortality, growth, damage, and crown condition
- Indicator plants, understory species, lichen communities; foliar chemistry, dendrochemistry
- Woody debris and soil carbon

In each of these indicators, a remote sensing information source was anticipated, although not yet fully developed, beyond forest stand typing and stand delineation. The remote-sensing contribution initially was conceived as a contextual map for each field site at which more detailed information would be acquired.

Australia Biodiversity Monitoring Program

An early initiative to inventory, and then monitor, *continental biodiversity* in Australia was described by Chapman and Busby (1994) with an information network comprised of databases on plant and animal specimens, land cover mapped with remote sensing (Wallace et al. 2004), and climate data and models. These ideas continued to evolve over the subsequent decade, and have culminated with the development of Australia's

Principles	Consideration
Define the problem and goals of monitoring	—based on explicit objectives related to ecosystem composition, structure, and function
Do not focus only on current perceptions	—avoid commitment to highly "specific" biodiversity indicators that may not have longer-term value and reduce confidence in data applicability and usefulness
Build on the past	—ensure value of existing data and provide certain sampling intensity and data pooling options for historical analyses
Make sure data are explicitly comparable	—avoid commitment to measures that are inadequately understood and techniques that are dependent on irrelevant advantages (e.g., efficiency)
Recognize the advantages of repeat measurements	—to detect change in trend, but also in validation and reconstruction or modeling efforts
Account for difficulties in establishing baselines	—recognize the difficulty or even impossibility of historical baselines in the face of pervasive human impacts and poor knowledge of previous dynamics
Collect interpretive data	—balance the need for cause-and-effect determinations and correlative studies
Ensure a long-term commitment	—identify resources and leadership to continue over the long-time periods necessary to identify trends
Ensure data storage and accessibility	—safeguard long-term investments and optimize potential knowledge gains

Source: With permission from Allen, R. B., P. J. Bellingham, and S. K. Wiser. 2003. *New Zealand Journal of Ecology* 27:207–220.

Table 6.4 Design Principles Underlying a Proposed New Zealand Biodiversity Monitoring Program

national environmental indicators in seven themes (biodiversity, the land, inland waters, estuaries and the sea, human settlements, the atmosphere, and natural and cultural heritage). Instead of a single overall biodiversity monitoring program, a number of different Australian environmental monitoring programs—some regional, some national, but all more or less loosely connected administratively—have been developed to provide necessary information on the key indicators of biodiversity. This is a similar pattern—reliance on existing local and regional programs—that has occurred in development of Canadian and New Zealand biodiversity monitoring initiatives.

Regional Biodiversity Monitoring

Early results following the development of a framework to monitor regional biodiversity in northern Australian rangeland ecosystems were outlined by Fisher and Kutt (2006) (http://savanna.cdu.edu.au/savanna_web/publications/downloads/Biodlc.pdf web site accessed January 22, 2009). A basic assumption was that comprehensive biodiversity monitoring programs, at local or regional scale, must include the direct assessment of selected biota. The overall goal of the *Tropical Savanna Biodiversity Monitoring Program* was to explore the link between biodiversity and *land condition* (a subjective measure or ranking of land units into *good, intermediate, poor* classes principally related to apparent vegetation productivity and resilience). Land condition was obtained from a land-cover change analysis of a time series of satellite imagery supported by aerial and ground inspection over the 10 years preceding the biodiversity field sampling:

- Biodiversity sampling, which included plants, ants and vertebrates, was undertaken at 216 sites in five land types in two important pastoral regions of northern Queensland and the Northern Territory:

> Biodiversity sampling occurred at 1 ha (100 × 100 m) sites, with groups of sites sampled over a 4 day period. Within a site, birds were censused during 8 diurnal and 2 nocturnal five-minute visits. Other vertebrates were sampled using 24 Elliott box traps (baited with a mixture of oats, peanut butter, honey and tuna or dog biscuits), four 20 litre pit buckets each with 10 m of drift fence, and 3 diurnal and 2 nocturnal, 15-minute searches. Ants were collected using 70 mm diameter pit-traps in a 3 × 5 array, with 10 m between pits, open for 48 hours. A complete floristic list for the site was collected, with cover and frequency of understory species estimated using 20–25 0.5 m 2 quadrats in a regular grid; these quadrats were also used to measure ground layer cover of vegetation, litter, rock, and bare soil. Additional "habitat" and "disturbance" variables were measured at each site, relating to vegetation structure, substrate, recent grazing pressure, and fire history.
>
> *(Fisher and Kutt 2006: pp. 4–5)*

- The response of biota to land condition was assessed in terms of species composition, total species richness, diversity and abundance (broken down into taxonomic and functional groups), and the abundance of individual species.
- Land condition was most strongly predictive for components of the biota whose ecology was closely linked to characteristics of the ground surface and density of ground layer vegetation (most notably ants). There was only a weak relationship between land condition and many aspects of biodiversity, and the response of biota to land condition was complex and highly variable between taxa, land types, and locations.
- The incorporation of additional habitat variables (such as litter cover, rock cover, tree canopy cover) substantially improved modeled relationships between land condition and biodiversity attributes. Across the five land types sampled, there was generally poor *surrogacy* between major taxonomic groups, in terms of site richness, site diversity, and assemblage fidelity.
- Land condition alone was inadequate to predict, and therefore, monitor biodiversity status in savanna rangelands. Nevertheless, improvements in land condition across rangeland landscapes were thought likely to have positive biodiversity consequences. The incorporation of additional habitat attributes into site-based condition assessment would greatly improve information content about potential biodiversity condition.
- Further investigation of five areas was recommended:
    - The use of additional habitat indicators, including structural complexity indices, and habitat complexity in the understory (litter cover, density of fallen logs, termite mounds).
    - *Benchmarking* to describe the expected richness and composition of various components of the biota in a range of land types and vegetation types in very good condition, through a combination of sampling *best-on-offer* sites and reference to existing and historical data in order to be able to test whether the fauna and flora actually remains intact (or near-intact) in sites that are rated in good "biodiversity condition" using habitat indicators.
    - The relationship between biodiversity condition and spatial configuration of patches in different land condition states across broader landscapes in order to be able to link biodiversity status to landscape-scale condition assessment derived from remote sensing.

- Continued baseline biodiversity survey at regional scales, using repeatable (and well-documented) methodology for large numbers of sites to reveal long-term trends in biodiversity status.
- Implementation of model biodiversity monitoring programs (that incorporate direct assessment of select biota) at multiple scales....

There remain very few examples in northern Australia, and the value of these as demonstration programs cannot be overstated..

(Fisher and Kutt 2006: p. 14)

National-Scale Biodiversity Monitoring

The overall context for national-scale biodiversity monitoring in Australia is based on a robust implementation of an environmental *condition-pressure-response model* that shows ecosystem responses to processes over larger areas and long-time periods (Lee et al. 2005). The environmental condition-pressure-response model is designed to identify key processes, such as landscape fragmentation and climate change that may influence changes in biodiversity over time at regional or national scales. For example, a set of 63 indicators were selected to monitor terrestrial biodiversity: of these, 17 indicators were related to condition, 12 were related to pressure, and 34 were related to responses to losses or perceived threats to biodiversity (Saunders et al. 1998). A complementary set of 53 indicators was recommended for inland waters: 6 relating to groundwater, 3 to human health, 13 to water quality, 12 to water quantity, 7 to physical change, 8 to biotic habitat quality, and 4 to effective management. Of these, 18 were indicators of environmental pressures, 19 of condition, and 16 of response. In any given ecosystem or region of Australia, there may or may not be wide applicability for a specific condition or response indicator. For example, a significant number of the response indicators relate to *biotic habitat quality*; in some regions of the country, habitat quality response indicators were developed using FROGWATCH records of frog populations in surface waters and wetlands, fish-kill records, observations of waterbird population size and breeding colonies, habitat loss, exotic pest number and rate of spread, wetland extent, and number and effectiveness of pest control programs (Fairweather and Napier 1998). Not all areas of the country have these records, or for that matter, these conditions and responses.

The various biodiversity assessments come together in another Australian program, called the *National Land and Water Resources Audit Australia*. This program was conceived to assess the condition and trend of wetlands, riparian zones, threatened species and ecosystems, and the processes that threaten various elements of biodiversity. Based on a biogeographic framework of bioregions and subregions, this program assembles the various regional survey results, and also conducts comprehensive mapping of biodiversity elements and quantitative and qualitative assessments. A number of products are produced, including landscape health assessment, rangeland biodiversity, rangeland function analysis framework, and rangeland ecosystem function indices. Components include a national vegetation information system, initially designed to support the Australian Greenhouse Office *National Carbon Accounting System* (Color Plate 6.2), a system to map catchments, rivers and estuaries, and an Australian-wide monitoring framework for condition and trend of rangelands (*Australian Collaborative Rangelands Information System* (ACRIS), and *Western Australian Rangeland System* (WARMS), comprised of 1600 monitoring sites since 1992) (Wallace et al. 2006). Much of the integration is provided by the *Australian Natural Resources Atlas*, which presents Audit products at a variety of spatial scales. A comprehensive overview of biodiversity—a one-year initiative with a recommended five-year update cycle—was provided by the *Australian Terrestrial Biodiversity Assessment 2002 (http://www.anra.gov.au/topics/vegetation/health/index.html#assess* web site accessed September15, 2008).

Circumpolar Biodiversity Monitoring Program

The final biodiversity monitoring program discussed in this book is an international-level effort overseen by the Conservation of Arctic Flora and Fauna (CAFF) working group of the Arctic Council. An emerging synthesis for a biodiversity and wildlife management monitoring program developed in the past 5 years to survey changes occurring and expected in the Arctic region (Petersen et al. 2004). A framework document and extensive web site (www.caff.is) has been developed to outline the future direction of the *Circumpolar Biodiversity Monitoring Program* (CBMP). This program was designed to coordinate and facilitate communication among existing biodiversity monitoring initiatives over almost 15 million km^2 of land and 13 million km^2 of ocean in eight Arctic countries. The program features national and regional networks, and thematic specialization of monitoring activities.

The Arctic, as a region, may be experiencing impacts of climate change at rates higher than other regions, and collectively is thought to represent an early warning system for anthropogenic effects of global scale. A biodiversity monitoring program for the Arctic region must consider the urgency and impact of this changing climate situation. For example, the *State of the Arctic Report Cards 2007* (www.arctic.noaa.gov/reportcard/) reported scientific evidence for, and political acknowledgement of, the following major phenomena:

- Arctic wind circulation patterns changes compared to circulation patterns in the past 100 years.
- Summer sea ice extent at a record minimum.

- Continued and consistent rates of ice loss for the Greenland ice sheet.
- North pole ocean temperatures may be returning to 1990s values, but currents are relatively warm around the edges of the Arctic Ocean.
- Permafrost temperatures are stabilizing in both North America and Eurasia, but permafrost melt remains a serious problem.
- Shrubs are moving northward into tundra areas, but treeline advance is much more variable and complicated by human activities, such as forest management.
- Mixed trends are apparent in Arctic vertebrate populations, for example, dramatic increases in some Arctic goose populations, and some barren ground caribou populations in the Canadian Northwest Territories have declined by as much as 80 percent.

These and other trends (e.g., elevated ultraviolet radiation levels expected to affect people, plants, and animals as climate changes) have helped stimulate the development of the key biodiversity monitoring initiatives. The Circumpolar Biodiversity Monitoring Program goals are to (CBMP 2007):

- Monitor all elements of Arctic ecosystems—including species, habitats, ecosystem structure (e.g., species assemblages, food webs), processes (e.g., predator-prey cycles, nutrient cycling), functions (e.g., NPP), and stressors to the ecosystems—to gain a meaningful understanding of Arctic biodiversity and changes in biodiversity over time.
- Focus on trends, including the recognition of the dynamic nature of Arctic ecosystems, and the importance of identifying change that is outside the realm of natural variability.
- Recognize the interplay between terrestrial, freshwater, and marine systems and the way it shapes Arctic ecology and the goods and services that Arctic biodiversity provides.
- Recognize the dependence of Arctic biodiversity on conditions outside the Arctic (e.g., high proportion of migratory species, significant impacts of pollutants originating from outside the Arctic).
- Include humans and their cultural diversity as an integral component of many ecosystems.
- Monitor the interaction between people and biodiversity, such as sustainable use and the ability of biodiversity to provide essential goods.

The CBMP program "is an innovative and important exploration of Arctic biodiversity, whose purpose is to synthesize and assess status and trends of biological diversity in the Arctic as a response to the United Nations CBD 2010 global target to halt or significantly reduce biodiversity loss and to enhance monitoring of biodiversity at the circumpolar level, fully utilizing traditional knowledge... to detect the impacts of global changes on biodiversity and to enable Arctic communities to effectively respond and adapt to these changes." (Petersen et al. 2004: p 4).

A range of program components that are scale-dependent (e.g., regional networks), and thematic network specializations have been proposed. The component networks of the CBMP currently in place include:

1. *Species and habitat monitoring networks*. These are surveys designed to sample marine, freshwater, and terrestrial vertebrates and higher plants, based on indicator species carefully selected for their capacity to show trends in ecosystem health and rarity. Three major networks have been established for caribou/reindeer, shorebirds and seabirds, and the International Tundra Experiment (ITEX). Another 40 individual species or habitat monitoring programs are being standardized and coordinated.
2. *Site networks*. Networks have been developed involving individual field sites at which long-term observations or experiments were established or are planned. Such sites are typically contained within or related to protected areas, and may already be monitoring particular species/environment relationships of interest—for example, as part of the seabird monitoring, colony establishment and dynamics are determined, and then a monitoring protocol developed to identify change.
3. *Ecosystem networks*. These biotic surveys were designed to examine selected Arctic ecosystems such as ice-edge food webs, wetlands, tree line forest systems expected to show large or rapid responses to climate change. The ecosystem networks now include landscape scale monitoring components, for example to detect changes in *ecosystem integrity* [e.g., Table 6.5, *see* Parks Canada (2008)]. An example ecological integrity monitoring framework may include separate, though linked, monitoring initiatives focused on related components, such as: plant and animal diversity, and ecological processes. Such an approach is strongly dependent on multiscale remote-sensing technology and methods.

Indicators selected for Arctic biodiversity monitoring are organized according to seven themes (species composition, ecosystem structure, habitat extent, habitat quality, ecosystem function and

Biodiversity (characteristic of region)	Ecosystem Function (resilient, evolutionary potential)	Stressors (unimpaired systems)
Species richness —change in species richness —numbers and extent of exotics	**Succession/ retrogression** —disturbance frequencies and size (fire, insects, flooding). —vegetation age class distributions	**Human–land-use patterns** —land-use maps, road and population densities
Population dynamics —mortality/natality rates of indicator species —immigration/ emigration of indicator species —population viability of indicator species	**Productivity** —remote or by site	**Habitat fragmentation** —patch size, inter-patch distance, forest interior
Trophic structure —size-class distribution of all taxa —predation levels	**Decomposition** —by site	**Pollutants** —water quality, petrochemicals —long-range transport of toxics
	Nutrient retention —Ca, N by site	**Climate** —weather data —frequency of extreme events

Source: With permission from Van Wieren, J. F. 2009. Ecological Integrity Monitoring in National Parks. Available online: http://www.monitoringthemoraine.ca/MonitoringAdvisoryCommittee/ewp3.ppt#6 [31 March 2009]. (*See also:* Parks Canada. 2008. Available online http://www.pc.gc.ca/progs/np-pn/eco/eco3_E.asp accessed March 27, 2009).

Table 6.5 An Example of a National Ecological Integrity Monitoring Framework with Components of Plant and Animal Diversity, Ecological Processes, and Stressors

services, human health and well-being, policy responses). Each theme is comprised of indicators, which in turn are a series of elements and subelements which can be combined to create an index of each theme for trend analysis. A partial example of the theme-indicator-element-subelement structure for three selected themes is shown in Table 6.6.

The seven themes are for the most part consistent with the CBD/Ramsar Convention and other global biodiversity initiatives (*see,* for example, the United Nations Environment Programme

Theme	Index	Indicator	Element	Subelement
Species composition	Arctic species trend index	Trends in abundance of key species and trends in species parameters (e.g., distribution, productivity, survival, body condition, etc.)	Terrestrial Fauna	Caribou invasive species Invertebrates Landbirds Predators Lemmings . . .
			Marine	Commercial species (e.g., salmon) Polar Bears Ringed Seals Walrus . . .
Habitat quality	Arctic habitat fragmentation index	Trends in patch size	Terrestrial	Human Footprint . . .
		Distribution of habitats	All ecosystems	
		Fragmentation of rivers systems	Aquatic	
		Extent of seafloor destruction	Marine	
Human health and well-being	Arctic human well-being index	Trends in availability of biodiversity for traditional food and medicine	Societal	
		Trends in use of TEK in research, monitoring, and management		
		Trends in incidence of pathogens and parasites in wildlife		
		Change in status of threatened species		

Source: Adapted from CBMP. 2007. *Circumpolar Biodiversity Monitoring Program Five-Year Implementation Plan Overview Document*. Meeting of Senior Arctic Officials, Tromso, Norway (available online http://arcticportal.org/uploads/fr/pM/frpMZcpixn5kDMMusJ0kAQ/CBMP-Imp-Plan—Overview-FINAL.pdf accessed March 27, 2009).

TABLE 6.6 A Partial Example of Index-Indicator-Element Structure for Three Selected Themes in the Circumpolar Biodiversity Monitoring Program

Biodiversity Indicators for National Use at http://www.unep-wcmc.org/collaborations/BINU/, and the 2010 Bioindicators Partnership Initiative http://www.twentyten.net/default.aspx, web sites accessed September 12, 2008). The CBMP indicators, through a rigorous scientific working group process, had to meet the following criteria: scientifically valid and ecologically relevant; appropriate to diverse audiences (e.g., local communities, decision makers, global public); sustainable monitoring capacity; amenable to target and threshold setting; and practical. The indicators represent information from all major Arctic biomes at various scales and reflect all known Arctic pressures, major trophic levels, biodiversity levels (genes, species, habitat) including humans, and critical ecosystem services and functions.

Over time, expectations are that selected indicators will enable further analysis and interpretation into the cause of change. Combining indicator data from the multitude of elements and subelements originating in the various networks will be the task of modeling systems to predict responses of Arctic biodiversity to climate change and other impacts and develop future scenarios. Additional issues to be completely addressed prior to full program implementation over the coming years include data sharing protocols, selection and implementation of data standards, development of comprehensive databases, communications and web portal design, and creation of pilot projects. An important component of the Circumpolar Biodiversity Monitoring Program is the *Traditional Ecological Knowledge (TEK) Network*, which can support, complement, and be integrated with scientific approaches and form a unique community-based monitoring initiative. Such an initiative is entirely consistent with a transdisciplinary approach to environmental complexity. Throughout the region, the TEK network will also serve to increase credibility and facilitate communication (Peterson et al. 2004).

CHAPTER 7

Conclusion

Mortals dwell properly when they learn to protect the earth, not to exploit or conquer or subordinate.

—J. G. Gray, 1952

This book set out to introduce specific elements of the problem-oriented, contextual, interdisciplinary framework that enables successful application of a rigorous, largely empirical, comprehensive and practical remote-sensing approach to issues in biodiversity and wildlife management. The goal was to present and synthesize a vast and growing literature to help answer the question: *In what ways can remote sensing deliver the information on spatially explicit and multitemporal ecosystem structure and processes required to support biodiversity and wildlife management*? The reason this question is of importance, even urgency, is that the planetary situation of resource management, wildlife, habitat, and biodiversity conservation appears dire (Wilson 2006). Habitat loss, fragmentation, invasive alien species, pollution, and overharvesting are complex processes leading to catastrophic species extinction rates. To reverse declines and preserve the ecological, economic, and spiritual values of wildlife and biodiversity will require enormous ingenuity and effort. A remote-sensing approach is without a doubt necessary, even crucial, contributor to such efforts and, ultimately, part of the required creativity essential to achieve a workable synthesis.

In the following sections, three of the significant dimensions that represent the scientific challenge for those attempting to understand remote-sensing approaches to solutions that are desirable in increasingly intense and urgent environmental management situations. These dimensions also can be viewed from the perspective of *scientific opportunities*—which together, help create a fertile environment to ensure additional progress and successful application of the remote-sensing approach:

1. The remarkable and challenging intellectual environment offered by interdisciplinary remote sensing and *transdisciplinarity*
2. The impressive array of *technological innovations* in remote sensing
3. The rich opportunity to acquire and implement *new insights in specific environmental remote-sensing applications*

Remote Sensing Across Disciplines—Revisited

Fifty years ago, the initial outlines of what would eventually become a *transdisciplinary remote-sensing approach* could be discerned among new initiatives to develop scientific images using visible and nonvisible wavelengths from aerial and spaceborne sensors, and to apply quantitative methods in image analysis. Conceptually, transdisciplinarity is a form of interdisciplinary science that is designed to encompass the relatively familiar and traditional methods-based interdisciplinarity, and then to go beyond a simple sharing of methods and instruments to identify and achieve a scientific objective that is not possible within a single disciplinary perspective or by individuals working in either a multidisciplinary or certain types of strictly interdisciplinary environments. Instead, transdisciplinarity gives rise to new ways of organizing teams of expertise to identify, articulate, and accomplish common goals that transcend single disciplinary, multidisciplinary, and possibly even many interdisciplinary perspectives and structures. Transdisciplinary remote-sensing projects are based on the development of a compelling vision, arching across methods and goals, which can help focus the energy needed to address the many perspectives. This vision will be informed by societal influences in understanding the values to be preserved and enhanced through team effort, and all participants will be able to appreciate the project goals and their individual contribution to achieving them.

Early remote-sensing expertise was often supplied by those interested in problems of sensor design and mission control; they were quickly joined by applications' specialists, the foresters, geographers, geologists, agronomists, meteorologists, and biologists, and others with impressive disciplinary expertise, who thought about new ways to use the imagery, identified effective platform and sensor characteristics, and created new applications. Enormous strides were made in advancing remote sensing within these disciplines, and as an overall interdisciplinary scientific approach, based on the theory of electromagnetic radiation interactions and quantitative analysis of spectral response patterns.

Today, remote sensing has matured beyond disciplinary applications, sensor design, and image acquisition issues, to provide a foundation for a new way of thinking about common and urgent environmental challenges that persist and are unlikely to be resolved with standard disciplinary approaches having significant limitations of spatial and temporal sampling and explanatory power. Such challenges are immediately obvious in biodiversity and wildlife management, and indeed in many other areas of environmental management and assessment, such as climate change analysis, urban planning, sustainable forestry, and ecosystem health studies. A common need is for researchers and managers to identify relevant environmental conditions through space and time, determine the processes of change, and assess the impact of change on fundamental ecosystem functioning in environmental processes and patterns over relevant scales. *Understanding and managing the distribution and potential changes to biodiversity or special species habitat over a large area and long-time periods represents a scientific goal that is inherently transdisciplinary; that is, the search for common underlying structures or patterns to explain features and attributes of natural systems requires transdisciplinary modes of knowledge production.*

A key feature of the remote-sensing approach to complex environmental science and management, then, is to recognize the transdisciplinary nature of the problem to be addressed and to develop effective ways of organizing transdisciplinary teams. A *conceptual framework* will provide structure to these activities. Some reasonably straightforward action items can be suggested to help ensure success in any transdisciplinary team endeavor, and which might be of value in applying any remote-sensing approach to biodiversity and wildlife management. For example:

- Create opportunities for both formal and informal team member discussion, and respectful intellectual engagement, such as ensuring visits to field sites together.
- Consider sharing of responsibilities and methods rather than promoting a "division of labor" based exclusively on expertise and disciplines.
- Emphasize learning and relearning skills, and "soft skills" such as communications, problem-solving, and team collaboration.
- Foster leadership skills, and a heterogeneous management organization, which encourages innovation and eliminate unproductive dualisms, such as technical/science, science/management, or applied/pure research.

Technological Innovation in Data and Image Processing

The scale and importance of environmental problems that remote sensing can help address, and the ways of organizing people and resources to accomplish goals, depend on the *fundamental technological innovations* that have marked the first five decades of remote-sensing technology development. Early remote-sensing applications in wildlife and biodiversity studies were based on high quality field- and aerial-based photography and insightful interpretations by skillful image analysts. These successes were soon joined by relatively coarse spatial resolution satellite imagery, which was scientifically exciting and stimulated many new applications, but had clear limitations in terms of image resolution and mission design. Computer technology continued to improve, with the result that, within a few decades of the initial development of remote sensing from origins in aerial photointerpretation projects, the majority of remote-sensing image analysis could be conducted in digital formats at astonishingly better image resolutions and quality.

In the 1980s, the multiconcept dominated remote sensing—*multispectral, multitemporal, multisensor, multiresolution data*—but such ideas were not yet integrated, and few had the resources or could marshal the sustained commitment necessary to fully implement a multiple remote-sensing approach in operational or research settings. With the tremendous parallel growth in GPS and GIS, an impressive geospatial integration is now effectively complete; remote sensing is frequently one component of data synergy within an *applications framework* that matches specific image qualities from a wide range of sources to well-defined information needs. Purpose-designed remote sensing has evolved to be a much more common and effective approach. This has also served to alleviate the widespread adoption of earlier remote-sensing experimental designs, which by necessity, tended to use available images even if these data were not particularly well-suited to the task at hand.

The pattern of improvements in remote-sensing data and image processing technology is likely to continue and even accelerate. Some of the most important innovations currently underway are related to spectral response pattern characteristics and to remote-sensing image analysis methods. For example:

- The continuing trend to increasingly *better image resolution*, including higher spatial, spectral, and radiometric resolution, resulting in greater detail in more frequent mapping and monitoring applications over larger areas.
- The improvement and expanding capability in *object-based image analysis* to complement and extend remote-sensing image information extraction based on pixel-based methods, such as spectral mixture analysis and texture analysis.
- The operational provision of new sensor capability—hyperspectral imaging, interferometric SAR, and LiDAR, for example—have stimulated new applications in three-dimensional sensing and tremendous *image data synergy* among different sensor packages and image characteristics.
- The impressive accumulation of technology and expertise in location sensing, and the continued development of GIS and spatial modeling techniques that rely on combinations of remote sensing and other spatial data, to estimate and predict environmental processes and patterns.

Applications Insights

Early remote-sensing applications in biodiversity and wildlife management relied on visual analysis of analog imagery, some digital analysis, and relatively simple map overlay techniques with habitat inferences or field observations of animal locations and behavior. This approach has now expanded to include the creation of a wide variety of habitat suitability and resource-selection maps, models of all kinds associated with individual animal and population characteristics, such as health and stress, and habitat and biodiversity indicators. The prediction of habitat quality and use can suggest future management scenarios, and has tremendous implications for research (such as clarifying the links and understanding the underlying mechanisms between climate, habitat, niche, and species). The continued development of these, and other complex modeling applications, will ensure an even greater relevance of remote-sensing approaches in the near future. *This trend is embedded within the urgent need to address the increasing demands for environmental information of all kinds.*

The applications that are now feasible, or will be implemented in the near future, are growing rapidly and are too numerous to list here. The indicator approach is a powerful and effective philosophy underlying many regional and thematic habitat and biodiversity monitoring applications. Based on the insights that have been gleaned from the literature, and decades of remote-sensing experience as outlined in this book a number of significant patterns in the development of remote-sensing applications in biodiversity and wildlife management can be identified. The insights gleaned in such applications are poised to stimulate even more widespread use of remote sensing in environmental assessment and monitoring at all relevant spatial and temporal scales. For example:

- *Characterization of habitat* with remote-sensing spectral response patterns, and relating macrohabitat and microhabitat patterns to observations of species presence/absence, abundance, or modeled distributions.
- More complex remote sensing-based predictions of *habitat suitability*, including those focusing on the relationships between habitat quality, biodiversity, phenology, and productivity; prediction of animal abundance, behavior, and health, special (e.g., threatened or endangered) species distribution, and species richness analysis.
- Improvements to habitat suitability models with the use of *historical pattern analysis*, establishment of reference conditions or intactness measures, and more complex modeling techniques that consider species distribution data (presence and absence), basic species biology and environmental

patterns developed synergistically from a remote-sensing image, ancillary data (e.g., topography), and existing (historical) map data.

- Improvements in the ability to detect and use the *structural properties of habitat*, food resources and quality, land cover and productivity relationships, phenology, chemical constituents of vegetation, and landscape energy flows and genetics (e.g., hybridization).
- *Changes to endangered or threatened species habitat*; monitoring habitat quality indicators of biodiversity coupled with a *natural habitat integrity* approach to generate a landscape-level quantitative assessment which can be readily updated; such a system, therefore, could serve as the basis for a specific habitat monitoring system to detect variations from expected or assumed *reference conditions*.
- Successful remote sensing of *invasive alien species*; based on improved methods of detection of biochemical, phenological, or structural characteristics or modifications of landscapes associated with a change in ecosystem functioning.
- *Measures of fragmentation and connectivity* obtained by remote-sensing methods create powerful diagnostic tools in the search to understand the impact of change on wildlife species; horizontal and vertical structure are increasingly required, and early success with indices of *structural complexity* suggests promising new applications development.
- Significant progress in the use of remote sensing to characterize *human use of landscapes*; for example, the mapping and field-based detection of human use components on linear features, such as roads, trails, seismic lines, pipelines, and other types of urban and transportation infrastructure, and clarifying with models and GIS data the link between remotely sensed land attributes (e.g., land cover, LAI) and land use.
- *Full geospatial systems support* to model wildlife habitat use and resource selection, with developments for components using digital cameras in field documentation of track sets, modules that facilitate the analysis and deployment of camera traps, and improved satellite image habitat modeling in a GIS environment with location-sensing (radio-collar) and bio-logging data.
- Remote-sensing approaches to *landscape epidemiology* improvements; for example, the use of multiple resolution satellite imagery in initiatives to map disease vectors and environmental correlates of disease, and ecosystem and animal health and environmental stress in relation to landscape change and population or behavioral characteristics.

- Improvements to mapping and modeling biodiversity, habitat and special species-environment relationships components within the overall framework of a *biodiversity monitoring system;* this may include regional and thematic system design goals; more powerful ways to remotely inventory of patterns of species richness, develop species-habitat relationships models, determine biophysical controls of species richness patterns, predict primary production, map spatial heterogeneity (for example, through optical/LiDAR synergy), and identify the effects of anthropogenic processes and understanding and quantifying energy/productivity relationships.
- At the regional, national, and international scale, the implementation of coarse-filter monitoring programs; regional examples, such as the Alberta Biodiversity Monitoring Institute or the Tropical Savanna Biodiversity Monitoring Program, and national initiatives, such as *BioSpace,* a proposed Canada-wide biodiversity monitoring system, are based on measures of productivity, disturbance, topography, and land cover, integrate the climatic and structural response of vegetation and faunal species; such approaches, over time, establish across large areas (e.g., over decades and continents), a unified scheme or foundation for biodiversity monitoring in changing landscapes.

Future

The capabilities of remote sensing to support applications in biodiversity and wildlife management have developed to an astonishing degree of sophistication since remote sensing was conceived in the late 1950s, and quickly expanded as a new field of science. The ability to predict habitat attributes, biodiversity, and the consequences of environmental change derives from the high quality imagery, processing methods, and increasingly complex models that integrate animal-based data sets (such as location acquired by GPS and health data) across a range of scales and ecological processes. The task of this book has been to highlight and synthesize these capabilities to illustrate the enormously powerful remote-sensing approach that may provide a foundation for biodiversity and wildlife management at multiple scales and over long-time periods. The success achieved to date may stimulate even greater effort to ensure that remote-sensing contributions are secured for future use.

But this synthesis has highlighted the existence of an even more profound challenge, one that was originally perceived as a fundamental reason to develop environmental remote sensing in the first place: *What is the ability of society to clarify and secure the common interest in managing Earth environmental processes (biodiversity and wildlife) sustainably?* This question may, in fact, represent the decisive issue of our times. Few would argue that progress on reconciling human needs with environmental resources is desperately needed: *How can remote sensing contribute to help resolve this urgent and complex societal challenge?*

This question will, perhaps, be accepted as a philosophical and practical challenge, not only to a new generation, but to all those involved in the development and continued use of remote sensing. The philosophical challenge is to ask: *What are you doing with remote sensing?* What are your values, are they aligned with a grand vision of remote sensing as a contributor to societal goals? Are you intent on capturing the critical moments that express your contribution? Are you leading by example? The opportunity is to answer: *We're saving species at risk; We're predicting environmental change; We're finding a link between health and societal organization; We're tackling poverty; We're fighting global hunger; We're conserving natural biodiversity; We're supporting world peace.* The early developers of remote sensing were committed to remote sensing because they were certain that this new approach would offer real strengths in resolving such challenges. Perhaps an initial priority for those building on the early achievements is to examine individual values, reflect on personal commitment, and seek alignment with the effort to address urgent societal needs by remote-sensing approaches.

Several practical dimensions to pursue can be suggested that might lead to greater innovation and creativity. First, the issues that prevent operational remote-sensing contributions to human societal needs must be identified and overcome; *the promise of remote sensing as an operational approach must become the reality of remote sensing*. Although many success stories are apparent in this book, rarely has a thorough evaluation of the steps necessary for *operational remote sensing* been completed. In too many apparently remote sensing-ready environmental situations, remote-sensing contributions are not actually realized—are deemed impractical, are overlooked, or are otherwise not available. Clearly, there are significant barriers to operational remote sensing that go beyond the fundamental technological and methodological limits, such as spatial resolution, that are relatively easily documented. Perhaps some of the complicating factors have more to do with education and human resources, a lack of clarity in objectives, or uncertain political and institutional commitment. But any such factors preventing the remote-sensing approach achieving full potential to help society to perceive and protect environmental values must be understood and quickly eliminated.

Second, and related to the first, the actual benefits of remote sensing must be quantified and enhanced. Few complete *cost-benefit analyses* of remote sensing have been reported. One unfortunate consequence is that, while certain specific examples of benefits are available, the immediate costs of remote sensing are more apparent, and are often considered in isolation of such benefits. This situation has not helped generate a greater awareness and appreciation for remote-sensing approaches. Apart from this, the overall benefits of remote sensing are not always simple to ascertain; for example, the trend of increasing remote-sensing image quality has been expected to stimulate the development of a broader perspective, perhaps leading to better environmental decision making. With remote sensing, people can *see* the connectedness of environmental patterns and processes; the context is apparent, linkages are better appreciated, artificial boundaries can be considered with the appropriate level of skepticism of their relevance in dealing with environmental processes that drive fundamental values, such as conservation of wildlife habitat and biodiversity. A confirmation of this important beneficial influence is urgently needed.

The impact of high quality remote-sensing imagery on the public *and* professionals can be dramatic, emotionally visceral, and lasting. Recall the astounding impact of those first photographs of *Earthrise* from space (Poole 2008), reflect on the compelling words of the astronauts on describing their often life-altering views of the Earth (http://www.spacequotations.com/earth.html web site accessed January 22, 2009), or perhaps evoke the thrill first experienced when examining exotic satellite imagery of a new or familiar location. Even if only a fraction of the power of such environmental images could be harnessed, a transformational change in perspective and new clarity in environmental thinking may be stimulated. The motivational use of remote sensing to inform and explain fundamental *human context* should be clarified for more effective contributions in future.

And finally, remote-sensing imagery (and GIS and GPS), ICT tools, and social networking have intensified discussion and engaged public participation in the critical environmental questions of the day like never before: *If people are able to contribute more frequently and meaningfully to the development of public policy on issues about which they care deeply, will things change? Can natural resource management be made more relevant and responsive to societal needs and values?* New patterns of participation and involvement are apparent in a host of social activities and creative endeavors, reducing barriers to informed discussion and decision making, and creating opportunities for a greater openness in all manner of public discourse, scientific communication, and management dialogue. In this new *e-action* environment, perhaps one of the most significant contributions to be made by remote-sensing science, in addition to the predictive scientific and technological developments highlighted in this book, will be in stimulating an increased awareness of the social and political context in which natural resource management operates and continues to evolve.

APPENDIX I

Annotated Bibliography of Selected Reference Books Documenting the Development of Remote Sensing in Environmental Applications

Development of a Remote-Sensing Approach

Swain, P. H., and S. M. Davis, eds., 1978. *Remote Sensing—The Quantitative Approach*. McGraw-Hill, New York

This book established the new science of remote sensing and introduced resource management scientists and analysts to the growing capacity for digital image processing that would soon become the dominant analytical approach. The basis for the new field was traced from the science and technology of aerial photointerpretation, and complex computer image analysis approaches were reviewed including classification and pattern recognition. A strong emphasis was placed on how sensors, computers, and human decision making together resolve real-world environmental problems.

Townshend, J. R. G., ed. 1981. *Terrain Analysis and Remote Sensing*. George Allen and Unwin, London

Terrain analysis encompasses applied land classification and aerial photointerpretation land-resource surveys that were the precursors to more specialized ecological land survey, forest inventory, and biophysical mapping systems. This book highlighted for the first time the intersection of aerial photography and remote-sensing technologies in terrain-mapping applications, many of which are now routinely using satellite remote sensing. The book presciently emphasized the small but emerging field of GIS. The complementary nature of aerial photography, field and remote-sensing data was envisioned, together with the need for effective information display to help resource management professionals and decision makers.

Land Cover and Forest-Mapping Applications

Howard, J. A. 1991. *Remote Sensing of Forest Resources*. Chapman and Hall, London

This book provided the background to understand and use remote sensing in forest resource mapping and practical forest management applications. Examples of forest change detection, forest-species mapping, fire detection and mapping, and growth modeling were provided, with many highlights from international projects and development work. The changing nature of remote sensing in forestry—from aerial photography to digital imagery of all types—was identified and documented to help users adopt new approaches.

Gholz, H., K. Nakane, and H. Shimoda, eds. 1997. *The Use of Remote Sensing in Forest Productivity Modeling*. Kluwer, Boston

The book dealt with carbon dynamics and covered the remote sensing of forest structure and chemistry for the purposes of productivity modeling and growth estimation. The focus was on collecting input variables (such as LAI and leaf N concentration) for light efficiency and mechanistic models, and model validation using remote sensing. Applications included stand and landscape/regional analyses, carbon-flux dynamics, and specific outcomes such as forest wood yield estimation and biomass prediction.

Wulder, M. A., and S. E. Franklin, eds. 2003, *Remote Sensing of Forest Environments: Concepts and Case Studies*. Kluwer Academic Publishers, Boston, MA.

A state-of-the-art synopsis of the current methods and applications employed in remote sensing of forests, this book suggested the importance of scale, image characteristics, and image acquisition design to successful use of environmental remote sensing. The book highlighted the attention that must be given to data preparation and processing, to extracting information from imagery, and in developing data fusion or integration models. Case studies using LiDAR, hyperspectral, and aerial videography were provided together with satellite-based regional land characterization and mapping examples.

Ecological Applications

Miller, R. I., ed. 1994. *Mapping the Diversity of Nature*. Chapman and Hall, London

This book presented field, GIS and remote-sensing aspects of biological species mapping techniques used in a variety of settings and case study applications. Elements of biodiversity and habitat studies were presented, and many strong statements supporting the importance of remote sensing in solving issues of habitat distribution, conservation status questions, and an assessment of new challenges were provided. The development of GIS as a tool to support conservation strategy was highlighted, and the book concluded by summarizing promising new ideas in videography, hyperspectral imaging, and some global scale mapping and modeling database issues.

Frohn, R. C., 1998. *Remote Sensing for Landscape Ecology*. CRC Press, Boca Raton

The introduction of new landscape metrics that work well with remote-sensing images helped stimulate the now wide-spread use of remote-sensing imagery to map fragmentation and landscape structure over long-time periods and large areas. Land-use change patterns in four areas including Brazil were examined and used to clarify the approach.

Majumdar, S. K., J. E. Huffman, F. J. Brenner, and A. I. Panah, eds. 2005. *Wildlife Diseases: Landscape Epidemiology, Spatial Distribution, and Utilization of Remote-Sensing Technology*. Pennsylvania Academy of Science

The book surveyed a field of enormous public interest, and included three significant remote-sensing chapters, which together highlight the intersection of remote sensing, environment and zoonotic disease-vectors. The authors point out that remote sensing is most useful if disease dynamics and distributions have clear and consistent relationships to environmental variables that can be mapped reliably. An increasing number of epidemiology researchers are applying higher level GIS analysis to wildlife disease studies. Improvements in related technologies, such as biologging, biotelemetry, and GPS, continue to signal a revolution in the nascent landscape epidemiology application.

Richardson, L.L., and E. F. LeDrew. 2006. *Remote Sensing of Aquatic Coastal Ecosystem Processes: Science and Management Applications*. Springer, Dordrecht

The book provided a comprehensive survey of aquatic ecosystem remote sensing and concludes with a chapter of recommendations for new science and changes to management practices. Specific chapters present the state-of-the-art in studies of coral reef ecosystems and coastal ecosystem change, and new sensor developments, including LiDAR and diagnostic and monitoring arrays, were highlighted.

Modeling and GIS Applications

Alexander, R., and A. Millington, eds. 2000. *Vegetation Mapping: from Patch to Planet*. John Wiley and Sons, Chichester UK. (*see also* Hunsaker, C., M. F. Goodchild, M. A. Friedl, and T. J. Case, eds. 2001. *Spatial Uncertainty in Ecology: Implications for Remote Sensing and GIS Applications*. Springer-Verlag, New York)

These two books examined vegetation and several types of thematic mapping using field and remote-sensing methods, and presented a number of case studies at local, regional, continental, and global scales. The emphasis was on the uses of generalized maps, the meaning of vegetation boundaries, the input to models, and historical developments of methods and techniques. Handling error, assessing accuracy, and understanding ecological and map uncertainty were treated in some detail based on early results of research into these topics.

Skidmore, A., ed. 2002. *Environmental Modeling with GIS and Remote Sensing*. Taylor and Francis, London

With a view to influencing environmental management through modeling, this book covered the rapidly developing GIS world with a presentation of case studies, theory, and applications aimed primarily at GIS and remote-sensing professionals. Specific chapters highlighted wildlife and biodiversity mapping and modeling, hazard and disaster management, and environmental impact assessment. The book concluded with an emphasis on the need for interdisciplinary expertise to adequately use GIS and remote-sensing tools, various communications challenges, data interpretation and quality issues, and the need for improved models.

Atkinson, P., G. M. Foody, S. E. Darby, and F. Wu, eds. 2005. *Geodynamics: Modeling Spatial Change and Process*. CRC Press, Boca Raton

Geocomputation is the application of computer power to solve geographical problems, and remote sensing and GIS present some of the more compelling instances in which to showcase new capabilities, including 3-D landscape modeling and cellular automata growth models. The editors of this book have highlighted the interdisciplinarity of geodynamics and the central "what-if" scenario paradigm that enables thoughtful management based on predictions. Remote sensing and spatially-distributed dynamic models were highlighted as a convergence of powerful, synergistic technologies. Scale, development of smart systems, and error-propagation were identified as critical issues in need of concerted development.

Aronoff, S. 2005. *Remote Sensing for GIS Managers*. ESRI Press. Redlands, CA

The "age of imagery is becoming quite interesting"—so begins this comprehensive reference text that most definitively has placed remote sensing firmly in the context of GIS, the environment in which most environmental analysts deploy imagery of all types. A survey of some key remote-sensing applications, in agriculture, urban planning, national security, forestry, geology and archaeology, was provided. Considerable background material on remote sensing was included, and a critical contribution was to provide helpful suggestions on the organizational issues that many users will encounter when using remote-sensing imagery.

APPENDIX II

Common and Scientific Names of Species Mentioned in the Text

Common Name	Scientific Name
African elephant	*Loxodonta africana*
American bison	*Bison bison*
American beech	*Fagus grandifolia*
American marten	*Martes americana*
Arctic musk ox	*Ovibos moschatus*
Asiatic golden cat	*Catopuma temminckii*
Bald cypress	*Taxodium distichum*
Balsam fir	*Abies balsamea*
Barred owl	*Strix varia*
Beaver	*Castoridae*
Beech	*Fagus sylvatica*
Black wildebeest	*Connochaetes gnou*
Black-capped vireo	*Vireo atricapillus*
Black-tailed prairie dog	*Cynomys ludovicianus*
Blue grama	*Bouteloua gracilis*
Blue tit	*Parus caeruleus*
Borneo clouded leopard	*Neofelis diardi*
Brushtail possum	*Trichosurus vulpecula*
Capercaillie	*Tetrao urogallus*

(*Continued*)

Common Name	Scientific Name
Caribou	*Rangifer tarandus*
Caucasian black grouse	*Tetrao mlokosiewiczi*
Chipping sparrow	*Spizella passerina*
Cholera	*Vibrio cholerae*
Clouded leopard	*Neofelis nebulosa*
Dholes	*Cuon alpinus*
Dutch elm disease	*Ceratocystis ulmi*
Eastern white pine	*Pinus strobus*
Elephant seal	*Mirounga*
Elk	*Cervus elaphus*
Emperor geese	*Chen canagica*
Emperor penguin	*Aptenodytes forsteri*
Engelmann spruce	*Picea engelmannii*
Fishing cat	*Prionailurus viverrinus*
Flamingo	*Phoenicopteridae*
Giant panda	*Ailuropoda melanoleuca*
Giraffe	*Giraffa camelopardalis*
Grassland bobolink	*Dolichonyx oryzivorus*
Great tit	*Parus major*
Grizzly bear	*Ursus arturus*
Guanaco	*Lama guanicoe*
Gypsy moth	*Lymantria dispar*
Hairy-nosed wombat	*Lasiorhinus latifrons*
Hairy woodpecker	*Picoides villosus*
Jack pine	*Pinus banksiana*
Javan hawk-eagle	*Spizaetus bartelsi*
Kaleej pheasant	*Lophura leucomelanos*
Kangaroo	*Macropus*
Leafy spurge	*Euphorbia esula*
Leopard cat	*Felix bengalensis*
Leopard	*Panthera pardus*
Lesser prairie chicken	*Tympanuchus pallidicinctus*

(*Continued*)

Common Name	Scientific Name
Lesser snow geese	*Anser caerulescens*
Loblolly pine	*Pinus taeda*
Marbled cat	*Parodfelis marmorata*
Midge	*Culicoides*
Mongolian gazelle	*Procapra gutturosa*
Mountain bongo antelope	*Tragelaphus euryceros isaaci*
Mountain caribou	*Rangifer tarandus*
Mountain laurel	*Kalmia latifolia*
Northern bobwhite	*Colinus virginianus*
Northern flying squirrel	*Glaucomys sabrinus*
Northern goshawk	*Accipiter gentiles*
Northern spotted owl	*Strix occidentalis caurina*
Norway spruce	*Picea abies*
Oak	*Quercus*
Palm civet	*Paguma larvata*
Pepperweed	*Lepidium latifolium*
Pine warbler	*Dendroica pinus*
Pond apple	*Anonna glabra*
Poplar box	*Eucalyptus populnea*
Puerto rican parrot	*Amazona vittata*
Red panda	*Aliurus fulgens*
Red spruce	*Picea rubens*
Red squirrel	*Tamiasciurus hudsonicus*
Red-eyed vireo	*Vireo olivaceus*
Reeve's muntjac	*Muntiacus reevesi micrurus*
Reindeer lichen	*Cladina*
Root-rot fungus	*Phytophthora cinnamomi*
Rosebay rhododendron	*Rhododendron maximum*
Ryegrass	*Lolium perenne*
Saltwater crocodile	*Crocodylus porosus*
Seagrass	*Thallasia testudinum*
Silver wattle	*Acacia dealbata*

(*Continued*)

Common Name	Scientific Name
Snapper	*Pagrus auratus*
Sonoran pronghorn	*Anilocarpra americana sonoriensis*
Spiny lobster	*Jasus edwardsii*
Spotted owl	*Strix occidentalis*
Spruce bark beetle	*Dendroctonus rufipennis*
Subalpine fir	*Abies lasiocarpa*
Sugar maple	*Acer saccharum*
Tibetan wild ass	*Equus kiang*
Tick	*Borrelia burgdorferi*
Trembling aspen	*Populus tremuloides*
Tupelo gum	*Nyssa aquatic*
Walrus	*Obenidae*
West India pine	*Pinus occidentalis*
Western hemlock	*Tsuga heterophylla*
White cypress pine	*Callitris glaucophylla*
White-breasted nuthatch	*Sitta carolinensis*
Wildebeest	*Connochaetes taurinus*
Willow	*Calix*
Wolf	*Canis lupus*
Wood stork	*Mycteria americana*
Yellow birch	*Betula alleghaniensis*

APPENDIX III

A Selection of Past, Current, and Planned Remote-Sensing Satellites

Satellite Program	Satellite Platform	Launch Date	End Date
ADEOS (Advanced Earth Observing Satellite)	ADEOS 2	14/12/2002	25/10/2003
	ADEOS 1	17/08/1996	30/06/1997
ALMAZ	ALMAZ 1	31/03/1991	17/10/1992
	COSMOS-1870	25/07/1987	30/07/1989
ALOS (Advanced Land Observing Satellite)	ALOS 1	24/01/2006	–
BIRD (Bi-Spectral Infrared Detection)	BIRD 1	22/10/2001	–
CBERS (China/Brazil—Earth Resources Satellite)	CBERS-4	01/07/2010	–
	CBERS-2b	28/07/2008	28/07/2008
	CBERS-3	01/07/2008	
	CBERS-2 (ZY-1B)	21/10/2003	–
	CBERS-1 (ZY-1)	14/10/1999	13/08/2003
CloudSat	CLOUDSAT 1	28/04/2006	–
COMPASS-M1	COMPASS-M1	04/12/2008	–
CORONA	KeyHole 1 to 4B	01/08/1960	25/05/1972

(*Continued*)

Satellite Program	Satellite Platform	Launch Date	End Date
COSMOS (Kometa Space Mapping System)	COSMOS-2373	29/09/2000	14/11/2000
COSMO-SkyMed (Constellation of Small Satellites for Mediterranean basin Observation)	COSMO-SKYMED 4	01/11/2010	–
	COSMO-SKYMED 3	25/10/2008	–
	COSMO-SKYMED 2	09/12/2007	–
	COSMO-SKYMED 1	07/09/2007	–
Digital Globe Constellation	WorldView 2	01/07/2008	–
	WorldView 1	18/09/2007	–
	QuickBird 2	18/10/2001	–
DMC (Disaster Monitoring Constellation)	Beijing-1	27/10/2005	–
	UK-DMCSat-1	27/11/2003	–
	BILSAT-1	27/09/2003	–
	AISAT-1	28/11/2002	–
	NigeriaSat-1	27/11/2003	–
DubaiSat-1	DubaiSat-1	TBD	
Earth Explorers	EarthCARE	01/07/2011	–
	Swarm	01/01/2010	–
	ADM-Aeolus (Earth Explorer Atmospheric Dynamics)	01/06/2009	–
	SMOS (Soil Moisture and Ocean Salinity)	01/05/2009	–
	CryoSat-2	01/05/2009	–
	GOCE (Gravity Field & Steady-State Ocean Explorer)	16/03/2009	–
EnMAP	EnMAP	01/01/2012	–
ENVISAT (Environmental Satellite)	ENVISAT	01/03/2002	–

(*Continued*)

Satellite Program	Satellite Platform	Launch Date	End Date
EO-1 (Earth Observing-1)	EO-1	21/11/2000	–
EOS (Earth Observing System)	GRACE	17/03/2002	–
	AQUA	04/05/2002	–
	TERRA	18/12/1999	–
EROS (Earth Remote Observation System)	EROS-B	25/04/2006	–
	EROS-A	05/12/2000	
	ERS-2	21/04/1995	–
	ERS-1	17/07/1991	10/03/2000
Etalon	Etalon-2	13/07/1989	–
	Etalon-1	26/01/1989	–
ETS-8	ETS-8	10/03/2007	–
FORMOSAT	FORMOSAT 3	16/04/2006	–
	FORMOSAT 2	21/05/2001	
	FORMOSAT 1	27/01/1999	17/06/2004
GeoEye	GeoEye-1	06/09/2008	–
GFO-1	GFO-1	22/04/1998	–
GFZ-1	GFZ-1	19/04/1995	08/6/1999
GOES (Geostationary Operational Environmental Satellites)	GOES 13	24/05/2006	–
	GOES 12	23/07/2001	–
	GOES 11	03/05/2000	–
	GOES 10	25/04/1997	–
	GOES 9	23/05/1995	–
	GOES 8	13/04/1994	05/05/2004
	GOES 7	26/02/1987	–
	GOES 6	28/04/1983	–
	GOES 5	22/05/1981	18/07/1990
	GOES 4	09/09/1980	–
	GOES 3	16/06/1978	–
	GOES 2	16/06/1977	–
	GOES 1	16/10/1975	07/03/1985
ICESat	ICESat	05/03/2003	–
Ikonos	IKONOS-2	24/09/1999	–

(*Continued*)

Satellite Program	Satellite Platform	Launch Date	End Date
INSAT (Indian National Satellite)	INSAT 3A	09/04/2003	
	KALPANA 1	12/09/2002	–
	INSAT 2D	03/06/1997	04/10/1997
	INSAT 2E	02/04/1999	–
	INSAT 2C	06/12/1995	–
	INSAT 2B	22/07/1993	07/11/2000
	INSAT 2A	09/07/1992	01/04/2002
	INSAT 1D	12/06/1990	14/05/2002
	INSAT 1C	21/07/1988	22/11/1989
	INSAT 1B	31/08/1983	01/08/1993
	INSAT 1A	10/04/1982	04/11/1982
IRS (Indian Remote-Sensing Satellites)	IRS P5 (CartoSat-1)	05/05/2005	–
	IRS P6 (ResourceSat-1)	17/10/2003	
	IRS P4 (OceanSat-1)	26/05/1999	–
	IRS 1D	29/09/1997	–
	IRS P3	21/03/1996	01/04/2004
	IRS 1C	28/12/1995	
	IRS P2	15/10/1994	01/09/1997
	IRS 1B	29/08/1991	29/08/2001
	IRS 1A	17/03/1988	01/01/1992
Jason (Joint Altimetry Satellite Oceanography Network)	JASON 1	07/12/2001	–
JERS (Japan Earth Resources Satellite)	JERS 1	11/02/1992	11/10/1998
KOMPSAT (Korea Multipurpose Satellite)	KOMPSAT 2	28/07/2006	–
	KOMPSAT 1	20/12/1999	–

(*Continued*)

Satellite Program	Satellite Platform	Launch Date	End Date
Landsat	Landsat 7	15/04/1999	–
	Landsat 5	01/04/1984	–
	Landsat 4	16/07/1982	15/06/2001
	Landsat 3	05/03/1978	31/03/1983
	Landsat 2	22/01/1975	05/02/1982
	Landsat 1	23/07/1972	06/01/1978
LDCM (Landsat Data Continuity Mission)	LDCM 1	01/07/2014	–
MACSat (Medium-Sized Aperture Camera Satellite)	RAZAKSAT 1	01/04/2007	–
METEOR-3	METEOR-3	24/10/1985	–
METEOR-3M	METEOR-3M 1	10/12/2001	05/04/2006
	METEOR-3M 2	01/07/2007	
MetOp (Operational Meteorological Satellites)	MetOp 1	19/10/2006	–
MFG (Meteosat First Generation)	Meteosat 7	03/09/1997	–
	Meteosat 6	20/11/1993	
	Meteosat 5	02/03/1991	–
	Meteosat 4	06/03/1989	01/11/1995
	Meteosat 3	15/06/1988	15/11/1995
	Meteosat 2	19/06/1981	06/12/1991
	Meteosat 1	23/11/1977	01/11/1979
Monitor-E (Monitor Experimental)	MONITOR-E 1	26/08/2005	–
MSG (Meteosat Second Generation)	MSG 4	01/07/2012	–
	MSG 3	01/04/2009	
	MSG 2	22/12/2005	–
	MSG 1	28/08/2002	–
MTI (Multispectral Thermal Imager)	MTI	12/03/2000	–
NPOESS (6X)	NPOESS (6X)	TBD	–

(*Continued*)

Satellite Program	Satellite Platform	Launch Date	End Date
OrbView	Orbview 5	09/06/2008	–
	GeoEye 1	16/03/2007	23/04/2007
	OrbView 3	26/06/2003	23/04/2007
	OrbView 2	01/08/1997	–
	OrbView 1	01/04/1995	
PLEIADES	PLEIADES 2	01/11/2009	–
	PLEIADES 1	01/11/2008	–
POES (Polar Orbiting Environmental Satellites)	NOAA 18	20/05/2005	
	NOAA 17	24/06/2002	
	NOAA 16	21/09/2000	–
	NOAA 15	13/05/1998	
	NOAA 14	30/12/1994	
	NOAA 12	14/05/1991	05/04/2001
	NOAA 11	24/09/1988	13/09/1994
	NOAA 10	17/09/1986	30/08/2001
	NOAA 9	12/12/1984	07/11/1988
	NOAA 8	28/03/1983	31/10/1985
	NOAA 7	23/06/1981	06/07/1986
	NOAA 6	27/06/1979	31/03/1987
PROBA (Project for On-Board Autonomy)	PROBA 1	22/10/2001	
RadarSat constellation	RadarSat 2	14/12/2007	–
	RadarSat 1	04/11/1995	–
RapidEye Constellation	RapidEye A - E	29/08/2008	–
RESURS 01	RESURS-01 4	10/07/1998	07/02/2002
	RESURS-01 3	04/11/1994	31/12/2000
	RESURS-01 2	20/04/1988	20/04/1995
	RESURS-01 1	03/10/1985	03/10/1988
Resurs-DK1 (Resurs—High Resolution 1)	RESURS DK1	15/06/2006	–
RESURS-F	RESURS-F 1	01/07/1988	01/07/1993
	RESURS-F 3	01/07/1993	01/07/1994
	RESURS-F 2	01/07/1987	01/07/1994

(*Continued*)

Satellite Program	Satellite Platform	Launch Date	End Date
RISAT (Radar Imaging Satellite)	RISAT 1	01/11/2007	–
SAC-C	SAC-C	21/11/2000	–
SAC-D	SAC-D (AQUARIUS)	31/03/2009	–
SAOCOM	SAOCOM 1 (2X)	31/12/2007	–
SciSat (Science Satellite)	SCISAT 1	13/08/2003	–
SEASAT	SEASAT	27/06/1978	10/10/1978
Sentinel	Sentinel 1A	30/06/2011	–
	Sentinel 1B	30/06/2013	–
	Sentinel 2A	30/11/2011	–
	Sentinel 2B	30/06/2013	–
SPOT (Système Pour l'Observation de la Terre)	SPOT 5	04/05/2002	
	SPOT 4	24/03/1998	–
	SPOT 3	26/09/1993	14/11/1996
	SPOT 2	22/01/1990	–
	SPOT 1	22/02/1986	01/11/2003
SSOT	SSOT	31/12/2010	–
TANDEM-X	TANDEM-X	31/12/2009	–
TerraSAR	TerraSAR X	31/10/2006	–
TerraSAR-X	Terrasar-X2	31/12/2011	–
THEOS (Thailand Earth Observation System)	THEOS	01/07/2007	–
TOPEX/Poseidon	TOPEX/ Poseidon	01/01/1992	15/12/2005
TOPSAT	TOPSAT 1	27/10/2005	–
TSINGHUA	TSINGHUA 1	28/06/2000	–
VENUS	VENUS	30/06/2009	–
X-SAT	X-SAT	01/03/2008	–

Source: Adapted from http://eo.belspo.be/Directory/Satellites.aspx, http://www.tbs-satellite.com/tse/online/mis_teledetection_res.html, http://ilrs.gsfc.nasa.gov/satellite_missions/list_of_satellites/, and http://www.space-risks.com/SpaceData/index.php?id_page=5&PHPSESSID=eeb322b0515c0d54654e63d319191235 accessed on March 4, 2009.

APPENDIX IV

A Selection of Past, Current, and Planned Sensor Payloads on Remote-Sensing Satellites

Satellite Platform	Sensor	Sensor Type	Spatial Resolution	
			Panchromatic (m)	Multispectral (m)
ADEOS 1	AVNIR	Imaging multispectral radiometers (vis/IR)	8	16
	NSCAT	Imaging microwave radars	50000	–
	OCTS	Imaging multispectral radiometers (vis/IR)	–	700
	POLDER	Multiple direction/ polarization instruments	6000	–
ADEOS 2	AMSR	Imaging microwave radars	3000	–
	GLI	Imaging multispectral radiometers (vis/IR)	250	–
	POLDER	Multiple direction/ polarization instruments	6000	–
	SeaWinds	Imaging microwave radars	50000	–

(*Continued*)

Satellite Platform	Sensor	Sensor Type	Spatial Resolution	
			Panchromatic (m)	Multispectral (m)
ALMAZ 1	EKOR-A	Imaging microwave radars	10	–
ALOS 1	AVNIR-2	Imaging multispectral radiometers (vis/IR)	–	10
	PALSAR	Imaging microwave radars	7	–
	PRISM	High resolution optical imagers	2,5	–
AISAT-1	ESIS	Imaging multispectral radiometers (vis/IR)	–	32
AQUA	AIRS	Atmospheric temperature and humidity sounders	–	2,3
	AMSR-E	Imaging multispectral radiometers (passive microwave)	4000	–
	AMSU	Atmospheric temperature and humidity sounders	40000	–
	CERES	Imaging multispectral radiometers (vis/IR)	10000	–
	HSB	Atmospheric temperature and humidity sounders	13500	–
	MODIS	Imaging multispectral radiometers (vis/IR)	–	250
Beijing-1	CMT	High resolution optical imagers	4	–
	ESIS	Imaging multispectral radiometers (vis/IR)	–	32
BILSAT-1	COBAN	Imaging multispectral radiometers (vis/IR)	–	120
	MSIS	Imaging multispectral radiometers (vis/IR)	–	26
	PanCam	High resolution optical imagers	12	–
BIRD 1	HORUS	High resolution optical imagers	14,6	–
	HSRS	Imaging multispectral radiometers (vis/IR)	–	372
	WAOSS-B	Imaging multispectral radiometers (vis/IR)	–	185

(*Continued*)

Satellite Platform	Sensor	Sensor Type	Spatial Resolution	
			Panchromatic (m)	Multispectral (m)
CBERS-1 (ZY-1)	HRCC	Imaging multispectral radiometers (vis/IR)	19,5	19,5
	IRMSS	Imaging multispectral radiometers (vis/IR)	78	78
	WFI	Imaging multispectral radiometers (vis/IR)	–	260
CBERS-2 (ZY-1B)	HRCC	Imaging multispectral radiometers (vis/IR)	19,5	19,5
	IRMSS	Imaging multispectral radiometers (vis/IR)	78	78
	WFI	Imaging multispectral radiometers (vis/IR)	–	260
CBERS-2b	CCD	Imaging multispectral radiometers (vis/IR)	–	1000
	HRC	High resolution optical imagers	8	–
	IRMSS	Imaging multispectral radiometers (vis/IR)	78	78
	WFI	Imaging multispectral radiometers (vis/IR)	–	260
CBERS-3	IRMSS-2	Imaging multispectral radiometers (vis/IR)	–	40
	MUXCAM	Imaging multispectral radiometers (vis/IR)	–	20
	PANMUX	Imaging multispectral radiometers (vis/IR)	5	–
	WFI-2	Imaging multispectral radiometers (vis/IR)	–	73
CBERS-4	IRMSS-2	Imaging multispectral radiometers (vis/IR)	–	40
	MUXCAM	Imaging multispectral radiometers (vis/IR)	–	20
	PANMUX	Imaging multispectral radiometers (vis/IR)	5	–
	WFI-2	Imaging multispectral radiometers (vis/IR)	–	73
CLOUDSAT 1	CPR	Cloud profile and rain radars	1,2	–
COSMOS-1870	EKOR	Imaging microwave radars	25	–

(*Continued*)

Satellite Platform	Sensor	Sensor Type	Spatial Resolution	
			Panchromatic (m)	Multispectral (m)
COSMOS-2373	KVR-1000	Film Camera	2	–
	TK-350	Film Camera	10	–
COSMO-SKYMED 1	SAR 2000	Imaging microwave radars	1	–
COSMO-SKYMED 2	SAR 2000	Imaging microwave radars	1	–
COSMO-SKYMED 3	SAR 2000	Imaging microwave radars	1	–
COSMO-SKYMED 4	SAR 2000	Imaging microwave radars	1	–
CryoSat-2	SIRAL	Radar altimeters	–	–
DubaiSat-1	Not defined	Imaging multispectral radiometers	–	–
ENVISAT	AATSR	Imaging multispectral radiometers (vis/IR)	–	1000
	ASAR	Imaging microwave radars	28	–
	GOMOS	Atmospheric chemistry instruments	1700	–
	LRR	Lidars	–	–
	MERIS	Imaging multispectral radiometers (vis/IR)	–	260
	MIPAS	Atmospheric chemistry instruments	–	3000
	MRW	Imaging multispectral radiometers (passive microwave)	20000	–
	RA-2	Radar altimeters	–	–
	SCIAMACHY	Atmospheric chemistry instruments	32000	–
EO-1	ALI	Imaging multispectral radiometers (vis/IR)	10	30
	Hyperion	Imaging multispectral radiometers (vis/IR)	–	30
	LAC	Imaging multispectral radiometers (vis/IR)	–	250

(Continued)

Satellite Platform	Sensor	Sensor Type	Spatial Resolution	
			Panchromatic (m)	Multispectral (m)
EROS-A	PIC	High resolution optical imagers	1,9	–
EROS-B	PIC-2	High resolution optical imagers	0,7	–
ERS-1	AMI SAR	Imaging microwave radars	10	–
	AMI WS	Imaging microwave radars	50000	–
	ATSR IRR	Imaging multispectral radiometers (vis/IR)	–	1000
	ATSR MWS	Imaging microwave radars	20000	–
	RA	Radar altimeters	16	–
ERS-2	AMI SAR	Imaging microwave radars	10	–
	AMI WS	Imaging microwave radars	50000	–
	ATSR IRR	Imaging multispectral radiometers (vis/IR)	–	1000
	ATSR MWS	Imaging microwave radars	20000	–
	GOME	Atmospheric chemistry instruments	40000	–
	MWR	Imaging multispectral radiometers (passive microwave)	20000	–
	RA	Radar altimeters	16	–
Etalon 1	Etalon 1	Retro-reflector array	–	–
Etalon 2	Etalon 1	Retro-reflector array	–	–
ETS-8	ETS-VIII	Laser retroreflector array	–	–
FORMOSAT 1	OCI	Imaging multispectral radiometers (vis/IR)	–	800
FORMOSAT 2	RSI	Imaging multispectral radiometers (vis/IR)	2	8
FORMOSAT 3	TIP	Atmospheric chemistry instruments	–	–
GEOEYE-1	GIS MS	High resolution optical imagers	–	1.64

(*Continued*)

Satellite Platform	Sensor	Sensor Type	Spatial Resolution	
			Panchromatic (m)	Multispectral (m)
GEOEYE-1	GIS PAN	High resolution optical imagers	–	0.41
GFO-1	GFO-1	Radar altimeters	–	–
GFZ-1	GFZ-1	Retroreflector array	–	–
GOES 8	GOES Imager	Imaging multispectral radiometers (vis/IR)	–	1000
GOES 9	GOES Sounder	Atmospheric temperature and humidity sounders	–	8000
	GOES Imager	Imaging multispectral radiometers (vis/IR)	–	1000
	GOES Sounder	Atmospheric temperature and humidity sounders	–	8000
	GOES Sounder	Atmospheric temperature and humidity sounders	–	8000
GOES 12	GOES Imager	Imaging multispectral radiometers (vis/IR)	–	1000
	GOES Sounder	Atmospheric temperature and humidity sounders	–	8000
GOES 13	GOES Imager	Imaging multispectral radiometers (vis/IR)	–	1000
	GOES Sounder	Atmospheric temperature and humidity sounders	–	8000
ICEsat	GLAS	Laser Altimeter	–	–
IKONOS-2	OSA	High resolution optical imagers	1	4
INSAT 1A	VHRR	Imaging multispectral radiometers (vis/IR)	–	2750
INSAT 1B	VHRR	Imaging multispectral radiometers (vis/IR)	–	2750
INSAT 1C	VHRR	Imaging multispectral radiometers (vis/IR)	–	2750
INSAT 1D	VHRR	Imaging multispectral radiometers (vis/IR)	–	2750
INSAT 2A	VHRR bis	Imaging multispectral radiometers (vis/IR)	–	2000

(*Continued*)

Satellite Platform	Sensor	Sensor Type	Spatial Resolution	
			Panchromatic (m)	Multispectral (m)
INSAT 2B	VHRR bis	Imaging multispectral radiometers (vis/IR)	–	2000
INSAT 2C	VHRR bis	Imaging multispectral radiometers (vis/IR)	–	2000
INSAT 2D	VHRR bis	Imaging multispectral radiometers (vis/IR)	–	2000
INSAT 2E	VHRR bis	Imaging multispectral radiometers (vis/IR)	–	2000
INSAT 3A	CCD	Imaging multispectral radiometers (vis/IR)	–	1000
	VHRR 2	Imaging multispectral radiometers (vis/IR)	–	2000
IRS 1A	LISS I	Imaging multispectral radiometers (vis/IR)	–	72,5
	LISS II	Imaging multispectral radiometers (vis/IR)	–	36,25
IRS 1B	LISS I	Imaging multispectral radiometers (vis/IR)	–	72,5
	LISS II	Imaging multispectral radiometers (vis/IR)	–	36,25
IRS 1C	LISS III	Imaging multispectral radiometers (vis/IR)	–	23,5
	PAN	High resolution optical imagers	5,8	–
	WiFS	Imaging multispectral radiometers (vis/IR)	–	188
IRS 1D	LISS III	Imaging multispectral radiometers (vis/IR)	–	23,5
	PAN	High resolution optical imagers	5,8	–
	WiFS	Imaging multispectral radiometers (vis/IR)	–	188
IRS 2B (Carosat 2A)	LISS IV	Imaging multispectral radiometers (vis/IR)	–	–
IRS P2	LISS IIM	Imaging multispectral radiometers (vis/IR)	–	32
IRS P3	MOS	Imaging multispectral radiometers (vis/IR)	–	523
	WiFS	Imaging multispectral radiometers (vis/IR)	–	188

(*Continued*)

Satellite Platform	Sensor	Sensor Type	Spatial Resolution	
			Panchromatic (m)	Multispectral (m)
IRS P4 (OceanSat-1)	MSMR	Imaging microwave radars	22000	–
	OCM	Imaging multispectral radiometers (vis/IR)	–	236
IRS P5 (CartoSat-1)	PAN-A	High resolution optical imagers	2,5	–
	PAN-F	High resolution optical imagers	2,5	–
IRS P6 (ResourceSat-1)	AWiFS	Imaging multispectral radiometers (vis/IR)	–	56
	LISS III	Imaging multispectral radiometers (vis/IR)	–	23,5
	LISS IV	Imaging multispectral radiometers (vis/IR)	–	5,8
JASON 1	JMR	Imaging multispectral radiometers (passive microwave)	–	–
	Poseidon-2	Radar altimeters	–	–
JASON 2	AMR	Atmospheric temperature and humidity sounders	–	–
	DORIS	Radar altimeters	–	–
	GPSP	Radar altimeters	–	–
	Poseidon-3	Radar altimeters	–	–
JERS 1	OPS	High resolution optical imagers	–	18,3
	SAR	Imaging microwave radars	18	–
KALPANA 1	VHRR 2	Imaging multispectral radiometers (vis/IR)	–	2000
KeyHole 1 to 4B	PAN	Film Camera	1,8	–
KOMPSAT 1	EOC	High resolution optical imagers	6,6	–
	OSMI	Imaging multispectral radiometers (vis/IR)	–	1000

(*Continued*)

Satellite Platform	Sensor	Sensor Type	Spatial Resolution	
			Panchromatic (m)	Multispectral (m)
KOMPSAT 2	MSC	High resolution optical imagers	1	4
LANDSAT 1	MSS	Imaging multispectral radiometers (vis/IR)	–	80
	RBV	Imaging multispectral radiometers (vis/IR)	–	80
LANDSAT 2	MSS	Imaging multispectral radiometers (vis/IR)	–	80
	RBV	Imaging multispectral radiometers (vis/IR)	–	80
LANDSAT 3	MSS	Imaging multispectral radiometers (vis/IR)	–	80
	RBV	Imaging multispectral radiometers (vis/IR)	–	80
LANDSAT 4	MSS	Imaging multispectral radiometers (vis/IR)	–	80
	TM	Imaging multispectral radiometers (vis/IR)	–	30
LANDSAT 5	MSS	Imaging multispectral radiometers (vis/IR)	–	80
	TM	Imaging multispectral radiometers (vis/IR)	–	30
LANDSAT 7	ETM+	Imaging multispectral radiometers (vis/IR)	15	30
METEOR-3	KLIMAT	Imaging multispectral radiometers (vis/IR)	450	–
	ScaRaB	Earth radiation budget radiometers	–	60000
	SM	Atmospheric temperature and humidity sounders	–	42000
	TOMS	Atmospheric chemistry instruments	–	47000

(*Continued*)

Satellite Platform	Sensor	Sensor Type	Spatial Resolution	
			Panchromatic (m)	Multispectral (m)
METEOR-3M 1	KGI-4C	Gravity, magnetic field, and geodynamic instruments	–	–
	KLIMAT	Imaging multispectral radiometers (vis/IR)	450	–
	MIVZA	Atmospheric temperature and humidity sounders	–	40000
	MR-2000-M1	Imaging multispectral radiometers (vis/IR)	700	–
	MSGI-MKA	Gravity, magnetic field, and geodynamic instruments	–	–
	MSU-E	Imaging multispectral radiometers (vis/IR)	–	33
	MTVZA	Imaging multispectral radiometers (passive microwave)	–	12000
	SAGE III	Atmospheric chemistry instruments	–	2000
	SFM-2	Atmospheric chemistry instruments	–	–
METEOR-3M 2	IKFS-2	Atmospheric temperature and humidity sounders	63000	–
	KGI-4C	Gravity, magnetic field, and geodynamic instruments	–	–
	KMSS	Imaging multispectral radiometers (vis/IR)	–	50
	MSGI-MKA	Gravity, magnetic field, and geodynamic instruments	–	–
	MSU-MR	Imaging multispectral radiometers (vis/IR)	–	1000
	MTVZA-GY	Imaging microwave radars	–	32000
	OBRC	Imaging microwave radars	400	–

(*Continued*)

Satellite Platform	Sensor	Sensor Type	Spatial Resolution	
			Panchromatic (m)	Multispectral (m)
Meteosat 1	MVIRI	Imaging multispectral radiometers (vis/IR)	–	2500
Meteosat 2	MVIRI	Imaging multispectral radiometers (vis/IR)	–	2500
Meteosat 3	MVIRI	Imaging multispectral radiometers (vis/IR)	–	2500
Meteosat 4	MVIRI	Imaging multispectral radiometers (vis/IR)	–	2500
Meteosat 5	MVIRI	Imaging multispectral radiometers (vis/IR)	–	2500
Meteosat 6	MVIRI	Imaging multispectral radiometers (vis/IR)	–	2500
MetOp 1	AMSU	Atmospheric temperature and humidity sounders	40000	–
	ASCAT	Scatterometers	25000	–
	AVHRR 3	Imaging multispectral radiometers (vis/IR)	–	1100
	GOME-2	Atmospheric chemistry instruments	40000	–
	GRAS	Atmospheric temperature and humidity sounders	100000	–
	IASI	Atmospheric temperature and humidity sounders	25000	–
	MHS	Atmospheric temperature and humidity sounders	16000	–
MONITOR-E 1	PSA	High resolution optical imagers	8	–
	RDSA	Imaging multispectral radiometers (vis/IR)	–	20
MSG 1	GERB	Earth radiation budget radiometers	–	40000
	SEVIRI	Imaging multispectral radiometers (vis/IR)	–	1000
MSG 2	GERB	Earth radiation budget radiometers	–	40000
	SEVIRI	Imaging multispectral radiometers (vis/IR)	–	1000

(*Continued*)

Satellite Platform	Sensor	Sensor Type	Spatial Resolution	
			Panchromatic (m)	Multispectral (m)
MSG 3	GERB	Earth radiation budget radiometers	–	40000
	SEVIRI	Imaging multispectral radiometers (vis/IR)	–	1000
MSG 4	GERB	Earth radiation budget radiometers	–	40000
	SEVIRI	Imaging multispectral radiometers (vis/IR)	–	1000
MTI	HXRS	Earth radiation budget radiometers	–	–
	MTI	Imaging multispectral radiometers (vis/IR)	–	5
NigeriaSat-1	ESIS	Imaging multispectral radiometers (vis/IR)	–	32
NOAA 6	AVHRR 1	Imaging multispectral radiometers (vis/IR)	–	1100
NOAA 7	AVHRR 2	Imaging multispectral radiometers (vis/IR)	–	1100
NOAA 8	AVHRR 2	Imaging multispectral radiometers (vis/IR)	–	1100
NOAA 9	AVHRR 2	Imaging multispectral radiometers (vis/IR)	–	1100
NOAA 10	AVHRR 2	Imaging multispectral radiometers (vis/IR)	–	1100
NOAA 11	AVHRR 2	Imaging multispectral radiometers (vis/IR)	–	1100
NOAA 12	AVHRR 2	Imaging multispectral radiometers (vis/IR)	–	1100
NOAA 14	AVHRR 2	Imaging multispectral radiometers (vis/IR)	–	1100
NOAA 15	ATOVS	Imaging multispectral radiometers (vis/IR)	–	–
	AVHRR 3	Imaging multispectral radiometers (vis/IR)	–	1100
NOAA 16	ATOVS	Imaging multispectral radiometers (vis/IR)	–	–
	AVHRR 3	Imaging multispectral radiometers (vis/IR)	–	1100

(*Continued*)

Satellite Platform	Sensor	Sensor Type	Spatial Resolution	
			Panchromatic (m)	Multispectral (m)
NOAA 17	ATOVS	Imaging multispectral radiometers (vis/IR)	–	–
	AVHRR 3	Imaging multispectral radiometers (vis/IR)	–	1100
NOAA 18	ATOVS	Imaging multispectral radiometers (vis/IR)	–	–
	AVHRR 3	Imaging multispectral radiometers (vis/IR)	–	1100
NPOESS	VIIRS	VIIRS Visible/Infrared Imager/Radiometer Suite	TBD	TBD
OrbView 1	OTD	Imaging multispectral radiometers (vis/IR)	10000	–
OrbView 2	SeaWiFS	Ocean Color instruments	–	1000
OrbView 3	OHIRIS	High resolution optical imagers	1	4
OrbView 5	Panchromatic	High resolution optical imagers	0.41	–
	Multispectral	High resolution optical imagers	–	1.64
PLEIADES 1	HiRI	High resolution optical imagers	0,7	2,8
PLEIADES 2	HiRI	High resolution optical imagers	0,7	2,8
PROBA 1	CHRIS	Imaging multispectral radiometers (vis/IR)	–	17
	HRC	High resolution optical imagers	8	–
	WAC	High resolution optical imagers	–	–
Quickbird 2	BGIS 2000	High resolution optical imagers	0,61	2,8
RadarSat 1	SAR	Imaging microwave radars	3	–
	SAR	Imaging microwave radars	8	–
RadarSat 2	SAR	Imaging microwave radars	100	–

(*Continued*)

Satellite Platform	Sensor	Sensor Type	Spatial Resolution	
			Panchromatic (m)	Multispectral (m)
RapidEye A - E	REIS	Imaging multispectral radiometers (vis/IR)	–	6,5
RAZAKSAT 1	MAC	Imaging multispectral radiometers (vis/IR)	2,5	5
RESURS DK1	ESI	High resolution optical imagers	1	2,5
RESURS-01 1	MSU-E	Imaging multispectral radiometers (vis/IR)	–	33
	MSU-S	Imaging multispectral radiometers (vis/IR)	–	240
	MSU-SK	Imaging multispectral radiometers (vis/IR)	–	170
RESURS-01 2	MSU-E	Imaging multispectral radiometers (vis/IR)	–	33
	MSU-SK	Imaging multispectral radiometers (vis/IR)	–	170
RESURS-01 3	MSU-E	Imaging multispectral radiometers (vis/IR)	–	33
	MSU-SK	Imaging multispectral radiometers (vis/IR)	–	170
RESURS-01 4	MSU-E1	Imaging multispectral radiometers (vis/IR)	–	35
	MSU-SK1	Imaging multispectral radiometers (vis/IR)	–	210
	ScaRaB	Earth radiation budget radiometers	–	60000
RESURS-F 1	KATE-200	Imaging multispectral radiometers (vis/IR)	–	15
	KFA-1000	Film Camera	–	4
RESURS-F 2	MK-4	Film Camera	–	5
RESURS-F 3	KF1-3000	Film Camera	2	–
RISAT 1	RISAT-SAR	Imaging microwave radars	3	–
SAC-C	HRTC	High-Resolution Technological Camera	–	35
	HSC	High Sensitivity Camera	–	–
	MMRS	Multispectral Medium Resolution Scanner	–	175

(*Continued*)

Satellite Platform	Sensor	Sensor Type	Spatial Resolution	
			Panchromatic (m)	Multispectral (m)
SAC-D (AQUARIUS)	HSC	High Sensitivity Camera	–	200
	NIRST	New InfraRed Scanner	–	350
	AQUARIUS	Not defined	–	–
SAOCOM 1 (2X)	SAR	Imaging microwave radars	–	–
SCISAT 1	ACE-FTS	Atmospheric chemistry instruments	–	3000
	MAESTRO	Atmospheric chemistry instruments	–	–
SEASAT	ALT	Radar altimeters	–	–
	SAR	Imaging microwave radars	25	–
	SASS	Scatterometers	50000	–
	SMMR	Imaging microwave radars	22000	–
	VIRR	Imaging multispectral radiometers (vis/IR)	2300	–
SENTINEL 1A	Not defined	Not defined	–	5
SENTINEL-1B	Not defined	Not defined	–	5
SENTINEL-2A	Not defined	Not defined	–	10
SENTINEL-2B	Not defined	Not defined	–	10
SMOS (Soil Moisture and Ocean Salinity)	MIRAS	Imaging microwave radars	–	–
SPOT 1	HRV	Imaging multispectral radiometers (vis/IR)	10	20
SPOT 2	HRV	Imaging multispectral radiometers (vis/IR)	10	20
SPOT 3	HRV	Imaging multispectral radiometers (vis/IR)	10	20
SPOT 4	HRVIR	Imaging multispectral radiometers (vis/IR)	10	20
	VEGETATION	Imaging multispectral radiometers (vis/IR)	–	1000
SPOT 5	HRG	Imaging multispectral radiometers (vis/IR)	5	10
	HRS	High resolution optical imagers	5	–
	VEGETATION	Imaging multispectral radiometers (vis/IR)	–	1000

(*Continued*)

Satellite Platform	Sensor	Sensor Type	Spatial Resolution	
			Panchromatic (m)	Multispectral (m)
SSOT	MULTISPECTRAL	Imaging multispectral radiometers (vis/IR)	–	5.8,10
	PAN	High resolution optical imagers	–	1.45,10
TANDEM-X	TSX-SAR	Imaging microwave radars	–	16
TERRA	ASTER	Imaging multispectral radiometers (vis/IR)	–	15
	CERES	Imaging multispectral radiometers (vis/IR)	10000	–
	MISR	Imaging multispectral radiometers (vis/IR)	–	275
	MODIS	Imaging multispectral radiometers (vis/IR)	–	250
	MOPITT	Atmospheric chemistry instruments	–	22000
TerraSAR X	TSX-SAR	Imaging microwave radars	1	–
TERRASAR-X2	Not defined	Not defined	–	–
THEOS	MS	Imaging multispectral radiometers (vis/IR)	–	15
	PAN	High resolution optical imagers	2	–
TOPEX/ Poseidon	Not defined	Microwave radiometer	–	–
TOPSAT 1	RALCam 1	Imaging multispectral radiometers (vis/IR)	2,5	5
TSINGHUA 1	MEIS	Imaging multispectral radiometers (vis/IR)	–	39
UK-DMCSat-1	ESIS	Imaging multispectral radiometers (vis/IR)	–	32
VENUS	Not defined	The multispectral camera	–	10
WorldView 1	WV60	High resolution optical imagers	0,5	–
WorldView 2	WV110	Imaging multispectral radiometers (vis/IR)	0,5	2
X-SAT	IRIS	Imaging multispectral radiometers (vis/IR)	–	10

Source: Adapted from http://eo.belspo.be/Directory/Satellites.aspx, http://www.tbs-satellite.com/tse/online/mis_teledetection_res.html, http://ilrs.gsfc.nasa.gov/satellite_missions/list_of_satellites/, and http://www.space-risks.com/SpaceData/index.php?id_page=5&PHPSESSID=eeb322b0515c0d54654e63d319191235 [4 March 2009].

References

ABMI Index Development Group. 2006. *The Alberta Biodiversity Monitoring Program Species Pyramid*. Alberta Biodiversity Monitoring Program, Alberta, Canada. Available online http://www.abmi.ca web site accessed March 12, 2009.

Adrados, C., C. Baltzinger, G. Janeau, and D. Pépin. 2008. Red deer *Cervus elaphus* resting place characteristics obtained from differential GPS data in a forest habitat. *European Journal of Wildlife Research* 54: 487–494.

Ahearn, S. 1988. Combining Laplacian images of different spatial frequencies (scales): implications for remote sensing analysis. *IEEE Transactions on Geoscience and Remote Sensing* 26: 826–831.

Ahern, F. J. 2007. *Ecosystem Status and Trends: Indicators Using Earth Observation by Satellite*. Available online http://www.eman-ese.ca/eman/reports/publications/TerreVista-En.pdf web site accessed September 10, 2008.

Albert, D. P., W. M. Gesler, and B. Levergood, eds. 2000. *Spatial Analysis, GIS, and Remote Sensing Applications in the Health Sciences*. CRC Press, Boca Raton, Florida.

Aldrich, R. C. 1979. *Remote Sensing of Wildland Resources: A State-of-the-Art Review*. USDA Forest Service General Technical Report RM-71. Rocky Mountain Forest and Range Experimental Station, Fort Collins, Colo.

Alisuakskas, R. T., J. W. Charlwood, and D. K. Kellett. 2006. Vegetation correlates of the history and density of nesting by Ross's Geese and Lesser Snow Geese at Karrak Lake, Nunavut. *Arctic* 59: 201–210.

Allen, R. B., P. J. Bellingham, and S. K. Wiser. 2003. Developing a forest biodiversity monitoring approach for New Zealand. *New Zealand Journal of Ecology* 27: 207–220.

Allen, T. F. H. 1998. The landscale "level" is dead: persuading the family to take it off the respirator. In Peterson, D. L. and V. Parker, eds., *Ecological Scale: Theory and Applications*. Columbia University Press, New York, U.S. 35–54.

Altmann, S. A. and J. Altmann. 2005. The transformation of behaviour field studies. In Lucas, J. R. and L. W. Simmons, eds., *Essays in Animal Behaviour*. Elsevier Academic Press, Burlington, MA. 57–79.

Amis, M. A., M. Rouget, A. Balmford, W. Thuiller, C. J. Kleynhans, J. Day, and J. Nel. 2007. Predicting freshwater habitat integrity using land-use surrogates. *Water SA* 33: 215–221.

Ammenberg, P., P. Flink, T. Lindell, D. Pierson, and N. Strömbeck. 2002. Bio-optical modelling combined with remote sensing to assess water quality. *International Journal of Remote Sensing* 23: 1621–1638.

Anderson, G.S. and B.J. Danielson. 1997. The effects of landscape composition and physiognomy on metapopulation size: the role of corridors. *Landscape Ecol* 12: 261–271.

Anderson, J. R., E. E. Hardy, J. T. Roach, and R. E. Witmer. 1976. *A Land Use and Land Cover Classification System for Use with Remote Sensor Data*. U.S. Geological Survey Professional Paper 964. Washington, DC.

Andréfouët, S. 2008. Coral reef habitat mapping using remote sensing: a user vs producer perspective. Implications for research, management, and capacity building. *Journal of Spatial Sciences* 53: 113–129.

Andréfouët, S., M. J. Costello, M. Rast, and S. Sathyendranath. 2008. Earth observation for marine and coastal biodiversity and ecosystems. *Remote Sensing of Environment* 112: 3297–3299.

Andrew, M. E., and S. L. Ustin. 2006. Spectral and physiological uniqueness of perennial pepperweed *(Lepidium latifolium). Weed Science* 54: 1051–1062.

Ardö, J. 1998. *Remote Sensing of Forest Decline in the Czech Republic.* Lund University Press, Lund, Sweden.

Aronoff, S. 1989. *Geographical Information Systems: A Management Perspective.* WDL Publications, Ottawa, Canada.

Aronoff, S. 2005. *Remote Sensing for GIS Managers.* ESRI Press, Redlands, U.S.

Arroyo-Rodriguez, V., A. Aguirre, J. Benitez-Malvido, and S. Mandujano. 2007. Impact of rainforest fragmentation on the population size of a structurally important palm species: *Astrocaryum mexicanum* at Los Tuxtlas, Mexico. *Biological Conservation* 138: 198–206.

Ashley, E. P. and J. T. Robinson. 1996. Road mortality of amphibians, reptiles and other wildlife on the Long Point Causeway, Lake Erie, Ontario. *Canadian Field-Naturalist* 110: 403–412.

Asner, G. P., S. Goetz, K. Bergen, N. Coops, W. Fan, M. Follows etal. 2008. *Biodiversity Studies in the NASA Remote Sensing Programs.* Available online http://cce.nasa.gov/cgi-bin/meeting_2008/mtg2008_ab_detagenda.pl web site accessed September 9, 2008.

Asner, G. P. and K. B. Heidebrecht. 2005. Desertification alters regional ecosystem-climate interactions. *Global Change Biology* 11: 182–194.

Asner, G. P., J. A. Hicke, and D. B. Lobell. 2003. Per-pixel analysis of forest structure: vegetation indices, spectral mixture analysis, and canopy reflectance modelling. In Wulder, M. A. and S. E. Franklin, eds., *Remote Sensing of Forest Environments: Concepts and Case Studies.* Kluwer Academic Publishers, Boston, MA. 209–254.

Anser, G. P., M. Palace, M. Keller, R. Pereira, Jr., J. N. M. Silva, and J. C. Zweede. 2006. Estimating canopy structure in an Amazon forest from laser range finder and IKONOS satellite observations. *Biotropica* 34: 483–492.

Avery, T. E. 1968. *Interpretation of Aerial Photographs.* 2d ed., Burgess, Minneapolis.

Avery, M. I. and R. H. Haines-Young. 1990. Population estimates for the dunlin *Calidris alpina* derived from remotely sensed satellite imagery of the Flow Country of northern Scotland. *Nature* 344: 860–862.

Badeck, F.-W., A. Bondeau, K. Böttcher, D. Doktor, W. Lucht, J. Schaber, and S. Sitch. 2004. Responses of spring phenology to climate change. *New Phytologist* 162: 295–309.

Bailey, R.G., R. D. Pfister, and J. A. Henderson. 1978. Nature of land and resource classification—a review. *Journal of Forestry* 76: 650–655.

Ball, M. 2008. A visual proliferation: imagery purchase and integration is getting easier. *GEOWorld Magazine.* Available online http://208.106.162.71/ME2/dirmod.asp?sid=119CFE3ACE2A48319AA7DE6A39B80D66&nm=News&type=Publishing&mod=Publications%3A%3AArticle&mid=8F3A7027421841978F18BE895F87F791&tier=4&id=8C0230330AA148B1BDE62FC442A26D99 web site accessed August 12, 2008.

Balmford, A., P. Crane, A. Dobson, R. E. Green, and G. M. Mace. 2006. The 2010 Challenge: data availability, information needs and extraterrestrial insights. *Philosophical Transactions of the Royal Society of London B.* 360: 221–228.

Band, L. E., D. L. Peterson, S. W. Running, J. Coughlan, R. Lammers, J. Dungan, and R. Nemani. 1991. Forest ecosystem processes at the watershed scale: basis for distributed simulation. *Ecological Modelling* 56: 171–196.

Bansal, V. 2007. Potential of GIS to find solutions to space related problems in construction industry. *Proceedings of World Academy of Science, Engineering and Technology* 26: 307–309.

Barber, D. G., P. R. Richard, K. P. Hocheim, and J. Orr. 1991. Calibration of aerial thermal imagery for walrus population assessment. *Arctic* 44, Suppl. 1: 58–65.

Barlow, J. and S. E. Franklin. 2007. Mapping hazardous slope processes using digital data. In Li, J., et al., eds., *Geomatics Solutions for Disaster Management*, Springer-Verlag, New York, U.S. 75–90.

Barnsley, M. 1984. Effects of off-nadir view angles on the detected spectral response of vegetation canopies. *International Journal of Remote Sensing* 5: 715–728.

Barnsley, M. 1999. Digital remotely sensed data and their characteristics. In Longley, P. A., et al., eds., *Geographical Information Systems*. 2d ed., John Wiley and Sons, New York, U.S. 451–466.

Barnsley, M. 2007. *Environmental Modeling: A Practical Introduction*. Taylor and Francis (CRC Press), Boca Raton, Florida.

Bass, B., R. Hansell, and J. Choi. 1998. Towards a simple indicator of biodiversity. *Environmental Monitoring and Assessment* 49: 337–347.

Bater, C. W. 2008. *Assessing Indicators of Forest Sustainability Using LiDAR Remote Sensing*. Unpublished M.Sc. Thesis, University of British Columbia.

Beauvais, G. P., D. A. Keinath, P. Hernandez, L. Master, and R. Thurston. 2006. *Element Distribution Modelling: A Primer*. Version 2.0. Wyoming Natural Diversity Database, University of Wyoming, Laramie, Wyoming.

Beck, L. R., B. M. Lobitz, and B. L. Wood. 2000. Remote sensing and human health: new sensors and new opportunities. *Emerging Infectious Diseases* 6: 217–227.

Beck, P. S. A., P. Jönsson, K.-A. Høgda, S. R. Karlsen, L. Eklundh, L., and A. K. Skidmore. 2007. A ground-validated NDVI dataset for monitoring vegetation dynamics and mapping phenology in Fennoscandia and the Kola peninsula. *International Journal of Remote Sensing* 28: 4311–4330.

Behera, M. D., S. P. S. Kushwaha, and P. S. Roy. 2005. Rapid assessment of biological richness in a part of the Eastern Himalaya: an integrated three-tier approach. *Forest Ecology and Management* 207: 363–384.

Bender, D. J., T. A. Contreras, and L. Fahrig. 1998. Habitat loss and population decline: a meta-analysis of the patch size effect. *Ecology* 79: 517–533.

Benson, B. J. and M. D. MacKenzie. 1995. Effects of sensor spatial resolution on landscape structure parameters. *Landscape Ecology* 10: 113–120.

Berber, M. A. and S. F. Gaber. 2004. Clouds and cloud shadows detection and removal from remote sensing images. *Proceedings of IEEE* 75–79.

Bergen, K. M., A. M. Gilboy, and D. G. Brown. 2007. Multi-dimensional vegetation structure modelling in avian habitat. *Ecological Informatics* 2: 9–22.

Bergen, K. M., T. Zhao, V. Kharuk, Y. Blam, D. G. Brown, L. K. Peterson, and N. Miller. 2008. Changing regimes: forested land cover dynamics in central Siberia 1974 to 2001. *Photogrammetric Engineering and Remote Sensing* 74: 787–798.

Bergen, S., E. Sanderson, P. Coppolillo, J. Berger, and C. Campagna. 2008. Monitoring large wildlife directly through high spatial resolution remote sensing. NASA Carbon Cycle & Ecosystems Joint Science Workshop: NASA Biodiversity and Ecological Forecasting Team Meeting. Available online http://cce.nasa.gov/meeting_2008/biodiversity_pres/may01_08_sanderson.ppt web site accessed August 16, 2008.

Betts, M. G., S. E. Franklin, and R. Taylor. 2003. Interpretation of landscape pattern and habitat change for local indicator species using satellite imagery and geographic information system data in New Brunswick, Canada. *Canadian Journal of Forest Research* 33: 1821–1831.

Bissonette, J. and I. Storch. 2003. *Landscape Ecology and Resource Management: Linking Theory and Practice*. Island Press, Washington, USA.

Blackburn, G. A. 1998. Spectral indices for estimating photosynthetic pigment concentrations: a test using senescent leaves. *International Journal of Remote Sensing* 19: 657–675.

Blackburn, G. A. and E. J. Milton. 1997. An ecological survey of deciduous woodlands using airborne remote sensing and geographical information systems (GIS). *International Journal of Remote Sensing* 18: 1919–1935.

Blais, J. A. R. and H. Esche. 2008. Geomatics and the new cyberinfrastructure. *Geomatica* 62: 11–22.

Blaschke, T., S. Lang and G. J. Hay, eds. 2008. *Object-based Image Analysis: Spatial Concepts for Knowledge-Driven Remote Sensing Applications*. Springer, Berlin.

Bolliger, J., H. H. Wagner, and M. G. Turner. 2007. Identifying and quantifying landscape patterns in space and time. In Kienast, F., et al., eds., *A Changing World: Challenges for Landscape Research*. Springer, Dordrecht, The Netherlands. 177–194.

Boone, J. D., K. C. McGwire, and S. C. St.-Jeor. 2005. Mapping the distribution of Sin Nombre Virus infections in deer mice using remote sensing and geographic information systems. In Majumdar, S. K., J. E. Huffman, F. J. Brenner, and A. I. Panah, eds., *Wildlife Diseases: Landscape Epidemiology, Spatial Distribution and Utilization of Remote Sensing Technology.* Pennsylvania Academy of Science, Easton. 448–458.

Boonstra, R., C. J. Krebs, S. Boutin, and J. M. Eadie. 1994. Finding mammals using far-infrared thermal imaging. *Journal of Mammalogy* 75: 1063–1068.

Borry, F. C., P. B. DeRoover, M. M. Leysen, R. R. DeWulf, and R. E. Goossens. 1993. Evaluation of SPOT and TM data for forest stratification: a case study for small-size poplar stands. *IEEE Transactions on Geoscience and Remote Sensing* 31: 483–490.

Bossler, J., J. Jenson, R. McMaster, and C. Rizos, eds. 2002. *Manual of Geospatial Science and Technology*. Taylor and Francis, London, UK.

Botkin, D. B., H. Saxe, M. G. Araújo, R. Betts, R. H. W. Bradshaw, T. Cedhagen, P. Chesson, et al., 2007. Forecasting the effects of global warming on biodiversity. *BioScience* 57: 227–236.

Bowden, L. W. and E. L. Pruitt, eds. 1975. *Manual of Remote Sensing—Volume II: Interpretation and Applications.* ASP, Falls Church, VA.

Boyce, M. S. and A. Haney 1997. *Ecosystem Management: Applications for Sustainable Forests and Wildlife Resources.* Yale University Press, New Haven, U.S.

Boyce, M. S. and L. McDonald. 1999. Relating populations to habitats using resource selection functions. *Trends in Ecology and Evolution* 14: 268–272.

Boyd, D. S. and F. M. Danson. 2005. Satellite remote sensing of forest resources: three decades of research development. *Progress in Physical Geography* 29: 1–26.

Bradley, B. A. and E. Fleishman. 2008. Can remote sensing of land cover improve species distribution modelling? *Journal of Biogeography* 35: 1158–1159.

Bradley, B. A., R. W. Jacob, J. F. Hermance, and J. F. Mustard. 2007. A curve fitting procedure to derive inter-annual phenologies from time series of noisy satellite NDVI data. *Remote Sensing of Environment* 106: 137–145.

Braun, C. E., ed. 2005. *Techniques for Wildlife Investigations and Management.* The Wildlife Society, Bethesda, MD.

Bray, C. 2006. *Satellite Linked, Wireless Camera Trap Network.* Unpublished B.Eng. Thesis, University of New South Wales.

Brewer, G. D. 1988. Policy sciences, the environment and public health. *Health Promotion* 2: 227–237.

Brewer, G. D. 1995. *Environmental Challenges: Interdisciplinary Opportunities and New Ways of Doing Business.* MISTRA Lecture, Foundation for Strategic Environmental Research, Stockholm, Sweden.

British Columbia Wildlife Habitat Rating Standards. 1999. *Sustainable Resource Management Terrestrial Information Branch for the Resources Inventory Committee.* Ministry of Environment, Lands and Parks Resources Inventory Branch, Victoria.

Brokaw, N. V. L., and R. A. Lent. 1999. Vertical structure. In Hunter, M. L., Jr., ed., *Maintaining Biodiversity in Forest Ecosystems.* Cambridge University Press, Cambridge, UK. 373–399.

Broschart, M. R., C. A. Johnston, and R. J. Naiman. 1989. Predicting beaver colony density in boreal landscapes. *Journal of Wildlife Management* 53: 929–934.

Brosofske, K. D. 2006. Spatial structure and plant diversity in hierarchical landscapes. In Chen, J., et al., eds., *Ecology of Hierarchical Landscapes: From Theory to Applications.* Nova Science Publishers, New York, U.S. 35–69.

Brown, H. E., M. A. Diuk-Wasser, Y. Guan, S. Caskey, and D. Fish. 2008. Comparison of three satellite sensors at three spatial scales to predict larval mosquito presence in Connecticut wetlands. *Remote Sensing of Environment* 112: 2301–2309.

Buckley-Beason, V. A., W. E. Johnson, W. G. Nash, R. Stanyon, J. C. Menninger, C. A. Driscoll, J. Howard et al., 2006. Molecular evidence for species-level distinctions in clouded leopards. *Current Biology* 16: 2371–2376.

Bucur, V. 2003. *Nondestructive Characterization and Imaging of Wood.* Springer-Verlag, Heidelberg, Germany.

Buerkert, A., F. Mahler, and H. Maraschner. 1996. Soil productivity management and plant growth in the Sahel: potential of an aerial monitoring technique. *Plant and Soil* 180: 29–38.

Buiten, H. J. and J. G. P. W. Clevers, eds. 1993. *Land Observation by Remote Sensing: Theory and Applications.* Gordon Breach, Amsterdam, The Netherlands.

Bunnell, F. L., M. Boyland, and E. Wind. 2002. *How Should We Spatially Distribute Dying and Dead Wood?* USDA Forest Service Gen. Tech. Rep. PSW-GTR-181.

Bunting, P. and R. Lucas. 2006. The delineation of tree crowns in Australian mixed species forests using hyperspectral Compact Airborne Spectrographic Imagery (CASI) data. *Remote Sensing of Environment* 101: 230–248.

Burbridge, A. A. 2004. *Threatened Animals of Western Australia.* DEC, Kensington, WA.

Burbridge, P. R. 2004. A critical review of progress towards integrated coastal management in the Baltic Sea region. *In Managing the Baltic Sea*, 63–75. In Schernewski G. and N. Löser.Warnemünde eds., Coastline Reports 2.

Burke, D. M. and E. Nol. 1998. Influence of food abundance, nest-site habitat, and forest fragmentation on breeding ovenbirds. *Auk* 115: 96–104.

Burnett, M., P. J. B. August, and K. Killingbeck. 1998. The influence of geomorphological heterogeneity on biodiversity. I. A patch-scale perspective. *Conservation Biology* 12: 363–370.

Burns, J. J. and S. J. Harbo. 1972. An aerial census of ringed seals, northern coast of Alaska. *Arctic* 25: 279–290.

Burroughs, W. E. 1986. *Deep Black: The Startling Truth Behind America's Top-Secret Spy Satellites.* Berkley Books, New York, U.S.

Burtenshaw, J. C., E. M. Oleson, J. A. Hildebrand, M. A. McDonald, R. K. Andrew, B. M. Howe, and J. A. Mercer. 2004. Acoustic and satellite remote sensing of blue whale seasonality and habitat in the Northeast Pacific. *Deep-Sea Research II* 51: 967–986.

Busby, J. R. 2002. Biodiversity mapping and modelling. In Skidmore, A., ed., *Environmental Modelling with GIS and Remote Sensing.* Taylor and Francis, London,UK. 145–165.

Butler, B. J., J. J. Swenson, and R. J. Alig. 2004. Forest fragmentation in the Pacific Northwest: quantification and correlations. *Forest Ecology and Management* 189: 363–373.

Bürgi, M., A. M. Hersperger, M. Hall, E. W. B. (Russell) Southgate, and N. Schneeberger. 2007. Using the past to understand the present land use and land cover. In Kienast, F., O. Wildi, and S. Ghosh, eds., *A Changing World: Challenges for Landscape Research.* Springer, Dordrecht, The Netherlands. 133–144.

Bütler, R. and R. Schlaepfer. 2004. Spruce snag quantification by coupling colour infrared aerial photos and a GIS. *Forest Ecology and Management* 195: 325–339.

Caenepeel, C. L. and C. Wyrick. 2001. Strategies for successful interdisciplinary projects: a California State Polytechnic University, Pomona, perspective. *International Journal of Engineering Education* 17: 391–395.

Caldas, C. H., D. Grau, and C. Haas. 2006. Using Global Positioning Systems to improve materials locating processes on industrial projects. *Journal of Construction Engineering and Management* 132: 741–749.

Calhoun, A. J. K. and P. deMaynadier. 2004. *Forestry Habitat Guidelines for Vernal Pool Wildlife.* MCA Technical Paper No. 6, Metropolitan Conservation Alliance, Wildlife Conservation Society, Bronx, New York, U.S.

Campbell, J. B. 2002. *Introduction to Remote Sensing.* Guilford Press, New York, U.S.

Campbell, J. B. 2005. Visual interpretation of aerial imagery. In Aronoff, S., ed., *Remote Sensing for GIS Managers.* ESRI, Redlands. 259–285.

Campbell, J. B. and L. Ran. 1993. CHROM: a C-program to evaluate the application of dark-object subtraction technique to digital remote sensing data. *Computers & Geosciences* 19: 1475–1499.

Canci, M., J. Wallace, E. Mattiske, and A. Malcolm. 2007. Evaluation of airborne multispectral imagery for monitoring the effect of falling groundwater levels on native vegetation. *International Archives of the Photogrammetry, Remote Sensing and Spatial Sciences* 34 (abs).

Cannon, R. W., F. L. Knopf, and L. R. Pettinger. 1982. Use of Landsat data to evaluate lesser prairie chicken habitats in western Oklahoma. *Journal of Wildlife Management* 46: 915–922.

Canter, L. W. 1993. The role of environmental monitoring in responsible project management. *Environmental Professional* 15: 76–87.

Carlson, K. M., G. P. Asner, R. F. Hughes, R. Ostertag, and R. E. Martin. 2007. Hyperspectral remote sensing of canopy biodiversity in Hawaiian lowland rainforests. *Ecosystems* 10: 536–549.

Cardille, J., M. Turner, M. Clayton, S. Gergel, and S. Price. 2005. METALAND: characterizing spatial patterns and statistical context of landscape metrics. *BioScience* 55: 983–988.

Case, T. and R. N. Fisher. 2001. Measuring and predicting species presence: coastal sage scrub case study. In Hunsaker, C. T., et al., eds., *Spatial Uncertainty in Ecology: Implications for Remote Sensing and GIS Applications.* Springer, New York, U.S. 47–71.

Castro-Esau, K. L., G. A. Sánchez-Azofeifa, and T. Caelli. 2004. Discrimination of lianas and trees with leaf-level hyperspectral data. *Remote Sensing of Environment* 90: 353–372.

Castro-Esau, K. L., G. A. Sánchez-Azofeifa, B. Rivard, S. J. Wright, and M. Quesada. 2006. Variability in leaf optical properties of Mesoamerican trees and the potential for species classification. *American Journal of Botany* 93: 517–530.

Catena, G., L. Palla, and M. Catalano. 1990. Thermal infrared detection of cavities in trees. *European Journal of Forest Pathology* 20: 201–210.

Catling, P. C. and N. C. Coops. 1999. Prediction of the distribution and abundance of small mammals in the eucalypt forests of south-eastern Australia from airborne videography. *Wildlife Research* 26: 641–650.

Catling, P. C. and N. C. Coops. 2004. *Identification and Functional Groups that Give Early Warning of Major Environmental Change (Indicator 1.2c): Part B. Report on the Efficacy of Videography (and Other High Spatial Resolution Imagery) for Habitat Mapping.* Forest and Wood Products Research and Development Corporation, Melbourne, Australia.

Cattet, M., N. A. Caulkett, and G. B. Stenhouse. 2003. Anesthesia of grizzly bears using xylazine–zolazepam–tiletamine. *Ursus* 14: 88–93.

Cattet, M., J. Boulanger, G. Stenhouse, R. A. Powell, and M. J. Reynolds-Hogland. 2008. An evaluation of long-term capture effects in ursids: implications for wildlife welfare and research. *Journal of Mammalogy* 89: 973–990.

CBMP. 2007. *Circumpolar Biodiversity Monitoring Program Five-Year Implementation Plan Overview Document.* Meeting of Senior Arctic Officials, Tromso, Norway. Available online http://arcticportal.org/uploads/fr/pM/frpMZcpixn5kDMMusJ0kAQ/CBMP-Imp-Plan-Overview-FINAL.pdf web site accessed September 12, 2008.

Champion, F. W. 1928. *With a Camera in Tiger-Land.* Doubleday Doran, New York, U.S.

Chapman, A. D. and J. R Busby. 1994. Linking plant species information to continental biodiversity inventory, climate and environmental monitoring. In Miller, R.I. ed., *Mapping the Diversity of Nature*. Chapman and Hall London, UK. 177–195.

Chavez, P. S. 1988. An improved dark-object subtraction technique for atmospheric scattering correction of multispectral data. *Remote Sensing of Environment* 24: 459–479.

Chastain, R. A., Jr. and P. A. Townsend. 2007. Use of Landsat ETM and topographic data to characterize evergreen understory communities in Appalachian deciduous forests. *Photogrammetric Engineering and Remote Sensing* 73: 563–576.

Chen, C. H., ed. 2007. *Image Processing for Remote Sensing*. Taylor and Francis (CRC Press), Boca Raton, Florida.

Chen, F., Z. Zhao, L. Peng, and D. Yan. 2005. Clouds and cloud shadows removal from high-resolution remote sensing images. *Proceedings of IEEE* 4256–4259.

Chen, J. M., J. Liu, J. Cihlar, and M. L. Goulden. 1999. Daily canopy photosynthesis model through temporal and spatial scaling for remote sensing applications. *Ecological Modelling* 124: 99–119.

Chen, J. T. and S. C. Saunders. 2006. Ecology of multiple ecosystems in space and time. In Chen, J. et al., eds., *Ecology of Hierarchical Landscapes: From Theory to Applications*. Nova Science Publications, New York, U.S. 1–34.

Chen, J. T., S. C. Saunders, K. D. Brosofske, and T. R. Crow, eds. 2006. *Ecology of Hierarchical Landscapes: From Theory to Applications.* Nova Science Publications, New York, U.S.

Chen, H., P. K. Varshney, and M. K. Arora. 2003a. Mutual information based image registration for remote sensing data. *International Journal of Remote Sensing* 24: 3701–3706.

Chen, H., P. K. Varshney, and M. K. Arora. 2003b. Performance of mutual information similarity measure for registration of multitemporal remote sensing images. *IEEE Transactions on Geoscience and Remote Sensing* 11: 2445–2454.

Chen, T., J. Wang, and K. Zhang. 2004. A wavelet transform based method for road centerline extraction. *Photogrammetric Engineering and Remote Sensing* 70: 1423–1432.

Cheng, P. 2007. Automated high-accuracy orthorectification and mosaicking of PALSAR data without ground control points. *Geoinformatics* September: 36–38.

Cheng, P. and C. Chaapel. 2006. DEM generation using Quickbird stereo data without ground control. *Geoinformatics* May: 36–38.

Cheng, K. S., H. C. Yeh, and C. H. Tsai. 2000. An anisotropic spatial modelling approach for remote sensing image rectification. *Remote Sensing of Environment* 73: 46–54.

Christ, A., J. Ver Hoef, and D. L. Zimmerman. 2008. An animal movement model incorporating home range and habitat selection. *Environmental and Ecological Statistics* 15: 27–38.

Christian, R. R., P. M. DiGiacomo, T. C. Malone, and L. Talaue-McManus. 2006. Opportunities and challenges of establishing coastal observing systems. *Estuaries and Coasts* 29: 871–875.

Christiansen, P. 2006. Sabertooth characters in the clouded leopard (*Neofelis nebulosa* Griffiths 1821). *Journal of Morphology* 267: 1186–1198.

Cihlar, J., J. M. Chen, and Z. Li. 1997. On the validation of satellite-derived products for land applications. *Canadian Journal of Remote Sensing* 23: 381–389.

Cihlar, J. and J. Howarth. 1994. Detection and removal of cloud contamination from AVHRR images. *IEEE Transactions on Geoscience and Remote Sensing* 32: 583–589.

Cihlar, J., R. Latifovic, J. Beabien, A. Trishchenko, J. Chen, and G. Fedosejevs. 2003. National scale forest information extraction from coarse resolution satellite data, part 1. In Wulder, M. A. and S. E. Franklin, eds., *Remote Sensing of Forest Environments: Concepts and Case Studies.* Kluwer Academic Publishers, Boston, MA. 337–357.

Cihlar, J., Q. Xiao, J. Chen, J. Beaubien, K. Fung, and R. Latifovic. 1998. Classification by progressive generalization: a new automated methodology for remote sensing multichannel data. *International Journal of Remote Sensing* 19: 2648–2704.

Clark, T. W., A. R. Willard, and C. M. Cromley, eds. 2000. *Foundations of Natural Resources Policy and Management.* Yale University Press, New Haven, U.S.

Clark, D. B., C. S. Castro, L. D. A. Alvarado, and J. M. Read. 2004. Quantifying the mortality of tropical rain forest trees using high-spatial-resolution satellite data. *Ecology Letters* 7: 52–59.

Clark, M. L., D. A. Roberts, and D. B. Clark. 2005. Hyperspectral discrimination of tropical rain forest tree species at leaf to crown scales. *Remote Sensing of Environment* 96: 375–398.

Clark, J. S., S. R. Carpenter, M. Barber, S. Collins, A. Dobson, J. A. Foley, D. M. Lodge, et al. 2001. Ecological forecasts: an emerging imperative. *Science* 293: 657–660.

Clarke, A. C. 1975. Satellites and saris. In *Future Space Programs 1975: Report of the Subcommittee on Space Science and Applications, Committee of Science and Technology.* US House of Representatives, 94th Congress, First Session. US Government Printing Office, Washington. Reprinted: Clarke, A. C. 1978. *The View from Serendip.* Pan Books, London.

Clawges, R., K. Vierling, L. Vierling, and E. Rowell. 2008. The use of airborne LiDAR to assess avian species diversity, density, and occurrence in a pine/aspen forest. *Remote Sensing of Environment* 112: 2064–2073.

Cliff, G., M. D. Anderson-Reade, A. P. Aitken, G. E. Charter, and V. M. Peddemors. 2006. Aerial census of whale sharks (*Rhincodon typus*) on the northern KwaZulu-Natal coast, South Africa. *Fisheries Research* 84: 41–46.

Clode, S., F. Rottensteiner, P. Kootsookos, and E. Zelniker. 2007. Detection and vectorization of roads from LiDAR data. *Photogrammetric Engineering and Remote Sensing* 73: 517–536.

Cloutis, E. A., D. R. Connery, D. J. Major, and F. J. Dover. 1996. Agricultural crop condition monitoring using airborne C-band synthetic aperture radar in southern Alberta. *International Journal of Remote Sensing* 17: 2565–2577.

Cohen, W. B. and S. N. Goward. 2004. Landsat's role in ecological applications of remote sensing. *Bioscience* 54: 535–545.

Cohen, W. B. and T. A. Spies. 1992. Estimating structural attributes of Douglas-fir/western hemlock forest stands from Landsat and SPOT imagery. *Remote Sensing of Environment* 41: 1–17.

Cohen, W. B., T. A. Spies, and M. Fiorella. 1995. Estimating the age and structure of forests in a multiownership landscape of western Oregon, USA. *International Journal of Remote Sensing* 16: 721–746.

Collins, J. B. and C. E. Woodcock. 1996. An assessment of linear change detection techniques for mapping forest mortality using multitemporal Landsat TM data. *Remote Sensing of Environment* 56: 66–77.

Colpaert, A., J. Kumpula, and M. Nieminen. 2003. Reindeer pasture biomass assessment using satellite remote sensing. *Arctic* 56: 147–158.

Colwell, R. N. 1961. Some practical applications of multiband spectral reconnaissance. *American Scientist* 49:1 (March).

Colwell, R. N. 1966. Aerial photography of the Earth's surface: its procurement and use. *Applied Optics* 5: 883–892.

Colwell, R. N., ed. 1983. *Manual of Remote Sensing*. 2d ed. ASPRS, Falls Church, VA.

Congalton, R. G., J. M. Stenback, and R. H. Barrett. 1993. Mapping deer habitat suitability using remote sensing and geographic information systems. *GeoCarto International* 3: 23–33.

Coops, N. C. 1999. Linking multi-resolution satellite-derived estimates of canopy photosynthetic capacity and meteorological data to assess forest productivity in a *Pinus radiata* (D. Don) stand. *Photogrammetric Engineering and Remote Sensing* 65: 1149–1156.

Coops, N. C. and P. C. Catling. 1997. Utilizing airborne multispectral videography to predict habitat complexity in eucalypt forests for wildlife management. *Wildlife Research* 24: 691–703.

Coops, N. C. and P. C. Catling. 2000. Estimating forest habitat complexity in relation to time since fire. *Austral Ecology* 23: 344–351.

Coops, N. C., R. H. Waring, M. A. Wulder, A. M. Pidgeon, and V. C. Radeloff. 2008a. Bird diversity: a predictable function of satellite-derived estimates of seasonal variation in canopy light absorbance across the United States. *Journal of Biogeography* in press.

Coops, N. C. and J. D. White. 2003. Modeling forest productivity using data acquired through remote sensing. In Wulder, M. A. and S. E. Franklin, eds., *Remote Sensing of Forest Environments: Concepts and Case Studies*. Kluwer Academic Publishers, Boston, MA. 411–432.

Coops, N. C., M. A. Wulder, D. S. Culvenor, and B. St.-Onge. 2004. Comparison of forest attributes extracted from fine spatial resolution multispectral and LiDAR data. *Canadian Journal of Remote Sensing* 30: 855–866.

Coops, N. C., M. A. Wulder, D. C. Duro, T. Han, and S. Berry. 2008b. The development of a Canadian dynamic habitat index using multi-temporal satellite estimates of canopy light absorbance. *Ecological Indicators* 8: 754–766.

Coops, N. C., M. A. Wulder, and D. Iwanicka. 2008c. Assessing the relative importance of seasonal variation in production and land cover for satellite-derived predictions of breeding bird distributions over Ontario, Canada. *Remote Sensing of Environment* in press.

Coops, N. C., M. A. Wulder, and J. C. White. 2007. Identifying and describing forest disturbance and spatial pattern: data selection issues and methodological implications. In Wulder, M. A. and S. E. Franklin, eds., *Understanding Forest Disturbance and Spatial Pattern: Remote Sensing and GIS Approaches.* Taylor and Francis, Boca Raton, Florida. 31–61.

Cooper, P. R., D. E. Friedman, and S. A. Wood. 1987. The automatic generation of digital terrain models by stereo. *Acta Astronica* 15: 171–180.

Coppin, P., I. Jonckheere, K. Nackaerts, B. Muys, and E. Lambin. 2004. Digital change detection methods in ecosystem monitoring: a review. *International Journal of Remote Sensing* 25: 1565–1596.

Corsi, F., de Leeuw, J. and Skidmore, A. 2000: Modeling species distribution with GIS. In Boitani, L. and Fuller, T.K., eds., *Research Techniques in Animal Ecology: Controversies and Consequences.* Columbia University Press, New York, U.S. 389–434.

Coulter, M. C., A. L. Bryan, Jr., H. E. Mackey, Jr., J. R. Jensen, and M. E. Hodgson. 1987. Mapping of wood stork foraging habitat with satellite data. *Colonial Waterbirds* 10: 178–180.

Crabbe, P., A. J. Holland, L. Ryszkowski, and L. Westra, eds. 2000. *Implementing Ecological Integrity, Restoring Regional and Global Environmental and Human Health.* Springer, Dordrecht, The Netherlands.

Crist, E. P. and R. C. Cicone. 1984. A physically based transformation of Thematic Mapper data—the TM Tasseled Cap. *IEEE Transactions on Geoscience and Remote Sensing* 22: 256–263.

Crist, E. P., R. Lauren, and R. C. Cicone. 1986. Vegetation and soils information contained in transformed Thematic Mapper data. *Proceedings of IGARSS'86 Symposium.* ESA Publication. Division SP–254, Zurich.

Crosse, W. 2005. The opportunities and constraints in using cost-effective satellite remote sensing for biodiversity monitoring. *Proceedings of the MTS/IEEE Oceans* 1: 844–849.

Culvenor, D. 2003. Extracting individual tree information: a survey of techniques for high spatial resolution imagery. In Wulder, M. A. and S. E. Franklin, eds., *Remote Sensing of Forest Environments: Concepts and Case Studies.* Kluwer Academic Publishers, Boston, MA. 255–278.

Curran, P. J. 1987. Remote sensing methodologies and geography. *International Journal of Remote Sensing* 8: 1255–1275.

Curran, P. J. 1989. Remote sensing of foliar chemistry. *Remote Sensing of Environment* 30: 270–278.

Curran, P. J. 1994. Imaging spectroscopy. *Progress in Physical Geography* 19: 247–266.

Curran, P. J., J. L. Dungan, and H. L. Gholz. 1990. Exploring the relationship between reflectance red edge and chlorophyll content in slash pine. *Tree Physiology* 7: 33–48.

Currie, D. J. 1991. Energy and large-scale patterns of animal- and plant-species richness. *The American Naturalist* 137: 27–49.

Curtis, A., S. Heath, and M. Hugh-Jones. 2005. GIS investigations of epizootics: the limitations of surveillance data. In Majumdar, S. K., J. E. Huffman, F. J. Brenner, and A. I. Panah, eds. 2005. *Wildlife Diseases: Landscape Epidemiology, Spatial Distribution and Utilization of Remote Sensing Technology.* Pennsylvania Academy of Science, Easton. 457–474.

Dai, X. and S. Khorram. 1999. A feature-based image registration algorithm using improved chain-code representation combined with invariant moments. *IEEE Transactions on Geoscience and Remote Sensing* 37: 2351–2362.

Daniels, R. J. R. 2003. Biodiversity of the Western Ghats: an overview. *ENVIS Bulletin* 4: 25–40.

Danson, F. M. 1987. Preliminary evaluation of the relationships between SPOT-1 HRV data and forest stand parameters. *International Journal of Remote Sensing* 8: 1571–1575.

Dare, P. M. 2005a. Current trends in low cost airborne remote sensing technology. *Proceedings of NARGIS*, Darwin. Available online http://www.spatialscientific.com.au/docs/Dare_PM_NARGIS_paper.pdf web site accessed August 5, 2008.

Dare, P. M., 2005b. The use of small environmental research aircraft (SERA's) for remote sensing applications. *International Journal of Geoinformatics* 1: 19–26.

Daszak, P., A. A. Cunningham, and A. D. Hyatt. 2001. Anthropogenic environmental change and the emergence of infectious diseases in wildlife. *Acta Tropica* 69: 91–98.

Davidson, C. 1998. Issues in measuring landscape fragmentation. *Wildlife Society Bulletin* 26: 32–37.

Davidson, E. A., G. P. Asner, T. A. Stone, C. Neill, and R. O. Figueiredo. 2008. Objective indicators of pasture degradation from spectral mixture analysis of Landsat imagery. *Journal of Geophysical Research 113*, G00B03, doi:10.1029/2007JG000622.

De Leuw, J., W. K. Ottichilo, A. G. Toxopeus, and H. H. T. Prins. 2002. Application of remote sensing and geographic information systems in wildlife mapping and modelling. In Skidmore, A., ed., *Environmental Modelling with GIS and Remote Sensing*. Taylor and Francis, London, UK. 121–144.

D'Erchia, F. 1997. Geographic information systems and remote sensing applications in ecosystem management. In Boyce, M. S. and A. Haney, eds. Ecosystem Management: Applications for Sustainable Forest and Wildlife Resources. Yale University Press, New Haven, U.S. 201–225.

De Sherbinin, A. M. 2005. *Remote Sensing in Support of Ecosystem Management Treaties and Transboundary Conservation.* CIESIN, Columbia University, Palisades, NY. Available online http://sedac.ciesin.columbia.edu/rs–treaties/laguna.html. web site accessed August 15, 2008.

De Wulf, R. R., R. E. Goosens, B. P. DeRoover, and F. C. Borry. 1990. Extraction of forest stand parameters from panchromatic and multispectral SPOT-1 data. *International Journal of Remote Sensing* 11: 1571–1588.

Delcourt, H. R., P. A. Delcourt, and W. Thompson. 1983. Dynamic plant ecology: the spectrum of vegetational change in space and time. *Quaternary Science Reviews* 1: 153–175.

Dendron Resources Inc. 1997. *Bioindicators of Forest Health and Sustainability.* Forest Resources Information Paper No. 138. Ontario Forest Resources Institute, Sault Ste. Marie, Ontario.

Diaz, R.J., M. Solan, and R.M. Valente. 2004. A review of approaches for classifying benthic habitats and evaluating habitat quality. *Journal of Environmental Management* 73: 165–181.

Díaz Varela, R. A., P. Ramil Rego, and M. S. Calvo Iglesias. 2006. Tracking environmental impacts and habitat fragmentation of coastal protected areas through object oriented analysis, identification and categorization of linear disturbances in Corrubedo Natural Park (NW Iberian Peninsula). *Proceedings of ISPRS Commission 4, OBIA*, p. 6.

Dierssen, H. M., R. C. Zimmerman, R. A. Leathers, T. V. Downes, and C. O. Davis. 2003. Ocean colour remote sensing of seagrass and bathymetry in the Bahamas Banks by high-resolution airborne imagery. *Limnology and Oceanography* 48: 444–455.

Dietz, J. M. 1997. Conservation of biodiversity in neotropical primates. In Reaka-Kudla, M. L., et al., eds., *Biodiversity II.* The National Academy of Sciences, Washington. U. S. 341–356.

Dique, D. S., H. J. Preece, J. Thompson, and D. L. de Villiers. 2004. Determining the distribution and abundance of a regional koala population in south-east Queensland for conservation management. *Wildlife Research* 31: 109–117.

Diner, D. J., G. P. Asner, R. Davies, Y. Knyazikhin, J. P. Muller, A. W. Nolin, B. Pinty, C. B. Schaaf, and J. Stroeve. 1999. New directions in Earth observing: scientific applications of multiangle remote sensing. *Bulletin of the American Meteorological Society* 80: 2209–2228.

Dobson, M. C. 2000. Forest information from synthetic aperture radar. *Journal of Forestry* 98: 41–43.

Dong, P. 2005. Development of a GIS/GPS–based emergency response system. *Geomatica* 59: 427–434.
Dottavio, C. L. and D. L. Williams. 1983. Satellite technology: an improved means for monitoring forest insect defoliation. *Journal of Forestry* 81: 30–34.
Dousse, T. and T. Wheeler. 2008. The future of remote sensing. *Position* 36 (Aug–Sept): 43–45.
Dowling, R. and A. Accad. 2003. Vegetation classification of the riparian zone along the Brisbane River, Queensland, Australia, using light detection and ranging (LiDAR) data and forward looking digital video. *Canadian Journal of Remote Sensing* 29: 556–563.
Dunning, J., B. Danielson, and H. Pulliam. 1992. Ecological processes that affect populations in complex landscapes. *Oikos* 65: 169–175.
Duro, D. C., N. C. Coops, M. A. Wulder, and T. Han. 2007. Development of a large-area biodiversity monitoring system driven by remote sensing. *Progress in Physical Geography* 31: 235–261.
Dutton, I. 2001. *An Examination of the Role of Monitoring in Protected Area Management.* Unpublished Ph.D. Thesis, University of Queensland, Brisbane, Australia.
Eastwood, P. 1967. *Radar Ornithology.* The Chaucer Press, London, UK.
Egbert, S. L., S. Park, K. P. Price, R.-Y. Lee, J. Wu, and M. D. Nellis. 2002. Using Conservation Reserve Program maps derived from satellite imagery to characterize landscape structure. *Computers and Electronics in Agriculture* 37: 141–156.
Egler, F. E. 1954. Vegetation science concept. I. Initial floristic composition: a factor in old field vegetation development. *Vegetatio* 4: 412.
Ehlers, M., 1997. Rectification and registration. In Star, J. L., et al., eds., *Integration of Geographic Information Systems and Remote Sensing.* Cambridge University Press, Cambridge, Cambridge, UK. 13–36.
Ehlers, M., G. Edwards, and Y. Bedard. 1997. Integration of remote sensing with geographic information systems: a necessary evolution. *Photogrammetric Engineering and Remote Sensing* 55: 1619–1627.
Ellner, S. P., E. McCauley, B. E. Kendall, C. J. Briggs, P. R. Hosseini, S. N Wood, A. Janssen, and others. 2001. Habitat structure and population persistence in an experimental community. *Nature* 412: 538–542.
El -Rabbany, A. 2006. *Introduction to GPS: the Global Positioning System.* 2d ed. Artech House Publications, Boston, MA.
Emch, M., J. W. Quinn, M. Peterson, and M. Alexander. 2005. Forest cover change in the Toledo District, Belize from 1975 to 1999: a remote sensing approach. *Professional Geographer* 57: 256–267.
Environmental Remote Sensing Group, 2003. *Determination of SRA Habitat Indicators by Remote Sensing.* CSIRO Land and Water Technical Report 28/03, Canberra.
Esche, H. A., S. E. Franklin, and M. A. Wulder. 2002. Assessing cloud contamination effects on K-means unsupervised classifications of Landsat data. *Proceedings of IGARSS'02*, Toronto, Ontario.
Essery, C. I. and A. P. Moore. 1992. The impact of ozone and acid mist on the spectral reflectance of young Norway spruce trees. *International Journal of Remote Sensing* 13: 3045–3054.
Estes, J. E. 1985. The need for improved information systems. *Canadian Journal of Remote Sensing* 11: 124–131.
Estes, J. E. and T. Foresman. 1996. Development of a remote sensing core curriculum. *Proceedings of IGARSS'96.* Lincoln, NE. 1: 820–822.
Estes, J. E. and J. L. Star. 1997. Research needed to improve remote sensing and GIS integration: conclusions and a look toward the future. In Star, J. L., et al., eds., *Integration of Geographic Information Systems and Remote Sensing.* Cambridge University Press, Cambridge, Cambridge, UK. 176–203.
Estes, L. D., G. S. Okin, A. G. Mwangi, and H. H. Shugart. 2008. Habitat selection by a rare forest antelope: a multi-scale approach combining field data and imagery from three sensors. *Remote Sensing of Environment* 112: 2033–2050.
Eva, H. D., E. E. de Miranda, C. M. D Bella, V. Gond, O. Huber, M. Sgrenzaroli, S. Jones, et al. 2002. *A Vegetation Map of South America.* Office for Official Publications of the European Communities, Luxembourg, U.S.

Evans, J. C. 2005. *With Respect for Nature: Living as a Part of the Natural World.* State University of New York Press, Albany, U.S.

Everitt, J. H., G. L. Anderson, D. E. Escobar, M. R. Davis, N. R. Spencer, and R. J. Andrascik. 1995. Use of remote sensing for detecting and mapping leafy spurge (*Euphorbia esula*). *Weed Technology* 9: 599–609.

Eyre, M. D., G. N. Foster, and M. L. Luff. 2005a. Exploring the relationship between the land cover and the distribution of water beetle species (*Coleoptera*) at the regional scale. *Hydrobiologica* 533: 87–98.

Eyre, M. D., J. C. Woodward, and R. A. Sanderson. 2005b. Assessing the relationship between grassland Auchenorrhyncha (*Homoptere*) and land cover. *Agriculture, Ecosystems and Environment* 109: 187–191.

Fahrig, L. 2003. Effects of habitat fragmentation on biodiversity. *Annual Review of Ecology, Evolution, and Systematics* 34: 487–515.

Fairweather, P. G. and G. M. Napier. 1998. *Environmental Indicators for National State of the Environment Reporting—Inland Waters.* State of the Environment (Environmental Indicator Reports), Department of the Environment, Canberra.

Faith, D. P. 2007. Biodiversity. *The Stanford Encyclopedia of Philosophy.* Available online http://plato.stanford.edu/entries/biodiversity/ web site accessed September 12, 2008.

Farnsworth, M. L., L. L. Wolfe, N. T. Hobbs, K. P. Burnham, E. S. Williams, D. M. Theobald, M. M. Connor, and M. W. Miller. 2005. Human land use influences chronic wasting disease prevalence in mule deer. *Ecological Applications,* 15: 119–126.

Faulkner, E. and D. Morgan. 2002. *Aerial Mapping.* 2d ed., CRC Press (Lewis Publishers), New York, U.S.

Ferguson, E. L., D. G. Jorde, and J. J. Sease. 1981. Use of 35 mm colour aerial photography to acquire mallard sex ratio data. *Photogrammetric Engineering and Remote Sensing* 47: 823–828.

Ferwerda, J. G. and A. K. Skidmore. 2007. Can nutrient status of four woody plant species be predicted using field spectrometry? *ISPRS Journal of Photogrammetry and Remote Sensing* 62: 406–414.

Findlay, C. S. and J. Bourdages. 2000. Response time of wetland biodiversity to road construction on adjacent lands. *Conservation Biology* 14: 86–94.

Fisher, A. and A. Kutt. 2006. *Biodiversity and Land Condition in Tropical Savanna Rangelands (Summary Report).* Tropical Savannas CRC, Darwin, Australia.

Fjeldså, J., D. Ehrlich, E. Lambin, and E. Prins. 1997. Are biodiversity "hotspots" correlated with current ecoclimatic stability? *Biodiversity and Conservation* 6: 401–422.

Flamm, R. O., E. C. G. Owen, C. F. W. Owen, R. S. Wells, and D. Nowacek. 2008. Aerial videogrammetry from a tethered airship to assess manatee life-stage structure. *Marine Mammal Science* 16: 617–630.

Fogel, D. N. and L. R. Tinney. 1996. *Image Registration Using Multiquadratric Functions, the Finite Element Method, Bivariate Mapping Polynomials, and the Thin Plate Spline.* NCGIA Technical Report 96-1. UCSB, Santa Barbara.

Foody, G. M. 1999. The continuum of classification fuzziness in thematic mapping. *Photogrammetric Engineering and Remote Sensing* 65: 443–451.

Foody, G. M. 2002. Status of land cover classification accuracy assessment. *Remote Sensing of Environment* 80: 185–201.

Forman, R. T. T. 1995. *Land Mosaics: The Ecology of Landscapes and Regions.* Cambridge University Press, Cambridge, UK.

Forman, R. T. T., D. Sperling, J. A. Bissonette, et al., eds. 2003. *Road Ecology: Science and Solutions.* Island Press, Washington DC.

Fortin, M.-J., B. Boots, F. Csillag, and T. K. Remmel. 2003. On the role of spatial stochastic models in understanding landscape indices in ecology. *Oikos* 102: 203–212.

Fortin, M. -J., and M. R. T. Dale. 2005. *Spatial Analysis: A Guide for Ecologists.* Cambridge University Press, Cambridge, UK.

Fournier, R. A., D. Mailly, J. -M. N. Walter, and K. Soudani. 2003. Indirect measurement of forest canopy structure from in situ optical sensors. In Wulder, M. A.

and S. E. Franklin, eds., *Remote Sensing of Forest Environments: Concepts and Case Studies.* Kluwer Academic Publishers, Boston, MA. 77–113.

Frair, J. L., S. E. Nielsen, E. H. Merrill, S. Lele, M. S. Boyce, R. H. M. Munro, G. B. Stenhouse, and H. L. Beyer. 2004. Approaches for removing GPS-collar bias in habitat selection studies. *Journal of Applied Ecology* 41: 201–212.

Francis, R. E., M. J. Morris, R. J. Myhre, and D. L. Noble. 1979. Inventory and analysis of leafy spurge (*Euphorbia esula*) sites...a feasibility study. *Proceedings of Northern Regional Leafy Spurge Conference,* Billings. Available online http://www.fs.fed.us/rm/sd/spurge_sites_feasibil.pdf web site accessed September 11, 2008.

Franklin, J. 1986. Thematic Mapper analysis of conifer forest structure and composition. *International Journal of Remote Sensing* 7: 1287–1301.

Franklin, J. 1995. Predictive vegetation mapping: geographic modelling of biospatial patterns in relation to environmental gradients. *Progress in Physical Geography* 19: 474–499.

Franklin, J., S. R. Phinn, C. E Woodcock, and J. E. Rogan. 2003. Rationale and conceptual framework for classification approaches to assess forest structure. In Wulder, M. A. and S. E. Franklin, eds., *Remote Sensing of Forest Environments: Concepts and Case Studies.* Kluwer Academic Publishers, Boston, MA. 279–300.

Franklin, S. E. 2001. *Remote Sensing for Sustainable Forest Management.,* Lewis Publishers (CRC Press) Boca Raton, Florida.

Franklin, S. E. and E. E. Dickson. 2001. Approaches for monitoring landscape composition and pattern using remote sensing. In Farr, D., et al., eds., *Monitoring Forest Biodiversity in Alberta: Program Framework,* Alberta Forest Biodiversity Monitoring Program Technical Report No. 3. Foothills Model Forest, Hinton, AB. 51–140.

Franklin, S. E., M. J. Hansen, and G. B. Stenhouse. 2002a. Quantifying landscape structure with vegetation inventory maps and remote sensing. *Forestry Chronicle* 78: 866–875.

Franklin, S. E., M. A. Lavigne, M. J. Deuling, M. A. Wulder, and E. R. Hunt, Jr. 1997. Estimation of forest leaf area using remote sensing and GIS data for modelling net primary production. *International Journal of Remote Sensing* 18: 3459–3471.

Franklin, S. E., A. J. Maudie, and M. B. Lavigne. 2001a. Using spatial co-occurrence texture to increase forest structure and species composition classification accuracy. *Photogrammetric Engineering and Remote Sensing* 67: 849–855.

Franklin, S. E., P. K. Montgomery, and G. B. Stenhouse 2005. Interpretation of land cover change using aerial photography and satellite imagery in the Foothills Model Forest of Alberta. *Canadian Journal of Remote Sensing* 31: 304–313.

Franklin, S. E., L. M. Moskal, M. Lavigne, and K. Pugh. 2000. Interpretation and classification of partially harvested forest stands in the Fundy Model Forest using multitemporal Landsat TM digital data. *Canadian Journal of Remote Sensing* 26: 318–333.

Franklin, S. E., D. R. Peddle, J. A. Dechka, and G. B. Stenhouse. 2002b. Evidential reasoning using Landsat TM, DEM, and GIS data in support of grizzly bear habitat analysis. *International Journal of Remote Sensing* 23: 4633–4652.

Franklin, S. E., R. H. Waring, R. McCreight, W. B. Cohen, and M. Fiorella. 1995. Aerial and satellite sensor detection and classification of western spruce budworm defoliation in a subalpine forest. *Canadian Journal of Remote Sensing* 21: 299–308.

Franklin, S. E. and M. A. Wulder. 2002. Remote sensing methods in medium spatial resolution satellite data land cover classification of large areas. *Progress in Physical Geography* 26: 173–205.

Franklin, S. E., M. A. Wulder, and G. R. Gerylo. 2001b. Texture analysis of IKONOS panchromatic imagery for Douglas-fir forest age class separability in British Columbia. *International Journal of Remote Sensing* 22: 2627–2632.

Fraser, D. E, J. F Gilliam, M. J. Daley, A. N. Le, and G. T. Skalski. 2001. Explaining leptokurtic movement distributions: intrapopulation variation in boldness and exploration. *American Naturalist* 158: 124–135.

Fraser, R. A. and J. J. Kay. 2004. Exergy analysis of ecosystems: establishing a role for thermal remote sensing. In Quattrochi, D. A. and J. C. Luvall, eds., *Thermal Remote Sensing in Land Surface Processing*. Routledge, New York. 283–360.

Frazer, G. W., R. A. Fournier, R. J. Hall, and J. A. Trofymow. 2001. A comparison of digital and film fisheye photography for analysis of forest canopy structure and gap-light transmission. *Agriculture and Forest Meteorology* 109: 249–255.

Frazer, G. W., J. A. Trofymow, and K. P. Lertzman.1997. *A Method for Estimating Canopy Openess, Effective Leaf Area Index, and Photosynthetically Active Photon Flux Using Hemispherical Photography and Computerized Image Analysis Techniques.* Canadian Forest Service Information Report BC-X-373. Pacific Forestry Centre, Victoria.

Frazer, T. K., R. M. RoSS, L. B. Quetin, and J. P. Montoya. 1997. Turnover of carbon and nitrogen during growth of larval krill, Euphausia superba Dana: a stable isotope approach. *Journal of Experimental Marine Biology and Ecology* 212: 259–275.

Friedl, M. A., D. K. McIver, J. C. F. Hodges, X. Y. Zhang, D. Muchoney, A. H. Strahler, C. E. Woodcock, et al. 2002. Global land cover mapping from MODIS: algorithms and early results. *Remote Sensing of Environment* 83: 287–302.

Friedl, M. A., C. E. Brodley, and A. H. Strahler. 1999. Maximizing land cover classification accuracies produced by decision trees at continental to global scales. *IEEE Transactions on Geoscience and Remote Sensing* 37: 969–977.

Fryxell, J. F., J. Greever, and A. R. E. Sinclair. 1988. Why are migratory ungulates so abundant? *American Naturalist* 131: 781–798.

Frohn, R. C. 1998. *Remote Sensing for Landscape Ecology: New Metric Indicators for Monitoring, Modelling and Assessment of Ecosystems.* Lewis Publications, Boca Raton, Florida.

Frohn, R. C. and Y. Hao. 2006. Landscape metric performance in analyzing two decades of deforestation in the Amazon Basin of Rondonia, Brazil. *Remote Sensing of Environment* 100: 237–251.

Fry, G. 2001. Multifunctional landscapes—toward transdisciplinary research. *Landscape and Urban Planning* 57: 159–168.

Fryxell, J. and J. Greever. 1988. Why are Migratory Ungulates So Abundant? *American Naturalist* 131: 781–798

Fullbright, T. E. and D. G. Hewitt, eds. 2008. *Wildlife Science: Linking Ecological Theory and Management Applications.* CRC Press, Boca Raton, Florida.

Fuller, R. M., G. B. Groom, and A. R. Jones. 1994. The land cover map of Great Britain: an automated classification of Landsat Thematic Mapper data. *Photogrammetric Engineering and Remote Sensing* 60: 553–562.

Fuller, R. M., G. B. Groom, S. Mugisha, P. Ipulet, D. Pomeroy, A. Katende, R. Bailey, and O. R. Ogutu. 1998. The integration of field survey and remote sensing for biodiversity assessment: a case study in the tropical forests and wetlands of Sango Bay, Uganda. *Biological Conservation* 86: 379–391.

Gardner, H. 2006. *Five Minds for the Future.* Harvard Business School Press, Boston, MA.

Gardner, R. H. and D. L. Urban 2007. Neutral models for testing landscape hypotheses. *Landscape Ecology* 22: 15–29.

Gates, D. M. 1967. Remote sensing for the biologist. *BioScience* 17: 303–307.

Gausman, H. 1977. Reflectance of leaf components. *Remote Sensing of Environment* 6: 1–9.

Gauthier, F. and A.Tabbagh,. 1994. The use of airborne thermal remote sensing for soil mapping: a case study in the Limousin region (France). *International Journal of Remote Sensing* 15: 1981–1989.

Gautier, F. and A. Tabbagh. 1994. The use of airborne thermal remote sensing for soil mapping: a case study in the Limousin Region (France). *International Journal of Remote Sensing* 15: 1981–1989.

Gavashelishvili, A. and V. Lukarevskiy. 2008. Modeling the habitat requirements of leopard Panthera pardus in west and central Asia. *Journal of Applied Ecology* 45: 579–588.

Gergel, S. E. 2007. New directions in landscape pattern analysis and linkages with remote sensing. In Wulder, M. A. and S. E. Franklin, eds., *Understanding Forest*

Disturbance and Spatial Pattern: Remote Sensing and GIS Approaches. Taylor and Francis, Boca Raton, Florida. 173–208.

Germade, E. I. I., J. Y. Jiya, B. E. B. Nwoke, E. O. Ogunba, H. Edegherre, J. I. Akoh, and A. Omojola. 1998. Human *Onchocerciasis*: current assessment of the disease burden in Nigeria by rapid epidemiological mapping. *Annals of Tropical Medicine and Parasitology* 92(Suppl.): 79–83.

Gerylo, G., R. J. Hall, S. E. Franklin, A. Roberts, and E. J. Milton. 1998. Hierarchical image classification and extraction of forest species composition and crown closure from airborne multispectral images. *Canadian Journal of Remote Sensing* 24: 219–232.

Ghitter, G. S., R. J. Hall, and S. E. Franklin. 1995. Variability of Landsat Thematic Mapper data in boreal deciduous and mixedwood stands with conifer understory. *International Journal of Remote Sensing* 16: 2989–3002.

Giardino, C., V. E. Brando, A. G. Dekker, N. Strombeck, and G. Candiani. 2007. Assessment of water quality in Lake Garda (Italy) using Hyperion. *Remote Sensing of Environment* 109: 183–195.

Gibbons, P. and D. Lindenmayer. 2002. *Tree Hollows and Wildlife Conservation in Australia.* CSIRO Publications, Collingwood, Australia.

Giles, R. H. Jr. 1978. *Wildlife Management.* W. H. Freeman and Co., San Francisco, California.

Gillespie, T. W. 2001. Remote sensing of animals. *Progress in Physical Geography* 25: 355–363.

Gillieson, D. S., T. J. Lawson, and L. Searle. 2006. *Applications of High Resolution Remote Sensing in Rainforest Ecology and Management.* Cooperative Research Centre for Tropical Rainforest Ecology and Management, CRC, Cairns.

Gillis, M. D. and D. G. Leckie. 1993. *Forest Inventory Mapping Procedures Across Canada.* Petawawa National Forestry Institute, Information Report PI-X-114. Canadian Forest Service, Ottawa.

Gilvear, D. J., P. Sutherland, and T. Higgins. 2008. An assessment of the use of remote sensing to map habitat features important to sustaining lamprey populations. *Aquatic Conservation: Marine and Freshwater Ecosystems* 18: 807–818.

Gleason, S. T. 2007. Sensing land and ice from low earth orbit. *GPSWorld.*Available online http://sidt.gpsworld.com/gpssidt/Innovation/Innovation-Reflecting-on-GPS/ArticleStandard/Article/detail/464786?contextCategoryId=285 web site accessed August 12, 2008.

Glenn, E. M., and W. J. Ripple. 2004. On using digital maps for assessing wildlife habitat. *Wildlife Society Bulletin* 32: 852–860.

Goetz, S., D. Steinberg, R. Dubayah, and B. Blair. 2007. Laser remote sensing of canopy habitat heterogeneity as a predictor of bird species richness in an eastern temperate forest, USA. *Remote Sensing of Environment* 108: 254–263.

Gholz, H. L., K. Nakane, and, H.Shimoda. 1997. *The Use of Remote Sensing in the Modeling of Forest Productivity.* Springer, Dordrecht, The Netherlands.

Gomez, C., F. A. Viscarra Rossel, and A. B. McBratney. 2008. Soil organic carbon prediction by hyperspectral remote sensing and field vis-NIR spectroscopy: an Australian case study. *Geoderma* 146: 403–411.

Gong, P., R. Pu, and B. Yu. 1997. Conifer species recognition: an exploratory analysis of *in situ* hyperspectral data. *Remote Sensing of Environment* 62: 189–200.

Gong, P. and B. Xu. 2003. Remote sensing of forests over time: change types, methods, and opportunities. In Wulder, M. A. and S. E. Franklin, eds., *Remote Sensing of Forest Environments: Concepts and Case Studies.* Kluwer Academic Publishers,, Boston, MA. 301–333.

Goodchild, M. F. 1992. Geographical information science. *International Journal of Geographical Information Systems* 6: 31–45.

Goodin, D. G., D. E. Koch, R. D. Owen, Y. -K. Chu, S. J. M. Hutchinson, and C. B. Jonsson. 2006. Land cover associated with hantavirus presence in Paraguay. *Global Ecology and Biogeography* 15: 519–527.

Goodwin, N. R. 2006. *Assessing Understory Structural Characteristics in Eucalypt Forests: An Investigation of LiDAR Techniques.* Unpublished Ph.D. Thesis, University of New South Wales.

Goodwin, N. R., N. C. Coops, and D. S. Culvenor. 2006. Assessment of forest structure with airborne LIDAR and the effects of platform altitude. *Remote Sensing of Environment* 103: 140–152.

Goosem, M. 2007. Fragmentation impacts caused by roads through rainforests. *Current Science* 93: 1587–1595.

Gordon, C. H., A. -M. E. Stewart, and E. Meijaard. 2007. Correspondence regarding Clouded leopards, the secretive top-carnivore of south-east Asian rainforests: their distribution, status and conservation needs in Sabah, Malaysia. *BMC Ecology* 7: 5.

Gordon, H. R., O. B. Brown, and M. M. Jacobs. 1975. Computed relationship between the inherent and apparent optical properties of a flat homogeneous ocean. *Applied Optics* 14: 417–427.

Gottschalk, T. K. and F. Huettmann. 2006. Thirty years of analyzing and modeling avian habitat relationships using satellite imagery data: A review. *Journal of Ornithology* 147: 175–195.

Gottschalk, T. K., K. Ekschmitt, S. Isfendiyaroglu, E. Gem, and V. Wolters. 2007. Assessing the potential distribution of the Caucasian black grouse *Tetrao mlokosiewiczi* in Turkey through spatial modeling. *Journal of Ornithology* 148: 427–434.

Gougeon, F. 1995. A crown-following approach to the automatic delineation of individual tree crowns in high spatial resolution aerial images. *Canadian Journal of Remote Sensing* 21: 274–284.

Gould, W. 2000. Remote sensing of vegetation, plant species richness, and regional biodiversity hotspots. *Ecological Applications* 10: 1861–1870.

Gower, J. F. R. 2006. *Remote Sensing of the Marine Environment: Volume 6.* Manual of Remote Sensing. 3d ed., ASPRS, Falls Church, VA.

Grabas, G., J. Ingram, and N. Patterson. 2006. Interdisciplinary approach to assessing the ecological integrity of the Great Lakes coast wetlands and evaluating restoration activities. *Proceedings of Annual Conference on Great Lakes Research* 49 (abs.). Windsor, ON.

Graham, R. and R. E. Read. 1986. *Manual of Aerial Photography*. Focal Press, London, UK.

Grant, E. H. C. 2005. Correlates of vernal pool occurrence in the Massachusetts, USA landscape. *Wetlands* 25: 480–487.

Grassman, L. I., Jr., M. E. Tewes, N. J. Silvy, and K. Kreetiyutanont. 2005. Ecology of three sympatric felids in a mixed evergreen forest in north-central Thailand. *Journal of Mammalogy* 86: 29–38.

Gray, J. G. 1952. Heidegger's "Being." *Journal of Philosophy* 49: 415–422.

Gregory, D. 1994. *Geographical Imaginations.* Blackwell Publishers, Cambridge, MA.

Green, E., P. Mumby, A. Edwards, and C. Clark. 1996. A review of remote sensing for the assessment and management of tropical coastal resources. *Coastal Management* 24: 1–40.

Green, K. 1999. Development of the spatial domain in resources management. In Morain, S., ed., *GIS Solutions in Natural Resource Management: Balancing the Technical-Political Equation.* Onward Press, Sante Fe.

Gregoire, J. -M. and R. Zeyen. 1986. An evaluation of ultralight aircraft capability for remote sensing applications in West Africa. *International Journal of Remote Sensing* 7: 1075–1081.

Gregory, A. F. 1971. Earth-observation satellites: a potential impetus for economic and social development. *World Cartography* XI: 1–15.

Gregory, S., K. Boyer, and A. Gurnell, eds. 2003. *The Ecology and Management of Wood in Rivers.* American Fisheries Society, Bethesda, MD.

Grierson, I. T. and J. A. Gammon. 2000. The use of aerial imaging in kangaroo management. *Proceedings of IEEE Transactions* Honolulu, HI. 1446–1448.

Grigg, G. C., A. R. Pople, and L. A. Beard. 1997. Application of an ultralight aircraft to aerial surveys of kangaroos on grazing properties. *Australian Wildlife Research* 24: 359–372.

Griffiths, M. and C. P. van Schalk. 1993. Camera-trapping: an new tool for the study of elusive rain forest animals. *Tropical Biodiversity* 1: 131–135.

Gruber, M. 2008. *UltraCamX, the New Digital Aerial Camera System by Microsoft Photogrammetry*. Available online http://www.ifp.uni–stuttgart.de/publications/phowo07/160Gruber.pdf web site accessed August 5, 2008.
Gruber, M. and S. Schneider. 2007. Digital surface models from UltracamX images. *International Archives of Photogrammetry, Remote Sensing and Spatial Information Sciences* 36: 47–52.
Gu, D., and A. Gillespie. 1998. Topographic normalization of Landsat TM images of forest based on subpixel sun-canopy-sensor geometry. *Remote Sensing of Environment* 64: 166–175.
Guild, L., B. Ganapol, P. Kramer, R. Armstrong, A. Gleason, J. Torres, L. Johnson, and T. Garfield. 2002. Clues to coral reef health: integrating radiative transfer modeling and hyperspectral data. *EOS Transactions AGU Fall Meeting Suppl.* 83: OS71A-0264.
Guindon, B. 2000. A framework for the development and assessment of object recognition modules from high-resolution satellite images. *Canadian Journal of Remote Sensing* 26: 334–348.
Guis, H., A. Tran, S. de La Rocque, T. Baldet, G. Gerbier, B. Barragué, F. Biteau-Coroller, et al. 2007. Use of high spatial resolution satellite imagery to characterize landscapes at risk for bluetongue. *Veterinary Research* 38: 669–683.
Guisan, A. and W. Thuiller. 2005. Predicting species distribution: offering more than simple habitat models. *Ecology Letters* 8: 993–1009.
Guisan, A., C. H. Graham, J. Elith, F. Huettmann, and the NCEAS Species Distribution Modelling Group. 2007. Sensitivity of predictive species distribution models to change in grain size. *Diversity and Distributions* 13: 332–340.
Gurney, C. M. 1981. The use of contextual information to improve land cover classification of digital remotely sensed data. *International Journal of Remote Sensing* 2: 379–388.
Gustafson, W. A. 2002. *Assessing Landsat TM Imagery for Mapping and Monitoring Prairie Dog Colonies*. Unpublished M.A. Thesis, University of Montana.
Gysel, L. W. and E. M. Davis, Jr. 1956. A simple automatic photographic unit for wildlife research. *Journal of Wildlife Management* 20: 451–453.
Haddad, N. M. 1999. Corridor and distance effects on interpatch movements: a landscape experiment with butterflies. *Ecological Applications* 9: 612–622.
Hager, H. A. 1998. Area-sensitivity of reptiles and amphibians: are there indicator species for habitat fragmentation? *Ecoscience* 5: 139–147.
Haines, A. M., L. I. Grassman, Jr., M. E. Tewes, and J. E. Jane ka. 2006. First ocelot (*Leopardus pardalis*) monitored with GPS telemetry. *European Journal of Wildlife Research* 52: 216–218.
Haldane, J. B. S. 1927. *Possible Worlds and Other Essays*. Chatto and Windus, London, UK.
Hall, F., G., D. B. Botkin, D. E. Strebel, K. D. Woods, and S. J. Goetz. 1991. Large-scale patterns of forest succession as determined by remote sensing. *Ecology* 72: 628–640.
Hall, R. J. 2003. The roles of aerial photographs in forestry remote sensing image analysis. In Wulder, M. A. and S. E. Franklin, eds., *Remote Sensing of Forest Environments: Concepts and Case Studies*. Kluwer Academic Publishers, Boston, MA. 47–75.
Hall, R. J., D. R. Peddle, and D. L. Klita. 2000. Mapping conifer understory within boreal mixedwoods from Landsat TM satellite imagery and forest inventory information. *Forestry Chronicle* 76: 75–90.
Hall, R. J., R. Skakun, and E. Arsenault. 2007. Remotely sensed data in the mapping of insect defoliation. In Wulder, M. A. and S. E. Franklin, eds., *Understanding Forest Disturbance and Spatial Pattern: Remote Sensing and GIS Approaches*. Taylor and Francis (CRC Press), Boca Raton, Florida. 85–111.
Hall, S. S. 1992. *Mapping the Next Millennium: The Discovery of New Geographies*. Random House, New York, U.S.
Hansen, A. J., S. L. Garman, B. Marks, and D. L. Urban. 1993. An approach for managing vertebrate diversity across multiple-use landscapes. *Ecological Applications* 3: 481–496.

Hansen, M. C., R. S. DeFries, J. R. G. Townshend, M. Carroll, C. Dimiceli, and R. A. Sohlberg. 2006. *Vegetation Continuous Fields MOD44B, 2001 Percent Tree Cover, Collection 4.* University of Maryland, College Park, Maryland.

Hansen, M. C., R. S. DeFries, J. R. G. Townshend, M. Carroll, C. Dimiceli, and R. A. Sohlberg. 2003. Global percent tree cover at a spatial resolution of 500 meters: first results of the MODIS Vegetation Continuous Fields algorithm. *Earth Interactions* 7: 1–15.

Hansen, M. C., S. V. Stehman, P. V. Potapov, T. R. Loveland, J. R. G. Townshend, R. S. DeFries, K. W. Pittman, et al. 2008. Humid tropical forest clearing from 2000 to 2005 quantified by using multitemporal and multiresolution remotely sensed data. *Proceedings of the National Academy of Sciences*, 105: 9439–9444.

Hansen, M. J., S. E. Franklin, C. Woudsma, and M. Peterson. 2001. Caribou habitat mapping and fragmentation analysis using Landsat MSS, TM and GIS data in the North Columbia Mountains, British Columbia, Canada. *Remote Sensing of Environment* 77: 50–65.

Hanson, P. J., J. S. Amthor, S. D. Wullscheger, K. B. Wilson, R. F. Grant, A. Hartley, D. Hui, et al. 2004. Oak forest carbon and water simulations: model intercomparisons and evaluations against independent data. *Ecological Monographs* 74: 443–489.

Haralick, R. M. 1979. Statistical and structural approaches to texture. *Proceedings of IEEE* 67: San Juan, Puerto Rico. 786–804.

Haralick, R. M., K. Shanmugan, and I. Dinstein. 1973. Texture features for image classification. *IEEE Transactions on Systems, Man and Cybernetics* 3: 610–621.

Hargis, C. D., J. A. Bissonette, and D. L. Turner. 1999. The influence of forest fragmentation and landscape pattern on American marten. *Journal of Applied Ecology* 36: 157–172.

Harris, L. D. 1984. *The Fragmented Forest: Island Biogeography Theory and the Preservation of Biotic Diversity.* University of Chicago Press, Chicago, IL.

Harvey, K. R. and G. Hill. 2003. Mapping the nesting habitat of saltwater crocodiles (*Crocodylus porosus*) in Melacca Swamp and the Adelaide River wetlands, Northern Territory: an approach using remote sensing and GIS. *Wildlife Research* 30: 365–375.

Hattenschwiler, S., A. V. Tiunov, and S. Scheu. 2005. Biodiversity and litter decomposition in terrestrial ecosystems. *Annual Review of Ecology, Evolution and Systematics* 36: 191–218.

Hay, K., 1958. Beaver census methods in the Rocky Mountain region. *Journal of Wildlife Management*, 22: 395–402.

Hayes, L and A. P. Cracknell. 1987. Georeferencing and registering satellite data for monitoring vegetation over large areas. *Pattern Recognition Letters* 5: 95–105.

He, Y., S. E. Franklin, and X. Guo. 2009. Narrow-linear and small-area forest disturbance detection and mapping from high spatial resolution imagery: a review. *Proceedings of ASPRS Annual Meeting*. Baltimore, MD (abs.).

Heath, G. R. 1956. A comparison of two basic theories of land classification and their adaptability to regional photointerpretation key techniques. *Photogrammetric Engineering* 22: 144–168.

Heberlein, T. A. 1988. Improving interdisciplinary research: integrating the social and natural sciences. *Society and Natural Resources* 7: 595–597.

Heinsch, F. A., M. Reeves, P. Votava, S. Kang, C. Milesi, M. Zhao and others. 2003. *User's Guide: GPP and NPP (MOD17A2/A3) Products*. NASA MODIS Land Algorithm. University of Montana, Missoula.

Hellweger, F. L., P. Schlosser, U. Lall, and J. K. Weissel. 2004. Use of satellite imagery for water quality studies in New York harbor. *Estuarine, Coastal and Shelf Science* 61: 437–448.

Helzer, C. J. and D. E. Jelinski. 1999. The relative importance of patch area and perimeter-area ratio to grassland breeding birds. *Ecological Applications* 9: 1448–1458.

Henry, P. -Y., S. Lengyel, P. Nowicki, R. Julliard, J. Clobert, T. Celik, B. Gruber, et al. 2008. Integrating ongoing biodiversity monitoring: potential benefits and methods. *Biodiversity and Conservation* 17: 3357–3382.

Hepinstall, J. A. and S. Sader. 1997. Using Bayesian statistics, Thematic Mapper satellite imagery, and breeding bird survey data to model bird species probability of occurrence in Maine. *Photogrammetric Engineering and Remote Sensing* 63: 1231–1237.

Herbreteau, V., G. Salem, M. Souris, J. -P. Hugot, and J. -P. Gonzalez. 2007. Thirty years of use and improvement of remote sensing, applied to epidemiology: from early promises to lasting frustration. *Health and Place* 13: 400–403.

Herold, M., L. Liu, and K. C. Clarke. 2003. Spatial metrics and image texture for mapping urban land use. *Photogrammetric Engineering and Remote Sensing* 69: 991–1001.

Heske, E. J., S. K. Robinson, and J. D. Brawn. 1999. Predator activity and predation on songbird nests on forest-field edges in east-central Illinois. *Landscape Ecology* 14: 345–354.

Hessburg, P. F., B. G. Smith, and R. B. Salter. 1999. Detecting change in forest spatial patterns. *Ecological Applications* 9: 1232–1252.

Hessburg, P. F., K. M. Reynolds, R. B. Salter, and M. B. Richmond. 2004. Using a decision support system to estimate departures of present forest landscape patterns from historical reference conditions. In Perera, A. H., et al., eds., *Emulating Natural Forest Landscape Disturbances: Concepts and Applications.* Columbia University Press, New York. 158–175.

Hill, R. A., S. A. Hinsley, and P. E. Bellamy. 2008. Integrating multiple datasets for the remote quantification of woodland bird habitat quality. *International Archives of Photogrammetry, Remote Sensing and Spatial Information Sciences* XXXV1-8/W2: 248–253.

Hill, R. A. and G. Thomson. 2005. Mapping woodland species composition and structure using airborne spectral and LiDAR data. *International Journal of Remote Sensing* 26: 3763–3779.

Hilty, J. A., W. H. Lidicker, Jr., and A. M. Merenlender. 2006. *Corridor Ecology: the Science and Practice of Linking Landscapes for Biodiversity Conservation.* Island Press, Washington.

Hird, J. N., and G. J. McDermid. 2009. Noise reduction of NDVI time series: an empirical comparison of selected techniques. *Remote Sensing of Environment* 113: 248–258.

Hirsch-Hadden, G., H. Hoffmann-Riem, S. Biber-Klemm, W. Grossenbacher-Mansuy, D. Joye, C. Pohl, U. Wiesmann, and E. Zemp, eds. 2008. *Handbook of Transdisciplinary Research.* Springer, Dordrecht, The Netherlands.

Hobbs, R. J. 1998. Managing ecological systems and processes. In Peterson, D. L. and V. Parker, eds., *Ecological Scale: Theory and Applications.* Columbia University Press, New York. 459–484.

Hoekman, D. H. 2007. Satellite radar observation of tropical peat swamp forest as a tool for hydrological modelling and environmental protection *Aquatic Conservation: Marine and Freshwater Ecosystems* 17: 265–275.

Hoekman, D. H. and M. A. M. Vissers. 2003. A new polarimetric classification approach evaluated for agricultural crops. *IEEE Transactions on Geoscience and Remote Sensing* 41: 2881–2889.

Hoffer, R. M., 1994. Challenges in developing and applying remote sensing to ecosystem management. In Sample, V. A., ed., *Remote Sensing and GIS in Ecosystem Management.* Island Press, Washington DC. 25–40.

Holben, B. N. and C. O. Justice. 1980. The topographic effect on spectral response from nadir-pointing sensors. *Photogrammetric Engineering and Remote Sensing* 46: 1191–1200.

Holden, H. and E. F. LeDrew. 1998. The scientific issues surrounding remote detection of submerged coral ecosystems. *Progress in Physical Geography,* 22: 190–221.

Holden, H. and E. F. LeDrew. 2001. Hyperspectral discrimination of healthy versus stressed corals using in situ reflectance. *Journal of Coastal Research* 17: 850–858.

Holdridge, L. R. 1967. *Life Zone Ecology.* Tropical Science Center, San Jose, Costa Rica.

Hollander, A. D., F. W. Davis, and D. M. Stoms. 1994. Hierarchical representations of species distributions, using maps, images and sighting data. In Miller, R. I., ed., *Mapping the Diversity of Nature.* Chapman and Hall, London. 71–88.

Holmgren, P. and T. Thuresson. 1998. Satellite remote sensing for forestry planning—a review. *Scandinavian Journal of Forest Research.* 13: 90–110.

Holmes, K.W., B. Radford, K. P. Van Niel, G. A.Kendrick, and S. L.Grove. 2008. Prediction of marine benthic biota distributions too deep for optical remote sensing. *Coastal Shelf Research* 28: 1800–1810.

Hoque, E., P. J. S. Hutzler, and H. Hiendl. 1992. Reflectance, colour, and histological features as parameters for the early assessment of forest damages. *Canadian Journal of Remote Sensing* 18: 105–110.

Hood, G. A., and S. E. Bayley. 2008. Beaver (*Castor canadensis*) mitigate the effects of climate on the area of open water in boreal wetlands in western Canada. *Biological Conservation* 141: 556–567.

Horn, R., A. Nottensteiner, and R. Scheiber. 2008. F-SAR—DLR's advanced airborne SAR system onboard DO228. *Proceedings of European Conference on Synthetic Aperture Radar (EUSAR).* Friedrichshafen, Germany. Available online http://elib.dlr.de/54519/01/EUSAR2008_F-SAR_Paper.pdf web site accessed August 7, 2008.

Hosking, E. and C. Newberry. 1944. *The Art of Bird Photography.* Country Life Ltd., London, UK.

Howard, J., K. Pelican, and D. Wildt. 2002. *Thailand Clouded Leopard and Fishing Cat Conservation Project.* Available online http://nationalzoo.si.edu/ConservationAndScience/ReproductiveScience/Thailand_Clouded_Leopard_and_Fishing_Cat_Conservation_Project.pdf web site accessed August 28, 2008.

Howard, J. A. 1991. *Remote Sensing of Forest Resources.* Chapman and Hall, London.

Howard, J. A. and I. J. Barton. 1973. Instrumentation for remote sensing solar radiation from light aircraft. *Applied Optics* 12: 2472–2476.

Hu, J. 2001. *Research on the Giant Panda.* Sichuan Publishing House of Science and Technology, Chengdu.

Huang, C., C. Homer, and L. Yang. 2003. Regional forest land cover characterization using medium spatial resolution satellite data. In Wulder, M. A. and S. E. Franklin, eds., *Remote Sensing of Forest Environments: Concepts and Case Studies.* Kluwer Academic Publishers, Boston, MA. 389–410.

Huang, C., X. Li, and L. Lu. 2008. Retrieving soil temperature profile by assimilating MODIS LST products with ensemble Kalman filter. *Remote Sensing of Environment* 112: 1320–1336.

Huber, T. P. and K. E. Casler. 1990. Initial analysis of Landsat TM data for elk habitat monitoring. *International Journal of Remote Sensing* 11: 907–912.

Hudson, W. D. 1991. Photo-interpretation of montane forests in the Dominican Republic. *Photogrammetric Engineering and Remote Sensing* 57: 79–84.

Huettmann, F. and S. Cushman, eds. 2009. *Spatial Information Management in Animal Science.* Springer-Verlag, Tokyo, in press.

Huettmann, F., S. E. Franklin, and G. B. Stenhouse. 2005. Predictive spatial modelling of landscape change in the Foothills Model Forest. *Forestry Chronicle* 81: 525–537.

Huguenin, R. L., M. A. Karaska, D. Van Blaricom, and J. R. Jensen. 1997. Subpixel classification of bald cypress and tupelo gum trees in Thematic Mapper imagery. *Photogrammetric Engineering and Remote Sensing* 63: 717–725.

Hunt, E. R., Jr. 2004. Hyperspectral remote sensing to detect flowering leafy spurge. *Proceedings of Monitoring Science and Technology Symposium* (abs.). Available online http://www.ars.usda.gov/research/publications/publications.htm?SEQ_NO_115=169974 web site accessed March 14, 2009.

Hunt, E. R., Jr., S. C. Piper, R. Nemani, C. D. Keeling, R. D. Otto, and S. W. Running. 1996. Global net carbon exchange and intraannual atmospheric CO_2 concentration predicted by an ecosystem process model and three-dimensional atmospheric transport model. *Global Biogeochemical Cycles* 10: 431–456.

Hunt, E. R., Jr., J. E. McMurtrey, A. E. P. Williams, and L. A. Corp. 2004. Spectral characteristics of leafy spurge (*Euphorbia esula*) leaves and flower bracts. *Weed Science* 52: 492–497.

Hunter, A., N. El -Sheimy, and G. Stenhouse. 2005. Up close and grizzly: GPS/ camera collar captures bear doings. *GPSWorld*, February: 24–31.

Hunter, M. L. Jr., ed. 1999. *Maintaining Biodiversity in Forest Ecosystems.* Cambridge University Press, Cambridge, UK.

Hutchinson, C. F. 1982. Techniques for combining Landsat and ancillary data for digital classification improvement. *Photogrammetric Engineering and Remote Sensing* 48: 123–130.

Hyde, P., R. Dubaya, W. Walker, J. B. Blair, M. Hofton, and C. Hunsaker. 2006. Mapping forest structure for wildlife habitat analysis using multi-sensor (LiDAR, SAR/InSAR, ETM+, QuickBird) synergy. *Remote Sensing of Environment* 102: 63–73.

Hyde, R. F. and N. J. Vesper. 1983. Some effects of resolution cell size on image quality. *Landsat Data User Notes* 29: 9–12.

Iiames, J. S., R. G. Congalton, A. N. Pilant, and T. E. Lewis. 2008. Leaf area index (LAI) change detection analysis on Loblolly Pine (*Pinus taeda*) following complete understory removal. *Photogrammetric Engineering and Remote Sensing* 74: 1389–1400.

Imhoff, M. L., P. Johnson, W. Holford, J. Hyer, L. May, W. Lawrence, and P. Harcombe. 2000. BioSAR™: an inexpensive airborne VHF multiband SAR system for vegetation biomass measurement. *IEEE Transactions on Geoscience and Remote Sensing* 38: 1458–1462.

Imhoff, M. L., T. D. Sisk, G. Milne, G. Morgan, and T. Orr. 1997. Remotely sensed indicators of habitat heterogeneity: use of synthetic aperture radar in mapping vegetation structure and bird habitat. *Remote Sensing of Environment* 60: 217–227.

Innes, J. L., and B. Koch. 1998. Forest biodiversity and its assessment by remote sensing. *Global Ecology and Biogeography Letters* 7: 397–419.

Inoue, Y., S. Morinaga, and A. Tomita. 2000. A blimp-based remote sensing system for low-altitude monitoring of plant variables: a preliminary experiment for agricultural and ecological applications. *International Journal of Remote Sensing* 21: 279–385.

Ito, T. Y., N. Miura, B. Lhagvasuren, D. Enkhbileg, S. Takasuki, A. Tsunekawa, and Z. Jiang. 2006. Satellite tracking of Mongolian gazelles (*Procapra gutturosa*) and habitat shifts in their seasonal ranges. *Journal of Zoology* 269: 291–298.

Iverson, L. R., E. A. Cook, and R. L. Graham. 1989. A technique for extrapolating and validating forest cover across large regions: calibrating AVHRR data with TM data. *International Journal of Remote Sensing* 10: 1805–1812.

Jackson, T. J. 1993. Measuring surface soil moisture using passive microwave remote sensing. *Hydrological Processes* 7: 139–152.

Jacquin, A., V. Cheret, J. -P. Denux, M. Gay, J. Mitchley, and P. Xofix. 2005. Habitat suitability modeling of Capercaillie (Tetrao uorgallus) using earth observation data. *Journal for Nature Conservation* 13: 161–169.

James, M. E. and S. N. V. Kalluri. 1994. The Pathfinder AVHRR land data set: An improved coarse resolution data set for terrestrial monitoring. *International Journal of Remote Sensing, Special Issue on Global Data Sets.* 15(17): 3347–3363.

Jano, A. P., R. L. Jeffries, and R. F. Rockwell. 1998. The detection of vegetational change by multitemporal analysis of Landsat data: the effects of goose foraging. *Journal of Ecology* 86: 93–99.

Jensen, J. R. 2005. *Introductory Digital Image Processing: A Remote Sensing Perspective.* 3d ed., Prentice Hall, Upper Saddle River, New Jersey.

Jensen, J. R. 2007. *Remote Sensing of the Environment: An Earth Resource Perspective.* 2d ed., Prentice Hall, Upper Saddle River, New Jersey.

Jianwen, M., L. Xiawne, C. Xue, and F. Chun. 2006. Target adjacency effect estimation using ground spectrum measurement and Landsat-5 satellite data. *IEEE Transactions on Geoscience and Remote Sensing* 44: 729–735.

John, R., J. Chen, N. Lu, K. Guo, C. Liang, Y. Wei, A. Noormets, K. Ma, and X.Han. 2008. Prediction of plant diversity based on remote sensing products in the semi-arid region of Inner Mongolia. *Remote Sensing of Environment* 112: 2018–2032.

Jones, L. A., J. S. Kimball, K. C. McDonald, S. T. K. Chan, E. G. Njoku, and W. C. Oechel. 2007. Satellite microwave remote sensing of boreal and arctic soil temperatures from AMSR-E. *IEEE Transactions on Geoscience and Remote Sensing* 45: 2004–2018.

Jones, L. L. C. and M. G. Raphel. 1993. *Inexpensive Camera Systems for Detecting Martens, Fisher, and Other Animals: Guidelines for Use and Standardization.* US Forest Service General Technical Report PNW-GTR-306. 16pp.

Jorgensen, E. E. 1997. Habitat suitability is a valuable concept. *Wildlife Society Bulletin* 25: 602–603.

Jorgensen A. F. and H. Nohr. 1996. The use of satellite images for mapping landscape and biological diversity. *International Journal of Remote Sensing* 17: 99–109.

Juntii, T. M. and M. A. Rumble. 2006. *Arc Habitat Suitability Index Computer Software.* General Technical Report RMRS-GTR-180WWW. USDA Rocky Mountain Research Station, Ft. Collins, CO.

Kalluri, S., P. Gilruth, and R. Bergman. 2003. The potential of remote sensing data for decision makers at the state, local and tribal level: experiences from NASA's Synergy program. *Environmental Science and Policy* 6: 487–500.

Kansas, J. L. 2003. *Effects of Mapping Scale, Disturbance Coefficients and Season on Grizzly Bear Habitat Effectiveness Models in Kananaskis Country, Alberta.* Unpublished M.Sc. Thesis. University of Calgary, Calgary, Alberta.

Kantety, R. V., E. Vansanten, F. M. Woods, and C. W. Wood. 1996. Chlorophyll meter predicts nitrogen status of tall fescue. *Journal of Plant Nutrition* 19: 881–889.

Kareiva, P. and U. Wennegren. 1995. Connecting landscape patterns to ecosystem and population processes. *Nature* 373: 299–302.

Karpouzli, E., and T. Malthus. 2003. Hyperspectral discrimination of coral reef benthic commuities. *Proceedings of IEEE* 2377–2379. Available online http://ieeexplore.ieee.org/iel5/9010/28604/01294447.pdf web site accessed August 6, 2008.

Kaplan, E. D. ed. 2005. *Understanding GPS: Principles and Applications.* 2d ed., Artech House Publishers, Boston, MA.

Kasischke, E. S. and N. L. Christensen, Jr. 1990. Connecting forest ecosystem and microwave backscatter models. *International Journal of Remote Sensing* 11: 1277–1298.

Kauffman, M. J., N. Varley, D. W. Smith, D. R. Strahler, D. R. MacNulty, and M. S. Boyce. 2007. Landscape heterogeneity shapes predation in a newly restored predator-prey system. *Ecology Letters* 10: 690–700.

Kayitakire, F., C. Hamel, and P. Defourny. 2006. Retrieving forest structure variables based on image texture analysis and IKONOS-2 imagery. *Remote Sensing of Environment* 102: 390–401.

Kelly, N. M., M. Fonseca, and P. Whitfield. 2001. Predictive mapping for management and conservation of seagrass beds in North Carolina. *Aquatic Conservation: Marine and Freshwater Ecosystems* 11: 437–451.

Kennedy, M. 2002. *The Global Positioning System and GIS.* 2d ed., Taylor and Francis, New York, U.S.

Kenward, R. 1987. *Wildlife Radio Tagging*. Academic Press, London, UK.

Keramitsoglou, I., H. Sarimveis, C. T. Kiranoudis, C. Kontoes, N. Sifakis, and E. Fitoka. 2006. The performance of pixel window algorithms in the classification of habitats using VHSR imagery. *ISPRS Journal of Photogrammetry and Remote Sensing* 60: 225–238.

Kern, S., G. Spreen, L. Kaleschke, S. De La Rosa Höhn, and G. Heygster. 2007. Polynya Signature Simulation Method polynya area in comparison to AMSR-E 89 GHz sea-ice concentrations in the Ross Sea and off the Adélie Coast, Antarctica, for 2002–2005: first results. *Annals of Glaciolology* 46: 409–418.

Kettig, R. L. and D. A. Landgrebe. 1976. Classification of multispectral image data by extraction and classification of homogeneous objects. *IEEE Transactions on Geoscience Electronics* 14: 19–26.

Killilea, M. E., A. Swei, R. S. Lane, C. J. Briggs, and R. S. Ostfeld. 2008. Spatial dynamics of lyme disease: a review. *EcoHealth* 5: 167–195.

King, D. J. 1991. Determination and reduction of cover type brightness variations with view angle in airborne multispectral video imagery. *Photogrammetric Engineering and Remote Sensing* 57: 1571–1577.

Knudby, A., E. F. LeDrew, and C. Newman. 2007. Progress in the use of remote sensing for coral reef biodiversity studies. *Progress in Physical Geography* 31: 421–434.

Koch, B. and E. Ivits. 2002. What can remote sensing provide for biodiversity assessment? *Proceedings of the ForestSat Symposium*, Heriot Watt University.

Koirala, R. A., and R. Shrestha. 1997. *Floristic Composition of Summer Habitat and Dietary Relationships between Tibetan argali (Ovis ammon hodgsonii), Naur (Pseudois nayaur), and Domestic Goat (Capra hircus) in the Damodar Kunda Region of Upper Mustang, Nepal, Himalaya.* Unpublished Manuscript. Agricultural University of Norway, Osloveien, Norway.

Kondratyev, K. Ya., O. B. Vasilyev, O. M. Pokrovsky, and G. A. Ivanyan. 1974. Remote sensing of natural formations from measurements of radiance coefficients. *Acta Astronautica* 1: 1415–1426.

Korpela, I. S. 2008. Mapping of understory lichens with airborne discrete-return LiDAR data. *Remote Sensing of Environment* 112: 3891–3897.

Kostelnick, J. C., D. L. Peterson, S. L. Egbert, K. M McNyser, and J. F. Cully. 2007. Ecological niche modelling of Black-tailed prairie dog habitats in Kansas. *Transactions of the Kansas Academy of Science* 110: 187–200.

Koukoulas, S. and G. A. Blackburn. 2005. Mapping individual tree locations, height and species in broadleaved deciduous forests using airborne LiDAR and multispectral remotely sensed data. *International Journal of Remote Sensing* 26: 431–455.

Kreig., R. A. 1970. Aerial photographic interpretation for land use classification in the New York State land use and natural resources inventory. *Photogrammetria* 26: 101–111.

Krishnan, P., J. D. Alexander, B. J. Butler, and J. W. Hummel. 1980. Reflectance technique for predicting soil organic matter. *Soil Science Society of America Journal* 44: 1282–1285.

Kumar, S., and K. B. Moore. 2002. The evolution of Global Positioning System (GPS) technology. *Journal of Science Education and Technology* 11: 59–80.

Kummer, D. M. 1992. Remote sensing and tropical deforestation: a cautionary note from the Philippines. *Photogrammetric Engineering and Remote Sensing* 58: 1469–1471.

Kunz, T. H. and M. B Fenton, eds. 2003. *Bat Ecology.* University of Chicago Press, Chicago.

Kushwaha, S. P. S. 2002. *Geoinformatics for Wildlife Habitat Characterisation*. Map India. Available online http://www.gisdevelopment.net/application/environment/wildlife/bewf0002.html web site accessed July 29, 2008.

Kushwaha, S. P. S., P. Padmanaban, D. Kumar, and P. S. Roy. 2005. Geospatial modelling of plant richness in Barsey Rhododendron Sanctuary in Sikkim Himalayas. *Geocarto International* 20: 63–68.

Lachapelle, G. 1991. Capabilities of GPS for airborne remote sensing. *Canadian Journal of Remote Sensing* 17: 305–312.

Lachapelle, G. 2004. GNSS-based indoor location technologies. *Journal of Global Positioning Systems* 3: 2–11.

Lai, Y. -C., K. J. -C. Pei, P. -J. Chiang, and H. Li -Ta. 2002. An auto-multivariate model of muntjacs habitat use for a geographic information system in southern Taiwan. *Proceedings of ACRS:* Available online http://www.gisdevelopment.net/aars/acrs/2002/adp/adp006.asp web site accessed August 28, 2008.

Lake, B. C., M. S. Lindberg, J. A. Schmutz, R. M. Anthony, and F. J. Broerman. 2006. Using videography to quantify landscale-level availability of habitat for grazers: an example with emperor geese in western Alaska. *Arctic* 59: 252–260.

Laliberte, A. S., and W. J. Ripple. 2003. Automated wildlife counts from remotely sensed imagery. *Wildlife Society Bulletin* 31: 362–371.

Lassau, S. A., and D. F. Hochuli. 2007. Testing predictions of beetle community patterns derived empirically using remote sensing. *Diversity and Distributions* 14: 138–147.

Landgrebe, D. A. 1978. Useful information from multispectral image data: another look. In Swain, P. H. and S. M. Davis, eds., *Remote Sensing: The Quantitative Approach*. McGraw-Hill, New York, U.S. 336–374.

Landgrebe, D. A. 1983. Land observation sensors in perspective. *Remote Sensing of Environment* 13: 391–402.

Landgrebe, D. A. 1997. The evolution of Landsat data analysis. *Photogrammetric Engineering and Remote Sensing* 63: 859–868.

Langlois, P. J., L. Fahrig, G. Merriam, and H. Artsob. 2001. Landscape structure influences continental distribution of hantavirus in deer mice. *Landscape Ecology* 16: 255–266.

Lassau, S. A., G. Cassis, P. K. J. Flemons, L. Wilkie, and D. F. Hochuli. 2005. Using high-resolution multi-spectral imagery to estimate habitat complexity in open-canopy forests: can we predict ant community patterns? *Ecography* 28: 495–504.

Lassau, S. A. and D. F. Hochuli. 2008. Testing predictions of beetle community patterns derived empirically using remote sensing. *Diversity and Distributions* 14: 138–147.

Lathrop, R. G., P. Montesano, J. Tesauro, and B. Zarate. 2005. Statewide mapping and assessment of vernal pools: a New Jersey case study. *Journal of Environmental Management* 76: 230–238.

Lattuca, L. R. 2001. *Creating Interdisciplinarity: Interdisciplinary Research and Teaching among College and University Faculty*. Vanderbilt University Press, Nashville.

Lauer, D. T., J. E. Estes, J. R. Jensen, and D. D. Greenlee. 1991. Institutional issues affecting the integration and use of remotely sensed data and geographic information systems. *Photogrammetric Engineering and Remote Sensing* 57: 647–654.

Lawrence, R. L., S. D. Wood, and R. L. Sheley. 2006. Mapping invasive plants using hyperspectral imagery and Breiman Cutler classifications (randomForest). *Remote Sensing of Environment* 100: 356–362.

Le Hegarat-Mascle, S., R. Seltz, L. Hubert-Moy, S. Corgne, and N. Stach. 2006. Performance of change detection using remotely sensed data and evidential fusion: comparison of three different cases of application. *International Journal of Remote Sensing* 27: 3515–3532.

LeDrew, E. F., and L. L. Richardson. 2006. Recommendations for scientists and managers for application of remote sensing to coastal waters. In Richardson, L. L. and E. F. LeDrew, ed. *Remote Sensing of Aquatic Coastal Ecosystem Processes: Science and Management Applications*. Springer, Dordrecht, The Netherlands. 307–315.

Lee, J. Y. and T. A. Warner. 2006. Segment based image classification. *International Journal of Remote Sensing* 27: 3403–3412.

Lee, R. J., J. Riley, I. Hunowu, and E. Maneasa. 2003. The Sulawasi palm civet: expanded distribution of a little known endemic viverid. *Oryx* 37: 378–381.

Lee, W, M. McGlone, and E. Wright, compilers. 2005. *Biodiversity and Monitoring: A Review of National and International Systems and a Proposed Framework for Future Biodiversity Monitoring by the Department of Conservation*. Landcare Research New Zealand Ltd., Wellington.

Leidner, A. 2007. Geo-Info CONOPS: developing the disaster response of the future. *GEOWorld Magazine*. Available online http://208.106.162.71/ME2/dirmod.asp?sid=C5C01050B7694BD09714467191307E13&nm=Resources+%26+Tools&type=Publishing&mod=Publications%3A%3AArticle&mid=8F3A7027421841978F18BE895F87F791&tier=4&id=645275FE03AA47A2BA1B9B6453C97D65 web site accessed August 12, 2008.

Leighly, J. 1955. What has happened to physical geography? *Annals of the Association of American Geographers* 45: 309–318.

Leimgruber, P., C. A. Christen, and A. Laborderie. 2005. The impact of Landsat satellite monitoring on conservation biology. *Environmental Monitoring and Assessment* 106: 81–101.

Lele, N., P. K. Joshi, and S. P. Agarwal. 2008. Assessing forest fragmentation in northeastern region (NER) of India using landscape matrices. *Ecological Indicators* 8: 657–663.

Leopold, A. 1933. *Game Management.* Scribner, New York, U.S.

Leopold, A. 1949. *A Sand County Almanac.* Oxford University Press, New York, U.S.

Letcher B.H., K. H. Nislow, J. A. Coombs, M. J. O'Donnell, and T. L. Dubreuil. 2007. Population response to habitat fragmentation in a stream-dwelling brook trout population. *PLoS ONE* 2: e1139.doi:10.1371/journal.pone.0001139.

Levick, S., K. Rogers., 2008. Patch and species specific responses of savanna woody vegetation to browser exclusion. *Biological Conservation* 141: 489–498.

Levin, N., A. Shmida, O. Levanoni, H. tamari, and S. Kark. 2007. Predicting mountain plant richness and rarity from space using satellite-derived vegetation indices. *Diversity and Distributions* 13: 692–703.

Leyequien, E., J. Verrelst, M. Slot, G. Schaepman. -Strub, I. M. A. Heitkonig, and A. Skidmore. 2007. Capturing the fugitive: applying remote sensing to terrestrial animal distribution and diversity. *International Journal of Applied Earth Observation and Geoinformation*, 9: 1–20.

Leyva, R. I., R. J. Henry, L. A. Graham, and J. M. Hill. 2002. *Use of LIDAR to Determine Vegetation Vertical Distribution in Areas of Potential Black-capped Vireo Habitat at Fort Hood, Texas.* Endangered Species Monitoring and Management 2002 Annual Report. The Nature Conservancy, Fort Hood, Texas, USA.

Li, X. and A. Strahler. 1985. Geometric-optical modelling of a conifer forest canopy. *IEEE Transactions on Geoscience and Remote Sensing* 23: 705–721.

Liang, P., M. Moghaddam, L. E. Pierce, and R. M. Lucas. 2005. Radar backscattering model for multilayer mixed-species forests. *IEEE Transactions on Geoscience and Remote Sensing* 43: 2612–2626.

Liang, S. 2004. *Quantitative Remote Sensing of Land Surfaces.* John Wiley and Sons, Hoboken, New Jersey.

Liang, S., ed. 2008. *Advances in Land Remote Sensing: System, Modeling, Inversion, and Application.* Springer, Dordrecht, The Netherlands.

Liebezeit, J. R. and S. Zack. 2008. Point counts underestimate the importance of arctic foxes as avian nest predators: evidence from remote video cameras in Arctic Alaskan oilfields. *Arctic* 61: 153–161.

Lillesand, T. M., R. W. Kieffer, and J. W. Chipman. 2004. *Remote Sensing and Image Interpretation.* 5d ed., John Wiley and Sons, New York, U.S.

Lillo-Saavedra, M., and C. Gonzalo. 2006. Spectral or spatial quality? A trade-off solution using the wavelet *à trous* algorithm. *International Journal of Remote Sensing* 27: 1453–1464.

Linke, J. 2003. *Using Landsat TM and IRS Imagery to Detect Seismic Cutlines: Assessing Their Effects on Landscape Structure and on Grizzly Bear (Ursus arctos) Landscape Use in Alberta.* Unpublished M.Sc. Thesis, University of Calgary, Calgary, AB.

Linke, J., M. G. Betts, M. B. Lavigne, and S. E. Franklin. 2007. Structure, function, and change of forest landscapes. In Wulder, M. A. and S. E. Franklin, eds., *Understanding Forest Disturbance and Spatial Pattern: Remote Sensing and GIS Approaches*. Taylor and Francis, Boca Raton, Florida. 1–30.

Linke, J. and S. E. Franklin. 2006. Interpretation of landscape structure gradients based on satellite image classification of land cover. *Canadian Journal of Remote Sensing* 32: 367–379.

Linke, J., S. E. Franklin, M. Hall-Beyer, and G. B. Stenhouse. 2008. Effects of cutline density and land-cover heterogeneity on landscape metrics in western Alberta. *Canadian Journal of Remote Sensing* 34: 390–404.

Linke, J., S. E. Franklin, G. B. Stenhouse, and F. Huettmann. 2005. Seismic lines, changing landscape metrics, and grizzly bear landscape use in Alberta, *Landscape Ecology* 20: 811–826.

Linke, J., G. J. McDermid, D.N. Laskin, A. J. McLane, A. Pape, J. Cranston, M. Hall-Beyer, and S. E. Franklin. 2009. Toward a framework for creating temporally- and categorically-dynamic land cover maps for reliable landscape monitoring. *Photogrammetric Engineering and Remote Sensing* in press.

Littaye, A., A. Gannier, S. Laran, and J. P. F. Wilson. 2004. The relationship between summer aggregation of fin whales and satellite-derived environment conditions in the northwestern Mediterranean Sea. *Remote Sensing of Environment* 90: 44–52.

Liu, D., M. Kelly, and P. Gong. 2006. A spatial-temporal approach to monitoring forest disease spread using multitemporal high spatial resolution imagery. *Remote Sensing of Environment* 101: 167–180.

Liu, J., J. M. Chen, J. Cihlar, and W. M. Park. 1997. A process-based boreal ecosystem productivity simulator using remote sensing inputs. *Remote Sensing of Environment* 62: 158–177.

Loeffler, E. and C. Margules. 1980. Wombats detected from space. *Remote Sensing of Environment* 9: 435–444.

Loveland, T. R., J. W. Merchant, D. O. Ohlen, and J. F. Brown. 1991. Development of a land-cover characteristics database for the conterminous U.S. *Photogrammetric Engineering and Remote Sensing*, 57: 1453–1463.

Lovelock, J. 2007. *The Revenge of Gaia: Earth's Climate Crisis and the Fate of Humanity.* Westview Press, London, UK.

Lu, D., P. Mausel, E. Brondízio, and E. Moran. 2004. Change detection techniques. *International Journal of Remote Sensing* 25: 2365–2407.

Lucas, K. L. and G. A Carter. 2008. The use of hyperspectral remote sensing to assess vascular plant species richness on Horn Island, Mississippi. *Remote Sensing of Environment* 3908–3915.

Lucas, N. S. and P. J. Curran. 1999. Forest ecosystem simulation models: the role of remote sensing. *Progress in Physical Geography* 23: 391–423.

Lucas, N. S., P. J. Curran, S. E. Plummer, and F. M. Danson. 2000. Estimating the stem carbon production of a coniferous forest using an ecosystem simulation model driven by the remotely sensed red edge. *International Journal of Remote Sensing* 21: 619–631.

Lucas, N. S., S. Shanmugam, and M. Barnsley. 2002. Sub-pixel habitat mapping of a coastal dune ecosystem. *Applied Geography* 22: 253–270.

Lugo, A. E. 1998. Biodiversity and public policy: the middle of the road. In Guruswamy, L. D. and J. A. McNeely, eds., *Protection of Global Biodiversity: Converging Strategies.* Duke University Press, Durham and London. 33–45.

Luoto, M., R. Virkkala, R. K. Heikkinen, and K. Rainio. 2004. Predicting bird species richness using remote sensing in boreal agricultural-forest mosaics. *Ecological Applications* 14: 1946–1962.

Luscombe, A. P., I. Ferguson, N. Shepperd, D. G. Zimick, and P. Naraine. 1993. The Radarsat synthetic aperture radar development. *Canadian Journal of Remote Sensing* 19: 298–310.

MacArthur, R. H. and E. O.Wilson. 1967. *The Theory of Island Biogeography.* Princeton UP, NJ.

Mack, R. N. 2005. Assessing biotic invasions in time and space: the second imperative. In Mooney, H. A. et al., eds. 2005. *Invasive Alien Species: A New Synthesis.* Island Press, Washington. 179–208.

Mackey, B., J. Bryan, and L. Randall. 2004. Australia's dynamic habitat template for 2004. *Proceedings of MODIS Vegetation Workshop II.* University of Montana, Missoula, Montana. (poster).

Mackinnon, J. and R. De Wulf. 1994. Designing protected areas for giant pandas in China. In Miller, R. I., ed., *Mapping the Diversity of Nature.* Chapman and Hall, London. 127–142.

Madden, M. 2004. Remote sensing and geographic information system operations for vegetation mapping of invasive exotics. *Weed Technology* 18: 1457–1463.

Madden, M., D. Jones, and L. Vilchek. 1999. Photointerpretation key for the Everglades vegetation classification system. *Photogrammetric Engineering and Remote Sensing* 65: 171–177.

Madhavan, B. B., T. Sasagawa, K. Tachibana, and K. K. Mishra. 2006. A high-level data fusion and spatial modelling system for change-detection analysis using high-resolution airborne digital sensor data. *International Journal of Remote Sensing* 27: 3571–3591.

Magurran, A. E. 2004. *Measuring Biological Diversity.* Blackwell Science Publishing, Oxford, UK.

Majumdar, S. K., J. E. Huffman, F. J. Brenner, and A. I. Panah, eds. 2005. *Wildlife Diseases: Landscape Epidemiology, Spatial Distribution and Utilization of Remote Sensing Technology.* Pennsylvania Academy of Science, Easton.

Maraj, R. and J. Seidensticker. 2006. *Assessment of a Framework for Monitoring Tiger Population Trends in India.* A Report to IUCN: World Conservation Union, and India's Project Tiger.

Marçal, A. R. S., J. S. Borges, J. A. Gomes, and J. F. Pinto da Costa. 2005. Land cover update by supervised classification of segmented ASTER images. *International Journal of Remote Sensing* 26: 1347–1362.

Marcus, T. and B. A. Burns. A method to estimate subpixel-scale coastal polynyas with satellite passive microwave data. *Journal of Geophysical Research* 100(C3): 4473–4487.

Marcus, W. A. and M. A. Fonstad. 2008. Optical remote mapping of rivers at sub-meter resolutions and watershed extents. *Earth Surface Processes and Landforms* 33: 4–24.

Martin, M. E., L. C. Plourde, S. V. Ollinger, M. -L. Smith, and B. E. McNeil. 2008. A generalizable method for remote sensing of canopy nitrogen across a wide range of forest ecosystems. *Remote Sensing of Environment* 112: 3511–3519.

Mather, P. M. and M. Koch. 2004. *Computer Processing of Remotely Sensed Images.* 3d ed., John Wiley and Sons, New York, U.S.

Matthews, B. J. H., A. C. Jones, N. K. Theodorou, and A. W. Tudhope. 1996. Excitation-emission-matrix fluorescence spectroscopy applied to humic acid bands in coral reefs. *Marine Chemistry* 55: 317–332.

McComb, B. C. 2008. *Wildlife Habitat Management: Concepts and Applications in Forestry.* CRC Press, Boca Raton, Florida.

McComb, J. and G. E. S. J. Hardy. 2003. *Phytophthora in Forests and Natural Ecosystems.* 2nd International IUFRO Working Party Meeting, Albany, WA, Murdoch University Print Room, Murdoch University, WA.

McComb, W., and D. Lindenmayer. 1999. Dying, dead, and down trees. In Hunter, M. I., Jr. ed., *Maintaining Biodiversity in Forest Ecosystems.* Cambridge University Press, Cambridge, UK. 335–372.

McCreight, R., C. F. Chen, and R. H. Waring. 1994. Airborne environmental analysis from an ultralight aircraft. *Proceedings, 1st International Airborne Remote Sensing Conference.* Ann Arbor, MI. 384–392.

McCullough, D. R., K. C. J. Pei, and Y. Wang. 2000. Home range, activity patterns, and habitat relations of Reeve's Muntjacs in Taiwan. *Journal of Wildlife Management* 64: 430–441.

McDermid, G. J. 2005. *Remote Sensing for Large-Area, Multi-Jurisdictional Habitat Mapping.* Unpublished Ph.D. Thesis, University of Waterloo, Waterloo, ON.

McDermid, G. J., S. E. Franklin, and E. F. LeDrew. 2005. Remote sensing for large-area habitat mapping. *Progress in Physical Geography* 29: 449–474.

McDermid, G., A. Pape, J. Linke, A. McLane, D. Laskin and S. E. Franklin. 2008. Object-based approaches to change analysis and thematic map update: challenges and limitations. *Canadian Journal of Remote Sensing* 34: 462–466.

McDermid, G. J., M. A. Wulder, N. C. Coops, S. E. Franklin, and N. E. Seitz. 2009. Critical remote sensing contributions to wildlife ecological knowledge and management. Huettmann, F. and S. Cushmann, eds., *Spatial Information Systems in Animal Science.* Springer-Verlag, Tokyo, in press.

McDonnell, R. A. 1996. Including the spatial dimension: using geographic information systems in hydrology. *Progress in Physical Geography* 20: 159–177.

McElhinny, C., P. Gibbons, C. Brack, and J. Bauhus. 2005. Forest and woodland stand structural complexity: Its definition and measurement. *Forest Ecology and Management* 218: 1–24.

McGarigal, K., S. Cushman, and S. Stafford. 2000. *Multivariate Statistics for Wildlife and Ecology Research.* Springer-Verlag, New York, U.S.

McGarigal, K. and B. J. Marks. 1995. *FRAGSTATS: Spatial Pattern Analysis Program for Quantifying Landscape Structure.* Reference Manual, Department of Forest Science, Oregon State University, Corvallis.

McGarigal, K., W. H. Romme, M. Crist, and E. Roworth. 2001. Cumulative effects of roads and logging on landscape structure in the San Juan Mountains, Colorado (USA). *Landscape Ecology* 16: 327–349.

McGill, M. J., D. L. Hlavka, and W. D. Hart. 2006. Applications of data from the Cloud Physics LiDAR.. http://ams.confex.com/ams/pdfpapers/85877.pdf

web site accessed August 5, 2008. *Proceedings of AMS Conference*, Madison, Wisconsin.

McGill, M. J., D. L. Hlavka, W. D. Hart, E. J. Welton, and J. R. Campbell. 2003. Airborne lidar measurements of aerosol optical properties during SAFARI 2000. *Journal of Geophysical Research* 108 (D13), 8493, doi:10.1029/2002JD002370.

McIntyre, N. E. 1995. Effects of patch size on avian diversity. *Landscape Ecology* 10: 85–99.

McKenzie, J. S., R. S. Morris, D. U. Pfeiffer, and J. R. Dymond. 2002. Application of remote sensing to enhance the control of wildlife-associated *Mycobacterium bovis* infection. *Photogrammetric Engineering and Remote Sensing*, 68: 153–159.

McLaren, I. A. 1966. An analysis of aerial census of ringed seals. *Journal of Fisheries Research Board of Canada*, 25: 769–773.

McLellan, B. N., and D. M. Shackleton. 1988. Grizzly bears and resource extracting industries: effects of roads on behaviour, habitat use and demography. *Journal of Applied Ecology* 25: 451–460.

McLellan, B. N. and D. M. Shackleton. 1989. Grizzly bears and resource extracting industries: habitat displacement in response to seismic exploration, timber harvesting and road maintenance. *Journal of Applied Ecology* 26: 371–380.

Mech, L. D. and S. M. Barber. 2002. *A Critique of Wildlife Radio-tracking and its Use in National Parks: a Report to the U.S. National Park Service.* U.S. Geological Survey, Northern Prairie Wildlife Research Center, Jamestown, N.D.

Meier, M. F. 1980. Remote sensing of snow and ice. *Hydrological Sciences* 25: 307–330.

Mena, J. B. and J. A. Malpica. 2005. An automatic method for road extraction in rural and semi–urban areas starting from high resolution satellite imagery. *Pattern Recognition Letters* 26: 1201–1220.

Meng, X., A. Dodson, T. Moore, and G. W. Roberts. 2007. Innovation: ubiquitous positioning, anyone, anything, anytime, anywhere. *GPSWorld*. Available online http://sidt.gpsworld.com/gpssidt/article/articleDetail.jsp?id=433155&sk=&date=&%0A%09%09%09&pageID=2 web site accessed August 12, 2008.

Menon, S. and K. S. Bawa. 1997. Applications of geographic information systems, remote sensing, and a landscape ecology approach to biodiversity conservation in the Western Ghats. *Current Science* 73: 134–145.

Metzger, M., R. Leemans, D. Schroter. 2005. A multidisciplinary multi–scale framework for assessing vulnerabilities to global change. *International Journal of Applied Earth Observation and Geoinformation* 7: 253–267.

Meyers, W. and L. Werth, 1990. Satellite data: management panacea or potential problem? *Journal of Forestry* 88: 10–13.

Mildrexler, D. J., M. Zhao, F. A. Heinsch, and S. Running. 2007. A new satellite based methodology for continental scale disturbance detection. *Ecological Applications* 17: 235–250.

Miller, R. I., ed. 1994. *Mapping the Diversity of Nature.* Chapman and Hall, New York, U.S.

Milner-Gulland, E. J., and J. M. Rowcliffe. 2007. *Conservation and Sustainable Use: a Handbook of Techniques.* Oxford University Press, Oxford.

Milton, E. J., E. M. Rollin, and D. R. Emery. 1995. Advances in field spectroscopy. In Danson, F. M. and S. E. Plummer, eds., *Advances in Environmental Remote Sensing.* John Wiley and Sons, London. 9–32.

Minar, J. and I. S. Evans. 2008. Elementary forms of land surface segmentation: the theoretical basis of terrain analysis and geomorphological mapping. *Geomorphology* 95: 236–259.

Minard, A. 2008. Cattle, deer graze along Earth's magnetic field. *National Geographic News*.Available online http://news.nationalgeographic.com/news/2008/08/080825–magnetic–cows.html?source=rss web site accessed August 28, 2008.

Moberg, G. P. and J. A. Mench. 2000. *The Biology of Animal Stress.* CABI Publishing, New York, U.S.

Moeur, M. and A. R. Stage. 1995. Most similar neighbour: an improved sampling inference procedure for natural resources planning. *Forest Science* 41: 337–359.

Moffiet, T., K. Mengersen, C. Witte, R. King, and R. Denham. 2005. Airborne laser scanning: exploratory data analysis indicates potential variables for classification of individual trees or forest stands according to species. *ISPRS Journal of Photogrammetry and Remote Sensing* 59: 289–309.

Molema G.J., J. Meuleman, J. G. Kornet, A. G. T. Schut, and J. J. M. H. Ketelaars. 2003. A mobile imaging spectroscopy system as tool for crop characterization in agriculture. In Werner, A. and A. Jarfe, eds. *Proceedings, 4th European Conference on Precision Agriculture.* ATB, Berlin. 499–500.

Molleman, F., A. Kop, P. M. Brakefield, P. J. De Vries, and B. J. Zwaan. 2006. Vertical and temporal patterns of biodiversity of fruit-feeding butterflies in a tropical forest in Uganda. *Biodiversity and Conservation* 15: 107–121.

Mooney, H. A., R. N. Mack, J. A. McNeely, L. E. Neville, P. J. Schei, and J. K. Waage, eds. 2005. *Invasive Alien Species: A New Synthesis.* Island Press, Washington.

Moore, W. and T. Polzin. 1990. ER-2 high altitude reconnaissance: a case study. *Forestry Chronicle* 66: 480–486.

Morley, L. R. 1992. Vignettes from the early days of remote sensing. *Proceedings of Multispectral Airborne Scanning for Forestry and Mapping Conference.* Canadian Forest Service, Ottawa (reprinted in 2007 *Geomatica* 61: 56–66.

Morrison, M. L., B. G. Marcot, and P. W. Mannan. 2006. *Wildlife–Habitat Relationships: Concepts and Applications.* 3d. Island Press, Washington.

Morsdorf, F., B. Kötz, E. Meier,, K. I. Itten, and B. Allgower., 2006. Estimation of LAI and fractional cover from small footprint airborne laser scanning data based on gap fraction. *Remote Sensing of Environment* 104: 50–61.

Moses, J. F., B. E. Weinstein, and J. L. Farnham. 2004. New NASA Earth Science data and data access methods. *Proceedings of the IEEE* 4418–4421.

Moskal, L. M. and S. E. Franklin. 2002. Multi-layer forest stand discrimination with spatial co-occurrence texture analysis of high spatial detail airborne imagery, *Geocarto International* 17: 53–65.

Moskal, L. M. and S. E. Franklin. 2004. Relationship between airborne multispectral image texture and aspen defoliation. *International Journal of Remote Sensing* 25: 2701–2711.

Muchoney, D. M. 2008. Earth observations for terrestrial biodiversity and ecosystems. *Remote Sensing of Environment* 112: 1909–1911.

Mumby, P. J., E. P. Green, A. J. Edwards, and C. D. Clark. 1999. The cost-effectiveness of remote sensing for tropical coastal resources assessment and management. *Journal of Environmental Management* 55: 157–166.

Murakami, Y., K. Nakayama, S.-I. Kitamura, H. Iwata, and S. Tanabe, eds. 2008. Interdisciplinary Studies on Environmental Chemistry, Vol. 1: Biological Responses to Chemical Pollutants. TerraPub, Tokyo.

Murden, S. B. and K. L. Risenhoover. 2000. A blimp system to obtain high–resolution, low-altitude aerial photography and videography. *Wildlife Society Bulletin* 28: 958–962.

Murtha, P. A. 1985. Photo-interpretation of spruce beetle–attacked spruce. *Canadian Journal of Remote Sensing* 11: 93–102.

Murtha, P. A. and R. Cozens. 1985. Colour infra-red photo-interpretation and ground surveys to evaluate spruce beetle attack. *Canadian Journal of Remote Sensing* 11: 177–187.

Musiega, D. E., S.-N. Kazadi, and K. Fukuyama. 2006. A framework for predicting and visualizing the East African wildebeest migration-route patterns in variable climate conditions using geographic information system and remote sensing. *Ecological Research* 21: 530–543.

Muybridge, E. 1887. *Animal locomotion: an Electro-photographic Investigation of Consecutive Phases of Animal Movement,* 1872–1884. University of Pennsylvania Press, Philadelphia.

Myers, N. 1997. The rich diversity of biodiversity issues. In Reaka-Kudla, M. L., et al., eds. *Biodiversity II.* The National Academy of Sciences, Washington. 125–138.

Myers, N., R. A. Mittermeier, C. G. Mittermeier, G. A. Da Fonseca, and J. Kent. 2000. Biodiversity hotspots for conservation priorities. *Nature* 403: 853–858.

Nagendra, H. 2001. Using remote sensing to assess biodiversity. *International Journal of Remote Sensing*, 22: 2377–2400.

Nagendra, H. and M. Gadgil. 1999a. Biodiversity assessment at multiple scales: linking remotely sensed data with field information. *Proceedings of the National Academy of Sciences* 96: 9154–9158.

Nagendra, H. and M. Gadgil. 1999b. Satellite imagery as a tool for monitoring species diversity: an assessment. *Journal of Applied Ecology* 36: 388–397.

Nagendra H. and G. Utkarsh. 2003. Landscape ecological planning through a multiscale characterization of pattern: studies in the Western Ghats, South India. *Environmental Monitoring and Assessment* 87: 815–833.

Naidoo, R., A. Balmford, R. Costanza, B. Fisher, R. E. Green, B. Lehner, T. R. Malcolm and T. H. Ricketts. 2008. Global mapping of ecosystem services and conservation priorities. *Proceedings of the National Academy of Sciences*, 105: 9495–9500.

Naiman, R. J. 1999. A perspective on interdisciplinary science. *Ecosystems* 2: 292–295.

Naveh, Z. 2007. *Transdisciplinary Challenges in Landscape Ecology and Restoration Ecology* Springer, Dordrecht, The Netherlands.

Neel, M., C. K. McGarigal, and S. A. Cushman. 2004. Behaviour of class–level landscape metrics across gradients of class aggregation and area. *Landscape Ecology* 19: 435–455.

Nelson, R. F., P. Hyde, P. Johnson, B. Emessiene, M. L. Imhoff, R. Campbell, and W. Edwards. 2007. Investigating RaDAR-LiDAR synergy in a North Carolina pine forest. *Remote Sensing of Environment* 110: 98–108.

Newman, C. M., E. F. LeDrew, and A. Lim. 2006. Mapping of coral reefs for management of marine protected areas in developing nations using remote sensing. In Richardson, L. L. and E. F. LeDrew, eds. *Remote Sensing of Aquatic Coastal Ecosystem Processes.* Springer, Dordrecht, The Netherlands. 251–278.

Nichol, J. and M. S. Wong. 2008. Habitat mapping in rugged terrain using multispectral Ikonos images. *Photogrammetric Engineering and Remote Sensing* 74: 1325–1334.

Nichols, W. F., K. T. Killingbeck, and P. V. August. 1998. The influence of geomorphological heterogeneity on biodiversity: II. A landscape perspective. *Conservation Biology* 12: 371–379.

Nielsen S.E. and Boyce M.S. 2002. Resource selection functions and population viability analyses. In: Stenhouse G.B. and Munro R.H. eds., *Foothills Model Forest Grizzly Bear Research Program.* 2001 Annual Report. Hinton, Alberta, 17–42.

Nielsen, S. E., E. M. Bayne, J. Schieck, J. Herbers, and S. Boutin. 2007. A new method to estimate species and biodiversity intactness using empirically derived reference conditions. *Biological Conservation* 137: 403–414.

Nielsen, S.E., J. Cranston, and G.B. Stenhouse. 2009. Identification of priority areas for grizzly bear conservation and recovery in Alberta, Canada. *Journal of Conservation Planning*, in press.

Niemann, O., and D. G. Goodenough. 2003. Estimation of foliar chemistry of western hemlock using hyperspectral data. In Wulder, M. A. and S. E. Franklin, eds., *Remote Sensing of Forest Environments: Concepts and Case Studies.* Kluwer Academic Publishers, Boston, MA. 447–467.

Nightingale, J. M., M. J. Hill, S. R. Phinn, and A. A. Held. 2007. Comparision of satellite-derived estimates of gross primary production for Australian old-growth tropical rainforest. *Canadian Journal of Remote Sensing* 33: 278–288.

Nilson, T., and J. Ross. 1997. Modeling radiative transfer through forest canopies: implications for canopy photosynthesis and remote sensing. In Gholz, H. K. et al., eds., *The Use of Remote Sensing in the Modeling of Forest Productivity.* Kluwer, Dordrecht. 23–60.

Noss, R. F. 1990. Indicators for monitoring biodiversity: a hierarchical approach. *Conservation Biology* 4: 355–364.

Obade, V. P. 2008. Wildlife habitat suitability mapping using remote sensing and geographical information science. *African Journal of Ecology* 46: 432–434.

Oberg, P., C. Rohner, and F. K. A. Schmiegelow. 2000. *GIS and Remote Sensing: Tools for Assessing Responses of Mountain Caribou to Different Types of Linear Developments in West-Central Alberta.* Report to the Research Subcommittee of the West-Central Alberta Caribou Standing Committee, Edmonton, AB.

Oswald, E. T. 1976. Terrain analysis from Landsat imagery. *Forestry Chronicle* 56: 274–282.

Oliver, C. D. and B. C. Larson. 1996. *Forest Stand Dynamics.* John Wiley and Sons, New York, U.S.

Oliver, R. E. and J. A. Smith. 1973. *Vegetation Canopy Reflectance Models.* Final Report, College of Forestry and Natural Resources. Colorado State University, Fort Collins.

Olsen, R. C. 2007. *Remote Sensing from Air and Space.* SPIE, New York, U.S.

Olthof, I., and D. J. King. 2000. Development of a forest health index using multispectral airborne digital camera imagery. *Canadian Journal of Remote Sensing* 26: 166–176.

OzEstuaries, 2003. *Coastal Indicator Knowledge and Information System I: Biophysical Indicators.* [Web document]. Geoscience Australia, Canberra. Available online www.ozestuaries.org/indicators web site accessed web site August 27, 2008.

Pacala, S. W. and D. Tilman,. 1994. Limiting Similarity in Mechanistic and Spatial Models of Plant Competition in Heterogeneous Environments. *The American Naturalist* 143: 222–257.

Panah, A.I. and R. P. Greene. 2005. Role of remote sensing, geographic information systems and telemetry in wildlife diseases investigation. In Majumdar, S. K., J. E. Huffman, F. J. Brenner, and A. I. Panah, eds., *Wildlife Diseases: Landscape Epidemiology, Spatial Distribution and Utilization of Remote Sensing Technology.* Pennsylvania Academy of Science, Easton. 434–447.

Pape, A. D. and S. E. Franklin. 2008. MODIS-based change detection for grizzly bear habitat mapping in Alberta. *Photogrammetric Engineering and Remote Sensing* 74: 973–986.

Park, S. and S. L. Egbert. 2008. Remote sensing-measured impacts of the Conservation Reserve Program (CRP) on landscape structure in southwestern Kansas. *GIScience and Remote Sensing* 45: 83–108.

Parks Canada. 2008. *Ecological Integrity Monitoring in National Parks.* Available online http://www.pc.gc.ca/progs/np–pn/eco/eco3_E.asp web site accessed March 27, 2009.

Parsons, D. M., N. T. Shears, R. C. Babcock, and T. R. Haggitt. 2004. Fine-scale habitat change in a marine reserve, mapped using radio-acoustically positioned video transects. *Marine and Freshwater Research* 55: 257–265.

Pasher, J. and D. J. King. 2009a. Modelling and mapping forest structural complexity using high resolution airborne imagery. *Remote Sensing of Environment,* in press.

Pasher, J. and D. J. King. 2009b. Modelling forest structural complexity as an indicator of biodiversity using airborne remote sensing. *Forest Ecology and Management,* in press.

Pauchard, A. and M. Maheu-Giroux. 2007. *Acacia dealbata* invasion across multiple scales: conspicuous flowering species can help us study invasion pattern and processes. Strand, H., et al. eds., *Sourcebook on Remote Sensing and Biodiversity Indicators.* Technical Series No. 32. Secretariate of the Convention on Biological Diversity, Montreal. 168–169.

Payne, N. F. and F. C. Bryant. 1998. *Wildlife Habitat Management of Forestlands, Rangelands, and Farmlands.* Krieger Publishers Malabar, Florida.

Payton, A., and M. L. Zoback. 2007. The inside track from academia and industry: crossing boundaries, hitting barriers. *Nature* 445: February 950.

Pearce, J. L. and M. S. Boyce. 2006. Modelling distribution and abundance with presence-only data. *Journal of Applied Ecology* 43: 405–412.

Pearson, O. P. 1960. Habits of *Microtus californicus* revealed by automatic photographic records. *Ecological Monographs,* 30: 231–249.

Peel, M., B. Finlayson, and T. McMahon. 2007. Updated world map of the Köppen-Geiger climate classification. *Hydrological Earth System Science* 11: 1633–1644.

Peddle, D. R. 1995. Knowledge formulation for supervised evidential classification. *Photogrammetric Engineering and Remote Sensing* 61: 409–417.

Peddle, D. R., S. E. Franklin, R. L. Johnson, M. B. Lavigne, and M. A. Wulder. 2003a. Structural change detection in a disturbed conifer forest using a geometric optical reflectance model in multiple-forward mode. *IEEE Transactions on Geoscience and Remote Sensing* 41: 163–166.

Peddle, D. R., P. M. Teillet, and M. A. Wulder. 2003b. Radiometric image processing. In Wulder, M. A. and S. E. Franklin, eds., *Remote Sensing of Forest Environments: Concepts and Case Studies.* Kluwer Academic Publishers, Boston, MA. 181–208.

Pei, K., 1998. An evaluation of auto-trigger cameras to record activity patterns of wild animals. *Taiwan Journal of Forest Science*, 13: 317–324.

Perera, A. H., L. J. Buse, and M. G. Weber, eds. 2004. *Emulating Natural Forest Landscape Disturbances: Concepts and Applications.* Columbia University Press, New York, U.S.

Peterson, D. L., V. T Parker,. eds. 1998. *Ecological Scale: Theory and Applications.* Columbia U. P. Press, New York, U.S.

Petersen, A., C. Zockler, and M. V. Gunnarsdottir. 2004. *Circumpolar Biodiversity Monitoring Program—Framework Document.* CAFF CBMP Report No. 1, CAFF International Secretariat, Akureyri, Iceland.

Peterson, D. L., and R. H. Waring. 1994. Overview of the Oregon Transect Ecosystem Research Project. *Ecological Applications* 4: 211–225.

Pettinger, L. R. 1982. *Digital Classification of Landsat Data for Vegetation and Land-cover Mapping in the Blackfoot River Watershed, Southeastern Idaho.* USGS Professional Paper 1219. Washington DC.

Phillips, L. B., A. J. Hansen, and C. H. Flather. 2008. Evaluating the species energy relationship with the newest measures of ecosystem energy: NDVI versus MODIS primary production. *Remote Sensing of Environment* 112: 3538–3549.

Philipson, W. , ed. 1997. *Manual of Aerial Photointerpretation.* 2d ed., ASPRS, Bethesda, MD.

Phinn, S. R., 1997. *Remote Sensing and Spatial Analytic Techniques for Monitoring Landscape Structure in Disturbed and Restored Coastal Environments.* Unpublished Ph.D. Thesis, UCSB and San Diego State University.

Phinn, S. R., K. Joyce, P. Scarth, and C. Roelfsema. 2006. The role of integrated information acquisition and management in the analysis of coastal ecosystem change. In Richardson, L. L. and E. F. LeDrew, eds., *Remote Sensing of Aquatic Coastal Ecosystem Processes.* Springer, Dordrecht, The Netherlands. 217–249.

Phinn, S. R., D. A. Stow, J. Franklin, L. A. K. Mertes, and J. Michaelsen. 2003. Remotely sensed data for ecosystem analyses: Combining hierarchy theory and scene models. *Environmental Management* 31: 429–441.

Pintea, L., M. E. Bauer, P. V. Bolstad, and A. Pusey. 2002. Matching multiscale remote sensing data to interdisciplinary conservation needs: the case of the chimpanzees in western Tanzania. *Proceedings, Pecora 15/Land Satellite Information IV/ ISPRS Commission I/FIEOS.* Denver, CO. http://www.isprs.org/commission1/proceedings02/contents_pecora.html web site accessed January 18, 2009.

Pohl, C. 2005. Transdisciplinary collaboration in environmental research. *Futures* 37: 1159–1178.

Polls, I. 1994. How people in the regulated community view biological integrity. *Journal of the North American Benthological Society* 13: 598–604.

Poole, R. 2008. *Earthrise: How Man First Saw the Earth.* Yale University Press, New Haven.

Potter, C., P. -N. Tan, M. Steinbach, S. Klooster, V. Kumar, R. Myneni, and V. Genovese. 2003. Major disturbance events in terrestrial ecosystems detected using global satellite data sets. *Global Change Biology* 9: 1005–1021.

Pouliot, D. A., and D. J. King. 2005. Approaches for optimal automated individual tree crown detection in young regenerating coniferous forests. *Canadian Journal of Remote Sensing* 31: 256–267.

Pouliot, D. A., D. J. King, and D. G. Pitt. 2006. Automated assessment of hardwood and shrub competition in regenerating forests using leaf-off airborne imagery. *Remote Sensing of Environment* 102: 223–236.

Prasad, S. N., S. P. Goyal, P. S. Roy, and S. Singh. 1994. Changes in wild ass (Equus heminous khur) habitat conditions in Little Runn of Kutch, Gujarat, from a remote sensing perspective. *International Journal of Remote Sensing* 15: 3155–3164.

Prenzel, B. G. and P. Treitz. 2006. Spectral and spatial filtering for enhanced thematic change analysis of remotely sensed data. *International Journal of Remote Sensing* 27: 835–854.

Pruitt, E. L. 1979. The Office of Naval Research and Geography. In *Seventy-five years of American Geography, Annals of the Association of American Geographers*, Special Issue No. 1. 69: 103–108.

Puissant, A., J. Hirsch, and C. Weber. 2005. The utility of texture analysis to improve per-pixel classification for high to very high spatial resolution imagery. *International Journal of Remote Sensing* 26: 733–745.

Purdy, A. 1972. *Trees at the Arctic Circle: Purdy Selected.* McClelland and Stewart, Toronto.

Purkis, S. J., N. A. J. Graham, and B. M. Riegl. 2008. Predictability of reef fish diversity and abundance using remote sensing data in Diego Garcia (Chagos Archipelago). *Coral Reefs* 27: 167–178.

Quattrochi, D. A. and J. C. Luvall, eds. 2004. *Thermal Remote Sensing in Land Surface Processing.* Routledge, New York.

Quigley, T. M., R. W. Haynes, and W. J. Hann. 2001. Estimating ecological integrity in the interior Columbia River basis. *Forest Ecology and Management* 153: 161–178.

Raclot, D., F. Colin, and C. Puech. 2005. Updating land cover classification using a rule-based decision system. *International Journal of Remote Sensing* 26: 1309–1321.

Ramanujan, K. 2004. New tools for conservation. *Earth Observatory*, October 5, 2004.

Rapport, D., R. Costanza, P. R. Epstein, C. Gaudet, and R. Levins. 1998. *Ecosystem Health.* Blackwell Science Publishers, Oxford, UK.

Raxworthy, C. J., E. Martinez-Meyer, N. Horning, R. A. Nussbaum, G. E. Schneider, M. A. Ortega–Huerta, and A. T. Peterson. 2003. Predicting distributions of known and unknown reptile species in Madagascar. *Nature* 427: 837–841.

Read, J. M., D. B. Clark, E. M. Venticinique, and M. P. Moreira. 2003. Application of merged 1-m and 4-m resolution satellite data to research and management in tropical forests. *Journal of Applied Ecology* 40: 592–600.

Reading, R. P., T. W. Clark, J. H. Seebeck, and J. Pearce. 1996. Habitat suitability index model for the eastern barred bandicoot, *Perameles gunnii. Wildlife Research* 23: 221–235.

Rees, W. G. 2001. *Physical Principles of Remote Sensing.* 2d ed., Cambridge University Press, Cambridge, UK.

Rees, W. G. 2005. *Remote Sensing of Snow and Ice.* CRC Press, Boca Raton Florida.

Reeves, R. G., ed. 1975. *Manual of Remote Sensing.* ASP, Falls Church, VA.

Reichardt, M. 2004. *The Havoc of Non-Interoperability: An Open GIS Consortium White Paper.* Available online http://www.opengeospatial.org/pressroom/papers web site accessed August 15, 2008.

Reid, N. J. 1987. Remote sensing and forest damage. *Environmental Science and Technology* 21: 428–429.

Remmel, T. K. and F. Csillag. 2003. When are two landscape pattern indices significantly different? *Journal of Geographical Information Systems* 5: 331–351.

Ren, B., M. Li, Y. Long, C. C. Grüter, and F. Wei. 2008. Measuring daily ranging distances of Rhinopithecus bieti via a global positioning system collar at Jinsichang, China: a methodological consideration. *International Journal of Primatology* 29: 783–794.

Rexstad, E., and K. P. Burnham. 1991. *User's Guide for Interactive Program CAPTURE Abundance Estimation of Close Populations.* Colorado State University, Fort Collins.

Reynolds, D. R., and J. R. Riley. 2002. Remote sensing, telemetric and computer-based technologies for investigating insect movement: a survey of existing and potential techniques. *Computers and Electronics in Agriculture* 35: 271–307.

Rhoten, D. 2004. Interdisciplinary research: trend or transition. *Items & Issues* 5: 6–12.

Ribaudo, M. O., D. L Hoag, M. E. Smith, and R. Heimlich. 2001. Environmental indices and the politics of the Conservation Reserve Program. *Ecological Indicators* 1: 11–20.

Richardson, L. L. 1996. Remote sensing of algal bloom dynamics. *BioScience* 46: 492–501.

Richardson, L. L. and E. F. LeDrew, eds., *Remote Sensing of Aquatic Coastal Ecosystem Processes: Science and Management Applications.* Springer, Dordrecht, The Netherlands.

Richter, R. 1990. A fast atmospheric correction algorithm applied to Landsat TM images. *International Journal of Remote Sensing* 11: 159–166.

Richter, R. 1997. Correction of atmospheric and topographic effects for high spatial detail satellite images. *International Journal of Remote Sensing* 18: 1011–1099.

Richter, R., M. Bachmann, W. Dorigo, and A. Mueller. 2006a. Influence of the adjacency effect on ground reflectance measurements. *IEEE Geoscience Remote Sensing Letters* 3: 565–569.

Richter, R., D. Schlaepfer, and A. Mueller. 2006b. An automatic atmospheric correction algorithm for visible/NIR imagery. *International Journal of Remote Sensing* 27: 2077–2085.

Ridd, M. K. and J. D. Hipple. 2006. *Remote Sensing of Human Settlements: Volume 5.* Manual of Remote Sensing. 3d ed., ASPRS, Falls Church, VA.

Riitters, K. H., R. V. O'Neill, C. T. Hunsaker, J. D. Wickham, D. H. Yankee, S. P. Timmons, K. B. Jones, and B. L. Jackson. 1995. A factor analysis of landscape pattern and structure metrics. *Landscape Ecology* 10: 23–39.

Riley, J. R. 1989. Remote sensing in entomology. *Annual Review of Entomology* 34: 247–271.

Ripple, W. J. and R. L. Beschta. 2006. Linking wolves to willows via risk-sensitive foraging by ungulates in the northern Yellowstone ecosystem. *Forest Ecology and Management* 230: 96–106.

Ripple, W. J., D. H. Johnson, K. T. Hershey, and E. C. Meslow. 1991. Old-growth and mature forests near spotted owl nests in western Oregon. *Journal of Wildlife Management* 55: 316–318.

Ripple, W. J., P. D. Lattin, K. T. Hershey, F. F. Wagner, and E. C. Meslow. 1997. Landscape composition and pattern around northern spotted owl nest sites in southwestern Orgeon. *Journal of Wildlife Management* 61: 151–158.

Risley, E. 1967. Developments in the application of Earth Observation satellites to geographic problems. *Professional Geographer* XIX: 130–132.

Robinove, C. J. 1981. The logic of multispectral classification and mapping of land. *Remote Sensing of Environment* 11: 231–244.

Robinson, G. 1990. Huddles. *Australian Geographic Journal* 1: 76–97.

Robinson, J., and T. Cully. 2003. Coastal Indicator Knowledge and Information System II: Socioeconomic Concerns and Values. [Web document]. Canberra : Geoscience Australia. Available online www.ozestuaries.org/indicators web site accessed August 27, 2008.

Robinson, R. 2008. Forest health surveillance in Western Australia: a summary of major activities from 1997 to 2006. *Australian Forestry* 71: 202–211.

Rocchini, D. 2007. Effects of spatial and spectral resolution in estimating ecosystem α-diversity by satellite imagery. *Remote Sensing of Environment* 111: 423–434.

Rocchini, D., A. Chiarucci, and S. A. Loiselle. 2004. Testing the spectral variation hypothesis by using satellite multispectral images. *Acta Oecologica* 26: 117–120.

Rogan, J. and D. Chen. 2004. Remote sensing technology for mapping and monitoring land cover and land use change. *Progress in Planning* 61: 301–325.

Rogan, J. and J. Miller. 2007. Integrating GIS and remotely sensed data for mapping forest disturbance and change. In Wulder, M. A. and S. E. Franklin, eds., *Understanding Forest Disturbance and Spatial Pattern: Remote Sensing and GIS Approaches.* Taylor and Francis, Boca Raton, Florida. 133–171.

Rogan, J. and S. R. Yool. 2001. Mapping fire-induced vegetation depletion in the Peloncillo Mountains, Arizona and New Mexico. *International Journal of Remote Sensing* 22: 3101–3121.

Ropert-Coudert, Y. and R. P. Wilson. 2005. Trends and perspectives in animal-attached remote sensing. *Frontiers of Ecology and Environment* 3: 437–444.

Rosenfeld, P. L. 1992. The potential of transdisciplinary research for sustaining and extending linkages between the health and social sciences. *Social Science and Medicine* 35: 1343–1357.

Rosenzweig, M. L. and Z. Abramsky. 1993. How are diversity and productivity related? In R. E. Ricklefs and D. Schluter, eds., *Species Diversity in Ecological Communities: Historical and Geographical Perspectives.* U of Chicago P, Chicago, Il, U.S. 52–65.

Rowcliffe, J.M., J. Field, S. T. Turvey, and C. Carbone. 2008. Estimating animal density using camera traps without the need for individual recognition. *Journal of Applied Ecology* DOI: 10.1111/j.1365-2664.2008.01473.x

Roy, P. S., H. Padalia, N. Chauhan, M. C. Porwal, S Gupta, S. Biswas, and R. Jagdale. 2005. Validation of geospatial model for biodiversity characterization at landscape level —a study in Andaman & Nicobar Islands, India. *Ecological Modelling* 185: 349–369.

Running, S. W. 1990. Estimating terrestrial primary productivity by combining remote sensing and ecosystem simulation. In Hobbs, R. and H. Mooney, eds., *Remote Sensing of Biosphere Functioning.* Springer-Verlag, New York, U.S. 65–86.

Running, S. W., and E. R. Hunt, Jr. 1993. Generalization of a forest ecosystem process model for other biomes, BIOME-BGC, and an application for global-scale models. In Ehlering, J. and C. Field, eds., *Scaling Physiological Processes: Leaf to Globe.* Academic Press, Toronto. 141–157.

Running, S. W., T. R. Loveland, L. L. Pierce, and E. R. Hunt, Jr. 1995. A remote sensing based vegetation classification logic for global land cover analysis. *Remote Sensing of Environment* 51: 39–48.

Running, S. W., R. R. Nemani, F. A. Heinsch, M. Zhao, M. Reeves, and H. Hashimoto. 2004. A continuous satellite–derived measure of global terrestrial primary production. *BioScience* 54: 547–560.

Running, S. W., R. R. Nemani, D. L. Peterson, L. E. Band, D. F. Potts, L. L. Pierce, and M. A. Spanner. 1989. Mapping regional forest evapotranspiration and photosynthesis by coupling satellite data with ecosystem simulation. *Ecology*, 70: 1090–1101.

Ryan, P. G., S. L. Petersen, G. Peters, and D. Grémillet. 2004. GPS tracking a marine predator: the effects of precision, resolution and sampling rate on foraging tracks of African Penguins. *Marine Biology* 145: 215–223.

Ryerson, R. A. 1989. Image interpretation concerns for the 1990s and lessons from the past. *Photogrammetric Engineering and Remote Sensing* 55: 1427–1430.

Ryerson, R. A., ed. 1998. *Manual of Remote Sensing.* 3d ed., ASPRS/Wiley, New York, U.S.

Sader, S. A., and S. Vermillion. 2000. Remote sensing education: an updated survey. *Journal of Forestry* 53: 133–144.

Sample, V. A., ed. 1994. *Remote Sensing and GIS in Ecosystem Management.* Island Press, Washington.

Sanchez, R. D. and G. L. Kooyman. 2004. *Advanced Systems Data for Mapping Emperor Penguin Habitat in Antarctica.* USGS Open-File Report 04-094. USGS, Washington DC.

Sánchez-Azofeifa, G. A., and K. L. Castro-Esau. 2005. Canopy observations on the hyperspectral properties of a community of tropical dry forest lianas and their host trees. *International Journal of Remote Sensing* 27: 2101–2109.

Sanderson, J. G. and M. Trolle. 2005. Monitoring elusive mammals. *American Scientist* 93: 148–155.

Sarkar, S. 2005. *Biodiversity and Environmental Philosophy: An Introduction.* Cambridge University Press, Cambridge, UK.

Sarkar, S., R. L. Pressey, D. P. Faith, C. R. Margules, T. Fuller, D. M. Stoms, A. Moffett, K. A. Wilson, K. J. Williams, P. H. Williams, and S. Andelman. 2006.

Biodiversity conservation planning tools: present status and challenges for the future. *Annual Review Environ. Resour.* 31: 123–159.
Sasamal, S. K., S. B. Chaudhury, R. N. Samal, and A.K. Pattanik. 2008. QuickBird spots flamingos off Nalabana Island, Chilika Lake, India. *International Journal of Remote Sensing* 29: 4865–4870.
Sattinger, I. J. 1966. *Peaceful Uses of Earth-observation Spacecraft: Volume I: Introduction and Summary.* The Institute of Science and Technology, Univesity of Michigan, Ann Arbor.
Saunders, D., C. Margules, and B. Hill. 1998. *Environmental Indicators for National State of the Environment Reporting—Biodiversity.* State of the Environment (Environmental Indicator Reports), Department of the Environment, Canberra.
Saura, S., O. Torras, A. Gil-Tenu, and L. Pascual-Hortal. 2008. Shape irregularity as an indicator of forest biodiversity and guidelines for metric selection. In Lafortezza, R., et al., eds., *Patterns and Processes in Forest Landscapes: Multiple Use and Sustainable Management*. Springer, Netherlands. 167–189.
Saxon, E., B. Baker, W. Hargrove, F. Hoffman, and C. Zganjar. 2005. Mapping environments at risk under different global climate change scenarios. *Ecology Letters* 8: 53–60.
Schaller, G. B. 1998. *Wildlife of the Tibetan Steppe.* University of Chicago Press, Chicago.
Scheiman, D. M., E. K. Bollinger, and D. H. Johnson. 2003. Effects of leafy spurge infestation on grassland birds. *Journal of Wildlife Management* 67: 115–121.
Schmidt, M., K. Konig, and J. V. Muller. 2008. Modelling species richness and life form composition in Sahelian Burkino Faso with remote sensing data. *Journal of Arid Environments* 72: 1506–1517.
Schnase, J. L., J. A. Smith, T. J. Stohlgren, S. Graves, and C. Trees. 2002. Biological invasions: a challenge in ecological forecasting. *Proceedings IEEE* 122–124.
Schrock, G. M. 2006. On-grid goal: seeking support for high-precision networks. *GPS World 17*: 34–40.
Schut, A. G. T. 2003. *Imaging Spectroscopy for Characterisation of Grass Swards.* Unpublished Ph.D. Thesis. Wageningen University, The Netherlands.
Schut A. G. T., J. J. M. H. Ketelaars, J. Meuleman, J. G. Kornet, and C. Lokhorst. 2002. Novel imaging spectroscopy for grass sward characterisation. *Biosystems Engineering* 82: 131–141.
Schut A. G. T, G. W. A. M. van der Heijden, I. Hoving, M. W. J. Stienezen, F. K. Van Evert and J. Meuleman. 2006. Imaging spectroscopy for on-farm measurement of grassland yield and quality. *Agronomy Journal* 98: 1318–1325.
Scott, J.M., B. Csuti, J. E. Estes, and H. Anderson. 1989. Status assessment of biodiversity protection. *Conservation Biology*, 3: 85–87.
Scott, J. M., B. Csuti, J. J. Jacobi, and J. E. Estes. 1987. Species richness: a geographic approach to protecting future biological diversity. *BioScience* 37: 782–788.
Scott, J. M., F. Davis, and B. Csuti. 1993. Gap analysis: a geographic approach to protection of biological diversity. *Journal of Wildlife Management*, No. 123: 1–41.
Sellers, P. J. 1985. Canopy reflectance, photosynthesis and transpiration. *International Journal of Remote Sensing* 6: 1335–1372.
Sellers, P. J., C. J. Tucker, G. J. Collatz, S. O. Los, C. O. Justice, D. A. Dazlich, and D. A. Randall. 1994. A global 1 x 1 NDVI data set for climate studies: 2. The generation of global fields of terrestrial biophysical parameters from the NDVI. *International Journal of Remote Sensing* 15: 3519–3545.
Sertel, E., S. H. Kutoglu, and S. Kaya. 2007. Geometric correction accuracy of different satellite sensor images: application of figure condition. *International Journal of Remote* Sensing 28: 4685–4692.
Seto, K. C., E. Fleishman, J. P. Fay and C. J. Betrus. 2004. Linking spatial patterns of bird and butterfly species richness with Landsat TM derived NDVI. *International Journal of Remote Sensing* 25: 4309–4324.
Shao, G., D. Liu, and G. Zhao. 2001. Relationships of image classification accuracy and variation of landscape statistics. *Canadian Journal of Remote Sensing* 27: 33–43.

Shao, G. and J. Wu. 2008. On the accuracy of landscape pattern analysis using remote sensing data. *Landscape Ecology* 23: 505–511.

Sharkov, E. A. 2003. *Passive Microwave Remote Sensing of the Earth: Physical Foundations.* Praxis Publishing, London, UK.

Sharma, B. D., J. Clevers, R. De Graaf, and N. R. Chapagain. 2004. Mapping *Equus kiang* (Tibetan Wild Ass) habitat in Surkhang, Upper Mustang, Nepal. *Mountain Research and Development* 24: 149–156.

Sharma, V. P., and A. Srivastava. 1997. Role of geographical information systems in malaria control. *Indian Journal of Medical Research* 106: 198–204.

Shedd, J. M., H. Devine, and D. Hurlbert. 2006. Mapping forest hurricane damage using automated feature extraction. *Proceedings of, 5th Southern Forestry and Natural Resources GIS Conference (Asheville, NC).* Warnell School of Forestry and Natural Resources, University of Georgia, Athens. 116–127.

Shiras, G. 1906. *Hunting Wild Game with Camera and Flashlight.* National Geographic Society, Washington.

Shugart, H. H. 1998. *Terrestrial Ecosystems in Changing Environments.* Cambridge University Press, Cambridge, UK.

Silbernagel, K., J. Chen, A. Noormets, and B. Song. 2006. Conducting sound ecological studies at the landscape scale: hypotheses, experiments, and challenges. In Chen, J., et al., eds., *Ecology of Hierarchical Landscapes: From Theory to Application.* Nova Science Publishing, New York, U.S. 283–299.

Silberstein, R. P., M. Sivapalan, and A. Wyllie. 1999. On the validation of a coupled water and energy balance model at small catchment scales. *Journal of Hydrology* 220: 149–168.

Simberloff, D. 1998. Flagship, umbrellas, and keystones: is single-species management passé in the landscape era? *Biological Conservation* 83: 247–257.

Simberloff, D., I. M. Parker, and P. N. Windle. 2005. Introduced species policy, management, and future research needs. *Frontiers in Ecology and Environment* 3: 12–20.

Simpson, J. J., and J. R. Stitt. 1998. A procedure for the detection and removal of cloud shadow from AVHRR data over land. *IEEE Transactions on Geoscience and Remote Sensing* 36: 880–897.

Sims, N. C., C. Stone, N. C. Coops, and P. Ryan. 2007. Assessing the health of *Pinus radiata* plantations using remote sensing data and decision tree analysis. *New Zealand Journal of Forestry Science* 37: 57–80.

Singh, A. 1989. Digital change detection techniques using remotely sensed data. *International Journal of Remote Sensing* 10: 989–1003.

Smit, I. P. J., C. C. Grant, and I. J. White. 2007. Landscape-scale sexual segregation in the dry season distribution and resource utilization of elephants in Kruger National Park, South Africa. *Diversity and Distribution* 13: 225–236.

Smith, J. A., and G. E. Carey. 2007. Those who are crossing the boundaries need less talk, more help and flexibility. *Nature* 447: June 638.

Smith, J. L. D., H. R. Mishra, and C. Wemmer.1992. Large mammal conservation in Asia: the role of land use planning and geographic information systems. In Wegge, P., ed., *Mammal Conservation in Developing Countries: A New Approach.* Proceedings of a Workshop held at the 5th Theriological Congress in Rome, Italy, August 1989. 64–71.

Smith, C. T., and R. J. Raison. 1998. *Utility of Montreal Process Indicators for Soil Conservation in Native Forests and Plantations in Australia and New Zealand.* Soil Science Society of America Spec. Publ. No. 53. Soil Science Society of America, Madison, WI. 121–135.

Snyder, M. N., S. J. Goetz, and R. K. Wright. 2005. Stream health rankings predicted by satellite derived land cover metrics. *Journal of the American Water Resources Association* 41: 659–677.

Solberg, A. H. S. 1999. Contextual data fusion applied to forest map revision. *IEEE Transactions on Geoscience and Remote Sensing* 37: 1234–1243.

Song, C., C. E. Woodcock, K. C. Seto, M. P. Lenney, and S. A. Macomber. 2001. Classification and change detection using Landsat TM data: when and how to correct for atmospheric effects. *Remote Sensing of Environment* 75: 230–244.

Song, J. and C. H. Caldas. 2008. Data processing for real-time construction site spatial modeling. *Automation in Construction* 17: 526–535.
Space Studies Board. 1998. *Development and Application of Small Spaceborne Synthetic Aperture Radars.* Commission on Physical Sciences, Mathematics, and Applications. National Research Council, Washington.
Spanner, M. A., L. L. Pierce, S. W. Running, and D. L. Peterson. 1990. Remote sensing of temperate coniferous forest leaf area index: the influence of canopy closure, understory vegetation, and background spectral response. *International Journal of Remote Sensing* 11: 95–111.
Sreekantha, M., D. Subash Chandran, D. K. Mesta, G. R. Rao, K. V. Guruaja, and T. V. Ramachandra. 2007. Fish diversity in relation to landscape and vegetation in central Western Ghats, India. *Current Science* 92: 1592–1602.
Stadt, J. J., J. Schieck, and H. A. Stelfox. 2006. Alberta Biodiversity Monitoring Program—monitoring effectiveness of sustainable forest management planning. *Environmental Monitoring and Assessment* 121: 33–46.
Stamnes, K., W. Li, H. Eide, T. Aoki, M. Hori, and R. Storvold. 2007. ADEOS-II/GLI snow / ice products—Part 1: scientific basis. *Remote Sensing of Environment* 111: 258–273.
Starks, P. J., S. W. Coleman, and W. A. Phillips. 2004. Determination of forage chemical composition using remote sensing. *Journal of Range Management* 57: 635–640.
Starks, P. J., and T. J. Jackson. 2002. Estimating soil water content in tallgrass prairie using remote sensing. *Journal of Range Management* 55: 474–482.
Ståhl, G., A. Ringall, and J. Fridman. 2001. Assessment of coarse woody debris—a methodological overview. *Ecological Bulletins* 49: 57–70.
Stenhouse, G. and K. Graham, eds. 2008. *Foothills Research Institute Grizzly Bear Program 2007 Annual Report*. Hinton, Alberta.
Stein, A. B., T. K. Fuller, and L.L. Marker. 2008. Opportunistic use of camera traps to assess habitat-specific mammal and bird diversity in northcentral Namibia. *Biodiversity and Conservation* 17: 3321–3630.
Steiner, D. 1974. Digital geometric picture correction using a piecewise zero-order transformation. *Remote Sensing of Environment* 3: 261–274.
Steininger, M., and N. Horning. 2007. The basics of remote sensing. In Strand, H. et al., eds., *Sourcebook on Remote Sensing for Biodiversity Indicators.* Technical Series No. 32. Secretariate of the Convention on Biological Diversity, Montreal.
Stenback, J. M., and R. G. Congalton. 1990. Using Thematic Mapper imagery to examine forest understory. *Photogrammetric Engineering and Remote Sensing* 56: 1285–1290.
Stenhouse, G. B. and K. Graham, eds. 2008. *Foothills Research Institute Grizzly Bear Program 2007 Annual Report.* Foothills Research Institute, Hinton, AB.
Stine, P. A. and C. T. Hunsaker. 2001. An introduction to uncertainty issues for spatial data used in ecological applications. In Hunsaker, C. T., et al., eds., *Spatial Uncertainty in Ecology: Implications for Remote Sensing and GIS Applications.* Springer, New York, U.S. 91–107.
St. -Louis, V., A. M. Pidgeon, V. C. Radeloff, T. J. Hawbaker, and M. K. Clayton. 2006. High–resolution texture as a predictor of bird species richness. *Remote Sensing of Environment* 105: 299–312.
Stöcklii, R. and P.L. Vidale. 2004. European plant phenology and climate as seen in a 20 year AVHRR land-surface parameter dataset. *International Journal of Remote Sensing* 25: 3303–3330.
Stockwell, D. and D. Peters. 1999. The GARP modelling system: problems and solutions to automated spatial prediction. *International Journal of Geographical Information Science* 13: 143–158.
Stokols, D., J. Fuqua, J. Gress, R. Harvey, K. Phillips, L. Baezconde-Garbanati, J. Unger, P. Palmer, M. A. Clark, S. M. Colby, G. Morgan, and W. Trochim. 2003. Evaluating transdisciplinary science. *Nicotine & Tobacco Research* 5: S21–S39.
Stoms, D. M. 1991. *Mapping and Monitoring Regional Patterns of Species Richness from Geographic Information.* Unpublished Ph.D. Thesis, University of California (Santa Barbara).

Stoms, D. M. 1994. Scale dependence of species richness maps. *Professional Geographer* 46: 346–358.

Stoms, D. M., and J. Estes. 1993. A remote sensing research agenda for mapping and monitoring biodiversity. *International Journal of Remote Sensing* 14: 1839–1860.

Stone, C., K. Old, G. Kile, and N. Coops. 2001. Forest health monitoring in Australia: national and regional commitments and operational realities. *Ecosystem Health* 7: 48–58.

Stone, J. S. 1992. *Vernal pools in Massachusetts: Aerial Photographic Identification, Biological and Physiographic Characteristics, and State Certification Criteria.* Unpublished M.Sc. Thesis, University of Massachusetts, Amherst.

St. -Onge, B., and F. Cavayas. 1995. Estimating forest stand structure from high resolution imagery using the directional variogram. *International Journal of Remote Sensing* 16: 1999–2021.

St. -Onge, B., J. Jumelet, M. Cobello, and C. Véga. 2004. Measuring individual tree height using a combination of stereophotogrammetry and LiDAR. *Canadian Journal of Forest Research* 34: 2122–2130.

St. -Onge, B., P. Treitz, and M. A. Wulder. 2003. Tree and canopy height estimation with scanning LiDAR. In Wulder, M. A. and S. E. Franklin, eds., *Remote Sensing of Forest Environments: Concepts and Case Studies.* Kluwer Academic Publishers, Boston, MA. 489–509.

Storlazzi, C., J. Logan, and M. Field. 2003. Quantitative morphology of a fringing reef tract from high resolution laser bathymetry: Southern Molokai, Hawaii. *Geological Society of America Bulletin* 115: 1344–1355.

Story, R., G. A. Yapp, and A. T. Dunn. 1976. Landsat patterns considered in relation to resource surveys. *Remote Sensing of Environment* 4: 281–303.

Stow, D., A. Hope, D. Richardson, D. Chen, C. Garrison, and D. Service. 2000. Potential of a colour infrared digital camera for inventory and mapping of alien plant invasions in South African shrublands. *International Journal of Remote Sensing* 21: 2965–2970.

Strand, H., R. Hoft, J. Strittholt, N. Horning, L. Miles, E. Fonight, and W. Turner, eds. 2007. *Sourcebook on Remote Sensing and Biodiversity Indicators.* Technical Series No. 32. Secretariate of the Convention on Biological Diversity, Montreal.

Strang, K. D. 2007. Examining effective technology project leadership traits and behaviours. *Computers in Human Behaviour* 23: 424–462.

Stutchbury, B. 2007. *Silence of the Songbirds: How We are Losing the World's Songbirds and What We Can Do to Save Them.* Walker and Company, New York, U.S.

Sunquist, F., 1997. Caught in the trap! Remote cameras with electronic beams are the newest tools for biologists investigating the mysteries in the wild. *International Wildlife* Nov–Dec, 1–6.

Swain, P. H. 1978. Fundamentals of pattern recognition in remote sensing. In Swain, P. H. and S. M. Davis, eds., *Remote Sensing: The Quantitative Approach.* McGraw–Hill, New York, U.S. 136–187.

Swain, P. H. and S. M. Davis, eds. 1978. *Remote Sensing: The Quantitative Approach.* McGraw–Hill, New York, U.S.

Syartinilia, S. T. 2008. GIS-based modelling of Javan Hawk–Eagle distribution using logistic and autologistic regression models. *Biological Conservation* 141: 756–769.

Taft, O. W., S. M. Haig, and C. Kiilsgaard. 2003. Use of radar remote sensing (RADARSAT) to map winter wetland habitat for shorebirds in an agricultural landscape. *Environmental Management* 32: 268–281.

Tatum, A. J., S. J. Goetz, and S. I. Hay. 2008. Fifty years of Earth-observation satellites: views from space have led to countless advances on the ground in both scientific knowledge and daily life. *American Scientist* 96: 390–399.

Teillet, P. M. and G. Fedosejevs. 1995. On the dark target approach to atmospheric correction of remotely sensed data. *Canadian Journal of Remote Sensing* 21: 374–387.

Teillet, P. M., D. N. H. Horler, and N. T. O'Neill. 1997. Calibration, validation, and quality control in remote sensing: a new paradigm. *Canadian Journal of Remote Sensing* 23: 401–413.

Théau, J., D. R. Peddle, and C. R. Duguay. 2004. Mapping lichen in a caribou habitat of Northern Quebec, Canada, using an enhancement classification method and spectral mixture analysis. *Remote Sensing of Environment* 94: 232–243.

Theil, C., P. Drezet, C. Weise, S. Quegan, and C. Schmullius. 2006. Radar remote sensing for the delineation of forest cover maps and the detection of deforestation. *Forestry* 79: 589–597.

Thompson, I. D. and P. Angelstam. 1999. Special species. In Hunter, M. L. Jr., ed., *Maintaining Forest Biodiversity in Forest Ecosystems.* Cambridge University Press, Cambridge, U.K. 434–459.

Throgmartin, W. W., A. L. Gallant, M. G. Knutson, T. J. Fox, and M. J. Suarez. 2004. Commentary: a cautionary tale regarding use of the National Land Cover Dataset 1992. *Wildlife Society Bulletin* 32: 970–978.

Tiner, R. W. 2002. Using remote sensing to assess natural habitat. *FieldNotes* Dec: 12.

Tiner, R. W. 2004. Remotely sensed indicators for monitoring the general condition of "natural habitat" in watersheds: an application for Delaware's Nanticoke River watershed. *Ecological Indicators* 4: 227–243.

Tompkins, E. L. 2005. *A Review of Interdisciplinary Environmental Centres of Excellence (Anonymized Version).* Report to MISTRA. Foundation for Strategic Environmental Research, Stockholm, Sweden.

Tou, J. T. and R. C. Gonzalez. 1974. *Pattern Recognition Principles.* Addison-Wesley, Reading MA.

Toutin, T. 2004. Geometric processing of remote sensing images: models, algorithms, and methods. *International Journal of Remote Sensing* 25: 1893–1924.

Townshend, J. R.G. 1981. The spatial resolving power of earth resources satellites. *Progress in Physical Geography* 5: 32–55.

Tran, A., N. Ponçon, C. Toty, C. Linard, H. Guis, J. -B. Ferré, D. L. Seen, F. Roger, S. de la Rocque, D. Fontenille, and T. Baldet. 2008. Using remote sensing to map larval and adult populations of *Anopheles hyrcanus* (Diptera: *Culicidae*) a potential malaria vector in Southern France. *International Journal of Health Geographics* 7: 9.

Travaini, A., J. Bustamante, A. Rodríguez, S. Zapata, D. Procopio, J. Pedrana, and R. M. Peck. 2007. An integrated framework to map animal distributions in large and remote regions. *Diversity and Distributions* 13: 289–298.

Truehaft, R. N., B. E. Law, and G. P. Asner. 2004. Forest attributes from radar interferometric structure and its fusion with optical remote sensing. *BioScience* 54: 561–571.

Tsai, F. and W. Philpot. 1998. Derivative analysis of hyperspectral data. *Remote Sensing of Environment* 66: 41–51.

Tuan, Y.–F. 1974. *Topophilia: A Study of Environmental Perception, Attitudes, and Values.* Columbia University Press, New York, U.S.

Tucker, C. J. 1979. Red and photographic infrared linear combinations for monitoring vegetation. *Remote Sensing of Environment* 8: 127–150.

Turner, M. G., R. H. Gardner, and R. V. O'Neill. 2001. *Landscape Ecology in Theory and Practice: Pattern and Process.* Springer-Verlag, New York, U.S.

Turner, R. 2007. An overview of airborne LiDAR applications in New South Wales state forests. *Proceedings, ANZIF Conference.* Coffs Harbour. Available online http://www.forestry.org.au/pdf/pdf–public/conference2007/papers/Turner%20Russell%20Lidar.pdf web site accessed August 18, 2008.

Turner, W., S. Spector, N. Gardiner, M. Fladeland, E. Sterling, and M. Steininger. 2003. Remote sensing for biodiversity science and conservation. *Trends in Ecology and Evolution* 18: 306–314.

Twedt, D. J., R. R. Wilson, and A. S. Keister. 2007. Spatial models of Northern Bobwhite populations for conservation planning. *Journal of Wildlife Management* 71: 1808–1818.

Tzeng, Y. C. and K. S. Chen. 2005. Image fusion of synthetic aperature radar and optical data for terrain classification with a variance reduction technique. *Optical Engineering* 44, 106202: 1–8.

Underwood, E. and S. Ustin. 2007. Trends in invasive species. In Strand, H., et al., eds., *Sourcebook on Remote Sensing and Biodiversity Indicators.* Technical Series No. 32. Secretariate of the Convention on Biological Diversity, Montreal. 163–179.

United States Fish and Wildlife Service. 1981. *Standards for the Development of Habitat Suitability Index Models.* United States Fish and Wildlife Service, Division of Ecological Services Manual 103.

Ustin, S. L., ed. 2004. *Remote Sensing for Natural Resource Management and Environmental Monitoring: Volume 4.* Manual of Remote Sensing. 3d ed., ASPRS, Falls Church, VA.

Ustin, S. L., D. A. Roberts, J. A. Gamon, G. P. Asner, and R. O. Green. 2004. Using imaging spectroscopy to study ecosystem processes and properties. *BioScience* 54: 523–534.

Van Bommel, F. P. J., I. M. A. Heitkönig, G. F. Epema, S. Ringrose, CD. Bonyongo, and E. M. Veenendaal. 2006. Remotely sensed habitat indicators for predicting distribution of impala (*Aepyceros melampus*) in the Okavango Delta, Botswana. *Journal of Tropical Ecology* 22: 101–110.

Van der Heijden G. W. A. M., J. G. P. W. Clevers, and A. G. T. Schut. 2007. Combining close-range and remote sensing for local assessment of bio-physical characteristics of arable land. *International Journal of Remote Sensing* 28: 5485–5502.

Van Horn, B. 1983. Density as a misleading indicator of habitat quality. *Journal of Wildlife Management* 47: 893–901.

Van Meter, R., L. L. Bailey, and E. H. C. Grant. 2008. Methods for estimating the amount of vernal pool habitat in the northeastern United States. *Wetlands* 28: 585–593.

Van Wieren, J. F. 2009. *Ecological Integrity Monitoring in National Parks.* Available online http://www.monitoringthemoraine.ca/MonitoringAdvisoryCommittee/ewp3.ppt#6 web site accessed March 31, 2009.

Van Zyl, J. J. 1993. The effect of topography on radar scattering from vegetated areas. *IEEE Transactions on Geoscience and Remote Sensing* 31: 153–160.

Vani, K., G. Priya, and V. Lakshmi. 2006. Detection and removal of cloud contamination from satellite images. *Proceedings of SPIE* 6408 (abs).

Vierling, L., M. Fersdahl, X. Chen, and P. Zimmerman. 2006. The Short Wave Aerostat-Mounted Imager (SWAMI): a novel platform for acquiring remotely sensed data from a tethered balloon. *Remote Sensing of Environment* 103: 255–264.

Vierling, K. T., L. A. Vierling, W. A. Gould, S. Martinuzzi, and R. M. Clawges. 2008. LiDAR: shedding new light on habitat characterization and modeling. *Frontiers in Ecology and the Environment* 6: 90–98.

Vincent, W. F., J. A. E. Gibson, and M. O. Jeffries. 2001. Ice–shelf collapse, climate change, and habitat loss in the Canadian high Arctic. *Polar Record* 201: 133–142.

Voutilainen, A., T. Pyhälahti, K. Kallio, J. Pulliainen, H. Haario, and J. P. Kaipio. 2006. A filtering approach for estimating lake water quality from remote sensing data. *International Journal of Applied Earth Observation and Geoinformation* 9: 50–64.

Wallace, C. S. A. and S. E. Marsh. 2005. Characterizing the spatial structure of endangered species habitat using geostatistical analysis of IKONOS imagery. *International Journal of Remote Sensing* 26: 2607–2629.

Wallace, C. S. A., J. M. Watts, and S. R. Yool. 2002. Characterizing the landscape structure of vegetation communities in the Mojave Desert using geostatistical techniques. *Computers & Geosciences* 26: 397–410.

Wallace, J., P. A. Caccetta, and H. T. Kiiveri. 2004. Recent developments in the analysis of spatial and temporal data for landscape qualities and monitoring. *Austral Ecology* 29: 100–107.

Wallace, J., G. Behn, and S. Furby. 2006. Vegetation condition assessment and monitoring from sequences and satellite imagery. *Ecological Management and Restoration* 7(Suppl.): 31–36.

Walker, R. E., D. E. Stoms, J. E. Estes, and K. D. Cayocca. 1992. Relationships between biological diversity and multi-temporal vegetation index data in

California. *Proceedings of Technical Papers of the ASPRS/ACSM Annual Meeting,* Albuquerque, NM. 562–571.
Walsh, P. D., K. A. Abernethy, M. Bermejo, R. Beyers, P. De Wachter, M. E. Akou, B. Huijbregts, et al. 2003. Catastrophic ape decline in western equatorial Africa. *Nature* 422: 10 April 611–614.
Walsh, S. J. 1980. Coniferous tree species mapping with Landsat data. *Remote Sensing of Environment* 9: 11–26.
Walsh, S. J. 1987. Variability of Landsat MSS spectral responses of forests in relation to stand and site characteristics. *International Journal of Remote Sensing* 8: 1289–1299.
Walsh, S. J., R. E. Bilsborrow, S. J. McGregor, B. G. Frizzelle, J. P. Messina, W. K. T. Pan, K. A. Crews-Mery, G. N. Taff, and F. Baquero. 2002. Integration of longitudinal surveys, remote sensing time series, and spatial analyses: approaches for linking people and place. In Fox, J., et al., eds., *People and the Environment: Approaches for Linking Household and Community Surveys to Remote Sensing and GIS*. Springer, Dordrecht, The Netherlands. 91–130.
Walsh, S. J., D. J. Weiss, D. R. Butler, and G. P. Malanson. 2004. An assessment of snow avalanche paths and forest dynamics using Ikonos satellite data. *Geocarto International* 19: 85–93.
Walter, V. 2004. Object-based classification of remote sensing data for change detection. *ISPRS Journal of Photogrammetry and Remote Sensing* 58: 225–238.
Wang, J. 1993. LINDA—a system for automated linear feature detection and analysis. *Canadian Journal of Remote Sensing* 19: 180–191.
Wang, Y., D. Moskovits, S. Packard, M. E. Ward, and R. Harari-Karemer. 1998. Remote sensing and GIS in regional biodiversity studies: a case study of Chicago Wilderness. *Proceedings of ASPRS Annual Meeting*. Tampa Bay, FLA. 431–439.
Wang, T., W. Xu, Z. Ouyang, J. Liu, X. Yi, Y. Chen, L. Zhao, and J. Huang. 2008. Application of ecological-niche factor analysis in habitat assessment of giant pandas. *Acta Ecologica Sinica* 28: 821–828.
Ward, J. T., 2008. From the management of single species to ecosystem management. In Fullbright, T. E. and D. G. Hewitt, eds.,. *Wildlife Science: Linking Ecological Theory and Management Applications.* CRC Press, Boca Raton, Florida. 225–240.
Wardell-–=Johnson, G. W. and P. Horwitz. 1996. Conserving biodiversity and the recognition of heterogeneity in ancient landscapes. *Forest Ecology and Management* 85: 219–238.
Wardell-Johnson, G. W., M. Williams, A. Mellican, and A. Annells. 2004. Floristic patterns and disturbance history in karri forest, south-western Australia: 1. Environment and species richness. *Forest Ecology and Management* 199: 449–460.
Waring, R. H., N. C. Coops, W. Fan, and J. M. Nightingale. 2006. MODIS enhanced vegetation index predicts tree species richness across forested regions in the contiguous USA. *Remote Sensing of Environment* 103: 218–226.
Waring, R. H., and S. W. Running. 1998. *Forest Ecosystems: Analysis at Multiple Scales.* Academic Press, San Diego.
Waring, R. H., and S. W. Running. 1999. Remote sensing requirements to drive ecosystem models at the landscape and regional scales. In Tenhunen, J. D. and P. Kabat, eds., *Integrating Hydrology, Ecosystem Dynamics, and Biogeochemistry in Complex Landscapes.* John Wiley and Sons, New York, U.S. 23–38.
Waring, R. H., J. B. Way, E. R. Hunt, Jr., L. Morrisey, K. J. Ranson, J. Weishempel, R. Oren and S. E. Franklin. 1995. Imaging radar for ecosystem studies. *BioScience* 45: 715–723.
Warren, A. J., and M. J. Collins. 2007. A pixel-based semi-empirical system for predicting vegetation diversity in boreal forest. *International Journal of Remote Sensing* 28: 83–105.
Washino, R. K. and B. J. Wood. 1994. Application of remote sensing to arthropod vector surveillance and control. *American Journal of Tropical Medicine and Hygiene* Suppl.: 134–144.
Watt, P. J. and D. N. M. Donoghue. 2005. Measuring forest structure with terrestrial laser scanning. *International Journal of Remote Sensing* 26: 1437–1446.

Weaver, K. F. 1969. Remote sensing—new eyes see the world. *National Geographic* 135 (Jan.): 47–73.
Weiers, S., M. Bock, M. Wissen, and G. Rossner. 2004. Mapping and indicator approaches for the assessment of habitats at different scales using remote sensing and GIS methods. *Landscape and Urban Planning* 67: 43–65.
Weisberg, S. B., J. A. Ranasinghe, D. M. Dauer, L. C. Schaffner, R. J. Diaz, and J. B. Frithsen. 1997. An estuarine benthic index of biotic integrity (B-IBI) for Chesapeake Bay. *Estuaries* 20: 158.
Welles, J. M. and S. Cohen. 1996. Canopy structure measurement by gap fraction analysis with commercial instrumentation. *Journal of Experimental Botany* 47: 1135–1142.
Weng, Q., ed. 2007. *Remote Sensing of Impervious Surfaces.* CRC Press, Boca Raton, Florida.
White, G. C. 1996. NOREMARK: population estimation from mark-resight surveys. *Wildlife Society Bulletin* 24: 50–52.
White, M.A., P. E. Thornton, and S. W. Running. 1997. A continental phenology model for monitoring vegetation responses to interannual climatic variability. *Global Biogeochemical Cycles* 11: 217–234.
White, T.H., Jr., G. G. Brown, and J. A. Collazo. 2006. Artificial cavities and nest site selection by Puerto Rican parrots: a multiscale assessment. *Avian Conservation and Ecology* 1: 3. Available online http://www.ace–eco.org/vol1/iss3/art5/
White, M. A., P. E. Thornton, and S. W. Running. 1997. A continental phenology model for monitoring vegetation responses to interannual climatic variability. *Global Biogeochemical Cycles* 11: 217–234.
Wickham, J. D., J. Wu, and D. F. Bradford. 1997. A conceptual framework for selecting and analyzing stressor data to study species richness at large spatial scales. *Environmental Management* 21: 247–257.
Wiens, J. A. 1989. Spatial scaling in ecology. *Functional Ecology* 3: 385–397.
Wiens, J. A. 1995. Landscape mosaics and ecological theory. In Hansson, L., et al., eds., *Mosaic Landscapes and Ecological Processes.* Chapman and Hall, London. 1–26.
Williams, A. E. P. and E. R. Hunt, Jr. 2004. Accuracy assessment for detection of leafy spurge with hyperspectral imagery. *Journal of Range Management* 57: 106–112.
Williams, S. E., and J. -M. Hero. 2001. Multiple determinants of Australian tropical frog biodiversity. *Biological Conservation* 98: 1–10.
Wilson, B. A. 1996. Estimating coniferous forest structure with SAR texture and tone. *Canadian Journal of Remote Sensing* 22: 382–389.
Wilson, B. A. and C. G. Ference. 2001. The influence of canopy closure on the detection of understory indicator plants in Kananaskis Country. *Canadian Journal of Remote Sensing* 27: 207–215.
Wilson, B. A., J. Luther, and T. Stuart. 1998. Spectral reflectance characteristics of Dutch Elm disease. *Canadian Journal of Remote Sensing* 24: 200–205.
Wilson, E. O. 1999. *Consilience: The Unity of Knowledge.* Vintage, New York, U.S.
Wilson, E. O. 2006. *The Creation: An Appeal to Save Life on Earth.* W. H. Norton, New York, U.S.
Wilson, E. O. and F. M. Peter, eds. 1988. *BioDiversity.* National Academy Press, Washington.
Wilting, A., F. Fischer, S. A. Bakar, and K. E. Linsenmair. 2006. Clouded leopards, the secretive top-carnivore of south-east Asian rainforests: their distribution, status and conservation needs in Sabah, Malaysia. *BMC Ecology* 6: 1–27.
Wilting, A., V. A. Buckley-Beason, H. Feldhaar, J. Gadau, S. J. O'Brien, and K. E. Linsenmair. 2007. Cloud leopard phylogeny revisited: support for species recognition and population division between Borneo and Sumatra. *Frontiers in Zoology* 4: 15 doi:10.1186/1742-9994-4-15.
Wolfram, S. 1984. Cellular automata as models of complexity. *Nature* 311: 419–424.
Womble, J. A. 2005. *Remote–Sensing Applications to Windstorm Damage Assessment.* Unpublished Ph.D. Thesis, Texas Tech University, Lubbock, TX.

Wong, A. and D. Clausi. 2007. ARRSI: Automatic registration of remote sensing images. *IEEE Transactions on Geoscience and Remote Sensing* 45: 1483–1493.

Wood, C. W., D. W. Reeves, R. R. Duffield, and K. L. Edmisten. 1992. Field chlorophyll measurements for evaluation of corn nitrogen deficiency status. *Journal of Plant Nutrition* 15: 487–500.

Woodcock, C. E., R. Allen, M. Anderson, A. Belward, R. Bindschadler, W. Cohen, F. Gao, et al. 2008. Free access to Landsat imagery. *Nature* 320: 1011.

Woodham, R. J. 1989. Determining intrinsic surface reflectance in rugged terrain and changing illumination. *Proceedings of International Geoscience and Remote Sensing Symposium/12th Canadian Symposium on Remote Sensing.* Canadian Aeronautics and Space Institute, Ottawa. 1–5.

Woodley, S. and J. Kay. 1993. *Ecological Integrity and the Management of Ecosystems.* CRC Press, Boca Raton, Florida.

Worboys, M. 1998. Imprecision in finite resolution spatial data. *GeoInformatica* 2: 257–279.

Wu, J., D. Wang and M. E. Bauer. 2005. Image-based atmospheric correction of Quickbird imagery of Minnesota croplands. *Remote Sensing of Environment* 99: 315–325.

Wulder, M. A., J. A. Dechka, M. A. Gillis, J. E. Luther, R. J. Hall, A. Beaudoin, and S. E. Franklin. 2003. Operational mapping of the land cover of the forested area of Canada with Landsat data: EOSD land cover program. *Forestry Chronicle* 79: 1075–1083.

Wulder, M. A. and S. E. Franklin, eds. 2007. *Understanding Forest Disturbance and Spatial Patterns: Remote Sensing and GIS Approaches.* Taylor and Francis (CRC Press), Boca Raton, Florida.

Wulder, M. A., S. E. Franklin, J. C. White, J. Linke, and S. Magnussen. 2006. An accuracy assessment framework for large area land cover classification products derived from medium resolution satellite remotely sensed data. *International Journal of Remote Sensing* 27: 663–683.

Wulder, M. A., R. J. Hall, N. C. Coops, and S. E. Franklin. 2004. High spatial resolution remotely sensed data for ecosystem characterization. *Bioscience* 54: 511–512.

Wulder, M. A., K. O. Niemann, and D. G. Goodenough. 2000. Local maximum filtering for the extraction of tree locations and basal area from high spatial resolution imagery. *Remote Sensing of Environment* 73: 103–114.

Wulder, M. A., J. C. White, G. J. Hay, and G. Castilla. 2008. Pixels to objects to information: spatial context to aid in forest characterization with remote sensing. In Blaschke, T., et al., eds., *Object-based Image Analysis: Spatial Concepts for Knowledge-Driven Remote Sensing Applications.* Springer, Berlin, Germany. 345–363.

Wunderle, A., S. E. Franklin, and X. Guo. 2007. Regenerating boreal forest structure estimation using SPOT-5 pan-sharpened imagery. *International Journal of Remote Sensing* 28: 4351–4364.

Wynne, R. H. and D. B. Carter. 1997. Will remote sensing live up to its promise for forest management? *Journal of Forestry* 95: 23–26.

Xiao, Q. and E. G. McPherson. 2005. Tree health mapping with multispectral remote sensing data at UC Davis, California. *Urban Ecosystems* 8: 349–361.

Xiao, X., M. Gilbert, J. Slingenbergh, F. Lei, and S. Boles. 2007. Remote sensing, ecological variables, and wild bird migration related to outbreaks of highly pathogenic H5N1 avian influenza. *Journal of Wildlife Diseases* 43: S40–S46.

Xu, W., Z Ouyang, A. Viña, H. Zheng, J. Liu, and Y. Xiao. 2006. Designing a conservation plan for protecting the habitat for giant pandas in the Qionglai mountain range, China. *Diversity and Distributions* 12: 610–619.

Yadav, V. and G. Malanson. 2007. Progess in soil organic matter research. *Progress in Physical Geography* 31: 131–154.

Yashuda, M. and K. Kawakami. 2002. New method of monitoring remote wildlife via the Internet. *Ecological Research* 17: 119–124.

Yukiko, K., I. Yoshikazu, and Y. Ryuzo. 2000. Automatic selection of GCPs for geometric correction and evaluation. *Proceedings of Japanese Conference on Remote Sensing* 28: 191–194.

Zaveh, N. 2007. *Transdisciplinary Challenges in Landscape Ecology and Restoration Ecology: An Anthology with Forewords by E. Lazlo and M Antrop, and Epilogue by E. Allen.* Springer, Dordrecht, The Netherlands.
Zhao, M., F. Heinsch, R. R. Nemani, and S. W. Running. 2005. Improvement of the MODIS terrestrial gross and net primary production global data set. *Remote Sensing of Environment* 95: 197–207.
Zhao, D., P. J. Starks, M. A. Brown, W. A. Phillips, and S. W. Coleman. 2008. Assessment of forage biomass and quality parameters of bermudagrass using promimal sensing of pasture canopy reflectance. *Grassland Science* 53: 39–49.
Zheng, D., E. R. Hunt, Jr., P. C. Doraiswamy, G. W. McCarty, and S. Ryu. 2006. Using remote sensing and models to understand the ecology of landscapes. In Chen, J., et al., eds., *Ecology of Hierarchical Landscapes: From Theory to Applications.* Nova Science Publishers, New York, U.S. 125–166.
Zheng, D., D. O. Wallin, and Z. Hao. 1997. Rates and patterns of landscape change between 1972 and 1988 in the Changbai Mountain area of China and North Korea. *Landscape Ecology* 12: 241–254.
Zhu, L. and R. Tateishi. 2006. Fusion of multisensor, multitemporal satellite data for land cover mapping. *International Journal of Remote Sensing* 27: 903–910.
Zimmerman, N. E., R. A. Washington-Allen, R. D. Ramsay, M. E. Schaepman, L. Mathys, B. Kotz, M. Kneubhler, and T. C. Edwards. 2007. Modern remote sensing for enviromental monitoring of landscape states and trajectories. In Kienast, F., O. Wildi, and S. Ghosh, eds., *A Changing World: Challenges for Landscape Research.* Springer, Dordrecht, The Netherlands. 65–91.
Zsilinsky, V. 1970. Supplemental aerial photography with miniature cameras. *Photogrammetria* 25: 27–38.

Index

A

B

C

D

E

I

J

K

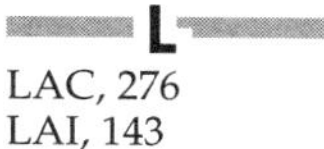

N

O

P

T

U

V

W

Y

Z